Multiwavelength Optical Networks

Network Theory and Applications

Volume 9

The titles published in this series are listed at the end of this volume.

Multiwavelength Optical Networks

by

Xiaohua Jia

Department of Computer Science,
City University of HongKong,
Hong Kong, SAR China

Xiao-Dong Hu

Institute of Applied Mathematics,
Academy of Mathematics and System Science,
Chinese Academy of Sciences, Beijing, P.R. China

and

Ding-Zhu Du

Department of Computer Science,
University of Minnesota,
Minneapolis, U.S.A.

KLUWER ACADEMIC PUBLISHERS
DORDRECHT / BOSTON / LONDON

A C.I.P. Catalogue record for this book is available from the Library of Congress.

ISBN 1-4020-0804-X

Published by Kluwer Academic Publishers,
P.O. Box 17, 3300 AA Dordrecht, The Netherlands.

Sold and distributed in North, Central and South America
by Kluwer Academic Publishers,
101 Philip Drive, Norwell, MA 02061, U.S.A.

In all other countries, sold and distributed
by Kluwer Academic Publishers,
P.O. Box 322, 3300 AH Dordrecht, The Netherlands.

Printed on acid-free paper

Printed in the Netherlands.

Contents

Preface

This book studies various of optimization problems arising from the development and applications of multiwavelength optical networks. It is divided into two parts.

The first part consists of the first four chapters. They comprise a self-contained treatment of some of the most significant results on three fundamental issues in multiwavelength optical networks. Most of them appeared in conference proceedings in the past few years, thus they are organized and presented in length at this book for the easy references of readers of the book. Chapter 1 first introduces some basic concepts and terminologies associated with multiwavelength optical networks, which will help the readers to understand the background and mathematical models of problems discussed in the rest of this book. It then describes the general mathematical approach that will be used for studying problems considered throughout the book. Chapter 2 discusses the routing problem for load balancing. Chapter 3 studies the wavelength assignment problem for minimization of wavelength usage. Chapter 4 focuses on the logical topology design problem for improving network performances.

The second part consists of the last five chapters. They present some recent work done by authors of this book and their coauthors. The obtained results are on some of the advanced topics in the design and application of multiwavelength optical networks. Chapter 5 studies the wavelength converter placement problem for reducing the number of wavelengths required in networks. Chapter 6 discusses the wavelength assignment and converter placement problems for minimization of overall blocking probabilities of networks. Chapter 7 addresses the wavelength assignment problems for minimization of wavelength conversions required in networks. Chapter 8 focuses on the routing and wavelength assignment problems for multicast connections in networks. Chapter 9 considers the constrained multicast routing and wavelength assignment problems in networks.

Finally, the authors of this book owe a great deal to many pioneers in this area. Their important work, especially those included in the first part of this book, have influenced and motivated our work at many points. Among them, without trying to be complete, are D. Banerjee, T. Erlebach, K. Jansen, C. Kaklamanis, V. Kumar, R. M. Krishnaswamy, K.-C. Lee, V. O. K. Li, B. Mukherjee, P. Persiano, P. Raghavan, R. Ramaswami, E. J. Schwabe, K. N. Sivarajan, E. Upfal, G. Wilfong and P. Winkler.

X. Jia

March 2002 *X.-D. Hu*

D.-Z. Du

Acknowledgments

Some part of this book was completed at Department of Computer Science, City University of Hong Kong, and supported by Research Grants of City University of Hong Kong under grant No. 7001121, 7000927, 7001035, 7000778 and by CERG Research Grants of Hong Kong under grant No. 9040442. Some part of this book was written in Institute of Applied Mathematics, Chinese Academy of Sciences, and supported by National 973 Information Technology and High-Performance Software Program of China under grant G1998030402. Some part of this book was finished in Department of Computer Science, University of Minnesota.

Many of our colleagues and students have contributed greatly to this book. We particularly thank X.-F. Du, S.-X. Gao, J. Gu, H.-J. Huang, M.-K. Lee, D.-Y. Li, L. Ruan, J.-H. Sun, Z. Sun, and M.-H. Zhang.

Finally, we thank our family members, old and young, whose love, patience, and encouragement have made this book possible. We affectionately dedicate this book to them.

March 2002

X. Jia
X.-D. Hu
D.-Z. Du

I

FUNDAMENTAL STUDY OF WDM NETWORKS

Chapter 1

INTRODUCTION

Nowadays as more and more users start to use data networks, and their usage patterns evolve to include more and more bandwidth-intensive networking applications such as data browsing on the worldwide web, java applications, video conferencing, etc., there emerges as acute need for very high-bandwidth transport network facilities, whose capabilities are much beyond those that current high-speed networks can provide. Wavelength optical networks are believed to be able to meet the information networking demands of the coming decades and constitute the backbones of the next generation of the Internet.

Wavelength optical networks are based on the current favorite multiplexing technology, optical *Wavelength Division Multiplexing* (WDM). By using WDM approach, all of the end-user equipment needs to operate only at the bit rate of a WDM channel, which can be chosen arbitrarily, e.g., peak electronic processing speed. Hence the major carriers today all devote significant effort to developing and applying WDM technologies in their business.

Research and development on WDM networks have matured considerably over the past ten years. There have been an explosive of articles and books on this topic as well as new conferences, workshops, journals, and magazines devoted to this topic.

This book studies various of optimization problems arising from the development and applications of multiwavelength optical networks. In this chapter, we will introduce some essential concepts, terminologies and methodologies that will be used throughout the book.

The rest of this chapter is organized as follows. Section 1 introduces some basic concepts and terminologies associated with multiwavelength optical networks. Section 2 describes the general mathematical model of multiwavelength optical networks and the general approach for solving problems to be discussed in the following chapters.

1. Multiwavelength Networks

Wavelength Division Multiplexing (WDM) is a technology of frequency division multiplexing in the optical frequency domain, which partitions the fiber bandwidth into a set of disjoint channels operating in different wavelengths. WDM provides an efficient way to utilize the tremendous bandwidth of a fiber. Thus WDM networks are believed to be the next generation networks that can meet the ever increasing bandwidth demand of end users.

In a WDM network, each network node is typically equipped with a small number of optical transmitters (lases) and optical receivers (filters), and some of these transceivers (transmitters or receivers) can be made dynamically tunable to operate on different wavelength channels. By tuning its transmitter(s) to one or more wavelength channels, a node can transmit into those channel(s); similarly, a node can tune its receiver(s) to receive from he appropriate channels. By employing high-speed tunable transceivers, WDM can be used for packet switching, as well as for wavelength routed circuit-switching.

Switches are referred broadly to the devices such as routers, cross-connects and add-drop multiplexer. In particular, an optical switch is the one that can switch an optical signal without converting it from the optical domain to the electronic domain, and then back to the optical domain, although the switch may still be controlled by electronic signals.

An *Add-Drop Multiplexer* (ADM) is an optical system that is used to modify the flow of traffic through a fiber at a routing node. An ADM passes through traffic on some of the wavelength channels without performing any optic-electronic-optic conversions, while letting other wavelength channels that carry traffic originating or terminating at the node be added or dropped.

A *Wavelength Router* (WR) is a more powerful system than an ADM. It takes in an optical signal at each of the wavelength channels at an input fiber, and routes it to a designated output fiber, independently from the other wavelengths. Because there is no wavelength conversion in a WR, the wavelength of a channel stays the same in the output fiber as it was in the input fiber. A WR with N input and N output ports, and the ability of handling ω wavelength channels, can be considered as ω independent $N \times N$ switches. These switches have to be preceded by a wavelength demultiplexer and followed by a wavelength multiplexer to implement a WR. They are sometimes also called *Wavelength Routing Switches* (WRS) or *wavelength crossconnects*.

A *Wavelength Converter* (WC) is an optical device that converts the wavelength that a channel is carrying at the router without intermediate optic-electronic-optic conversions. See a simple example [3] as shown in Fig. 1.1. In Fig. 1.1(a) without wavelength conversion, the connection from node 1 to node 3 using wavelength w_a will be blocked since w_a is used by the connection from node 2 to node 3; As contrast, in Fig. 1.1(b), the wavelength converter equipped at node 2 can convert the wavelength w_a of the connection from node 1 to node 3 to the free wavelength w_b, and thus it avoids wavelength conflict. Clearly, placing wavelength converters at routing nodes can reduce the blocking probability experienced by connection requests.

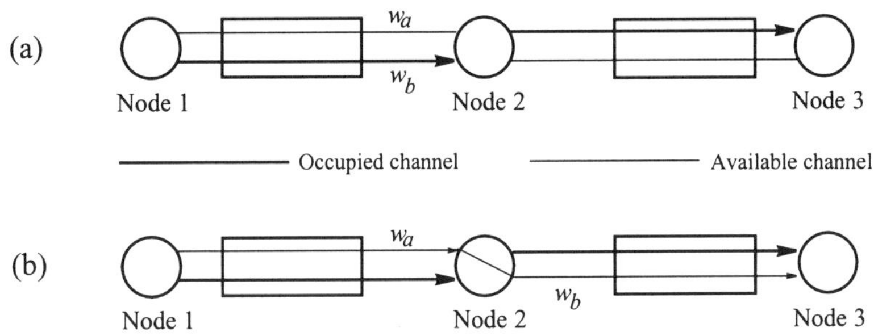

Figure 1.1. (a) Connection is blocked without wavelength conversion. (b) Connection is established through wavelength conversion.

A possible wavelength-convertible switch architecture of WDM networks [3] is shown in Fig. 1.2. The switching node provides the basic cross-connect function of transit optical signals by wavelength multiplexers and demultiplexers, optical switch, and wavelength converters. At first, an aggregate optical signal received from an incoming fiber is wavelength demultiplexed into separate wavelengths of signals each supporting a channel. And then each separate wavelength is switched to one of the output ports connected to the designated output fiber by the nonblocking optical switch. The output signal of the optical switch may have its wavelength changed by the wavelength converter. Finally, various wavelengths of channels are combined to form an aggregate signal coupled to the outgoing optical fiber.

The above described wavelength convertible switch is not very cost efficient since some channels may not need wavelength conversions. An more effective method is to share wavelength converters. Two structures are proposed in [3] depending on how the wavelength converters are shared. In the *share-per-node structure* as shown in Fig. 1.3, all the converters at the switching node are collected in the converter bank which can be accessed by all the incoming channels. Demultiplexed wavelengths are feeded to the first optical switch. Those wavelengths which do not need wavelength conversions, e.g., w_a, will be directed by this optical switch to the proper outgoing link; Otherwise, they, e.g., w_b, will be directed to the converter bank in which available converters can change their wavelengths. Converted wavelengths, e.g., w_c, are switched to the appropriate outgoing link by the second optical switch. A *Wavelength Converter Bank* (WCB) contains a few wavelength converters each of which has identical characteristics and can convert the wavelength of an incoming channel from any input to any output wavelength in the discrete wavelength set.

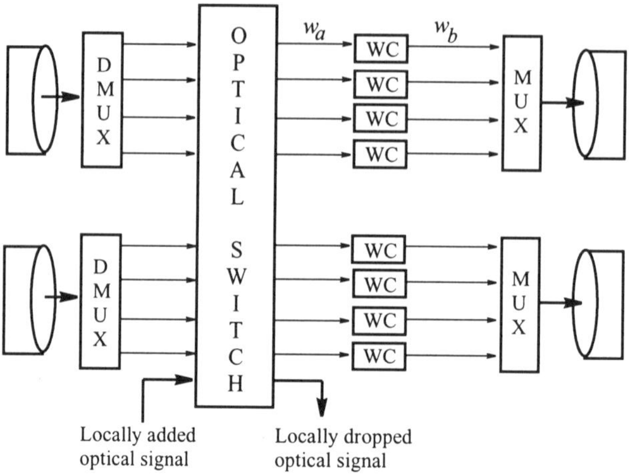

Figure 1.2. The wavelength convertible switch architecture.

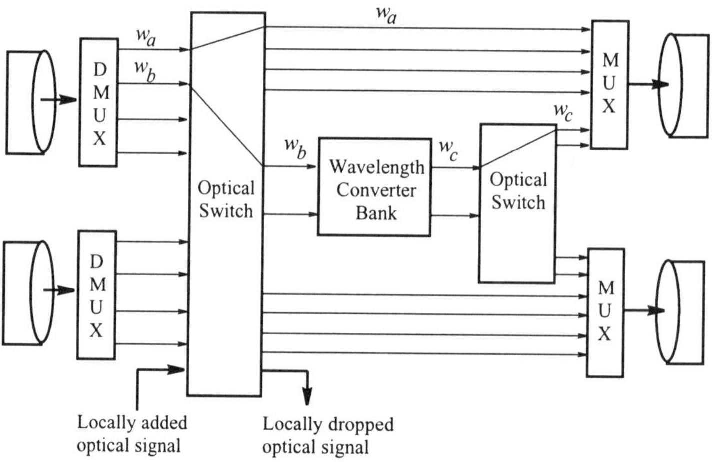

Figure 1.3. The share-per-node switch architecture.

In the *share-per-link structure* of Fig. 1.4, each outgoing link has a dedicated converter bank which can be accessed only by those channels going out on that link.

A WDM optical network consists of switching nodes interconnected by fiber links. Each fiber link supports a set of wavelength channels and each switching node is able

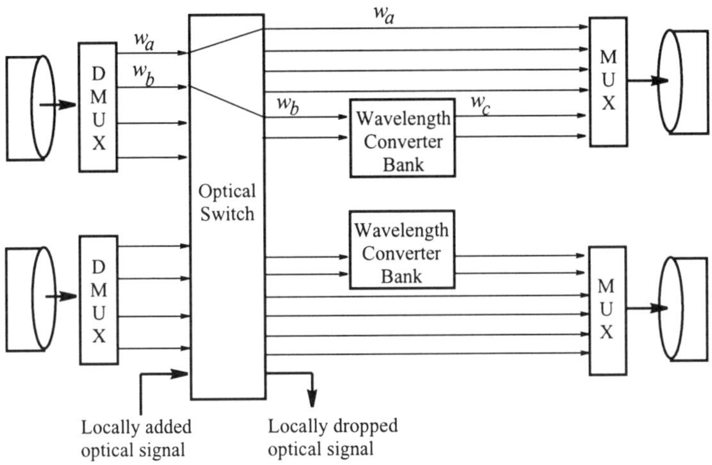

Figure 1.4. The share-per-link switch architecture.

to route a signal coming in on one wavelength at an input port to any other output port. In WDM optical networks, one-to-one connections are supported by lightpaths. A *lightpath* is an all-optical channel which may be used to carry circuit-switched traffic. A lightpath may span multiple fiber links, and all links along the path are assigned the same wavelength. This is called the *wavelength continuity* constraint in wavelength-routed networks.

In a *single-hop network* [4], the nodes must communicate with one another in one hop, that is, the connection between a pair of nodes should use the same wavelength throughout the route of the connection. This requires a significant amount of dynamic coordination between the nodes, since, during the data transmission at least, one of the transmitters of the sending node and one of the receivers of the destination node must be tuned to the same wavelength. In order for the single-hop network to be efficient, the transceivers must be very rapidly tunable. Thus the transceiver tuning times play a vital role in determining the performance and characteristics of single-hop networks.

In a *multihop network* [5], the communication between a pair of nodes consists of one or several lightpaths. Wavelength conversion is required at the joint of two lightpaths if they use different wavelengths. This is realized by placing wavelength converters at switching nodes. In order for the multihop network to be efficient, the processing complexity at the nodes must be small because the high-speed environment allows very little processing time. Thus simple routing mechanisms play an important role.

From a performance perspective, single-hop and multihop systems appears to be no clear winner, thus they are currently equally attractive and will be studied in the book.

In this book we will consider two types of communication channels: *bidirectional* (duplex) and *unidirectional.* In a bidirectional channel, data can be transmitted in both directions of the channel. The *wavelength conflict rule* for duplex channels is that channels over the same link must use different wavelengths on the link. In a unidirectional channel, data can be transmitted only in one direction from the source to the destination. Thus the *wavelength conflict rule* for unidirectional channels is that channels over the same link and in the same direction must use different wavelengths. That is, two unidirectional channels over the same link but in opposite directions can use the same wavelength.

A large part of this book (Chapters 2-7) focuses on unicast (one-to-one) traffic in WDM networks, while a small part of this book (Chapters 8-9) concentrates on multicast (one-to-many) traffic in WDM networks. *Multicasting* is the ability of an application at a node to send a single message to the communication network and have it delivered to multiple recipients at different locations. Multicasting in optical domain is realized by employing light-trees. A *light-tree* [8] is a point-to-multipoint all-optical channel which may interconnect multiple fiber links forming a tree. At the branching nodes in the tree, lights are split to reach multiple outgoing links. The multicast-capable wavelength-routing switch architecture is similar to the share-per-node switch architecture as shown in Fig. 1.3 with wavelength convert bank replaced by power splitters. In theory optical multicasting has some improved characteristics over electronic multicasting since "splitting light" is conceptually easier than "copying data" in electronic buffer.

2. Mathematical Model and Approach

A WDM network is modelled as a connected graph $G(V, E)$. V is the vertex-set of G representing the set of nodes in the network. E is the edge-set of graph G corresponding to point-to-point optical fiber links between nodes in the network, each of them carries two oppositely-directed fibers for data transmissions in the two directions of the link. In this book, the graph-theoretic term "vertex" and networking term "node", the graph-theoretic term "edge" and networking term "link" are interchangeable, respectively.

For some problems to be discussed in this book, each edge $e \in E$ is associated with two weight functions $c(e)$ and $d(e)$, where $c(e)$ represents the communication cost of e and $d(e)$ the delay of e. $c(e)$ can be interpreted as the monetary cost of using link e. $d(e)$ represents the time for data transmission between the two endpoints of link e, which includes switching and propagation delays. Both functions c and d are additive over the links of a path $P(u, v)$ between two nodes u and v, i.e.,

$$c(P(u,v)) \equiv \sum_{e \in P(u,v)} c(e) \text{ and } d(P(u,v)) \equiv \sum_{e \in P(u,v)} d(e).$$

Each edge $e \in E$ may be also associated with a wavelength function $w(e)$ that represents the set of wavelengths available on e.

In this book we will study many optimization problems arising from the development and applications of WDM networks by using the same approach. Our methodology essentially consists of the following six steps.

Step 1 Establish a mathematical models of the problems under discussion, most of which will be formulated as combinatorial optimization problems.

Step 2 Study the complexity of established mathematical problems, that is, how difficult it is to find optimal solutions to these problems.

Step 3 Propose efficient algorithms to find the optimal or approximate solutions to these problems.

Step 4 Analyze the performances of proposed algorithms, that is, how close the objective values of obtained solutions are to that of the optimal solutions.

Step 5 Run the proposed algorithms by implementing them in real or simulated network environments.

Step 6 Analyze the obtained solutions from simulation to investigate the effectiveness of the proposed algorithms and examine the correctness of the proposed mathematical models.

The above approach has two parts, theoretical study and analysis consisting of Steps 1-4, the simulation study and analysis consisting of Steps 5-6. In the following we will explain in detail how each of these steps will be carried out.

At **Step 1**, all of the problems addressed in this book will be formulated, under some reasonable assumptions on communications models in WDM networks, as combinatorial optimization problems. In general, a combinatorial optimization problem can be characterized by three components as follows.

PROBLEM 1.1 *Minimization/Maximization Problem* Π

Instance A set Ω of input instances Is specifying problem Π.
Solution The set $S(I)$ of all feasible solutions for an instance $I \in \Omega$.
Objective Minimizing/Maximizing the value function of solutions $f : S(I) \rightarrow \Re$, that is, to find an optimal solution $\sigma_{opt}(I) \in S(I)$ such that $f(\sigma_{opt}(I)) \leq f(\sigma)$ or $f(\sigma_{opt}(I)) \geq f(\sigma), \ \forall \sigma \in S(I)$.

A large number of the combinatorial optimization problems including those we will study in this book are $\mathcal{NP}$-hard. The computational complexity theory tells us that it is impossible to find efficient algorithms for such problems unless $\mathcal{P} = \mathcal{NP}$, which is believed to be not true. Here an algorithm is called *efficient* if its running time is upper bounded by a polynomial of the size of problem input.

At **Step 2**, we will give $\mathcal{NP}$-hardness proofs for our problem formulated at Step 1. For this purpose, we consider the decision version of Problem Π. In general, a decision problem is a question which can be answered by 'Yes' or 'No'. The decision version of Problem Π is to decide, given an instance $I \in \Omega$ and a bound of real number B, if there is a solution in $S(I)$ whose value function is at most B for minimization problems or at least B for maximization problem. The proof consists of two steps. (1) Select a known $\mathcal{NP}$-complete problem Π', which is a decision problem, among hundreds of candidates (refer to [1]); (2) Construct a *polynomial reduction* from the selected problem Π' to the decision version of our problem Π. That is, to produce an instance I of Problem Π and set a bound B, which are based on the instance I' of the selected problem, in such a way that Problem Π' has a positive answer if and only if the decision version of Problem Π has an one. It is essential that the construction can be finished in polynomial-time in the size of input I of Problem Π'.

However, $\mathcal{NP}$-hardness only means that we are unable to find algorithms which can find *exactly* the optimal solution to all instances of the problem in time which is polynomial in the size of the input. When we relax this rather stringent requirement, we may be possible to find *approximately* the optimal solution to all instances of the problem in time which polynomial in the size of the input, in other words, we may solve the problem reasonably well in an efficient way.

At **Step 3**, we will design approximation algorithms for our problems proved $\mathcal{NP}$-hard at Step 2. Some of the basic combinatorial algorithms will be used, such as shortest path algorithm, minimum spanning tree algorithm, coloring algorithm, etc.

The quality of an approximation algorithm is usually measured in terms of the ratio between the value of its solution and that of the optimal solution. In our discussion we will denote the value of the optimal solution as $Opt(I) \equiv f(\sigma_{opt}(I))$, which sometimes, for the simplicity of presentation, will be abused to denote the optimal solution as well. Similarly, we will denote by $A(I)$ the solution produced by algorithm A and its value. The following two definitions formalize the concept of *worst-case performance analysis* of an approximation algorithm.

DEFINITION 1.2 *Let A be an approximation algorithm for Problem 1.1. The performance ratio $R_A(I)$ of Algorithm A on an input instance I is defined as follows.*

$$R_A(I) \equiv \begin{cases} \frac{A(I)}{OPT(I)}, & \text{for minimization problem;} \\ \\ \frac{OPT(I)}{A(I)}, & \text{for maximization problem.} \end{cases}$$

The ratio is defined differently for maximization and minimization problems so as to have a uniform measure for the quality of the solution produced by an algorithm. The ratio is always at least 1 and the algorithm produces a better approximation if the ratio is closer to 1. We now define the worst-case ratio for an approximation algorithm.

DEFINITION 1.3 *For an approximation algorithm A for Problem 1.1, its performance ratio in worst-case is* $R_A \equiv \{r \mid R_A(I) < r, \forall\, I \in \Omega\}$.

Algorithm A is called an α-approximation algorithm if its performance ratio in worst-case is $R_A = \alpha$.

At **Step 4**, we will give worst-case performance analysis of our algorithms designed at Step 3. Assume without loss of generality that Problem Π is a minimization problem. Then the analysis consists of two parts. One part is a proof of a lower bound on the $OPT(I)$ in terms of some parameters ϕ. Another part is a proof of an upper bound on $A(I)$ in terms of parameters ϕ. To obtain the bound on the performance ratio $R_A(I)$, we just need to eliminate ϕ from these two inequalities.

At **Step 5**, after the theoretical study on the proposed algorithms are finished (which is based on the analysis of approximation ratios of worst-case performance), we will also apply experimental study on them to evaluate their average performances. For this purpose we will simulate the proposed algorithms on both randomly generated and typical real-world network environments.

Typical real-world network is obtained through modification of NSFnet. Fig. 1.5 shows the physical topology of NSFnet [6], where fourteen cities of United States are connected into the backbone of the network. Table 1.1 gives the modified traffic between two cities for a 15-minute duration. The traffic data in this table are used as the probabilities of requesting for connections between two cities (after normalization). Table 1.2 shows the traffic routes between cities, which were produced by using the shortest path algorithm.

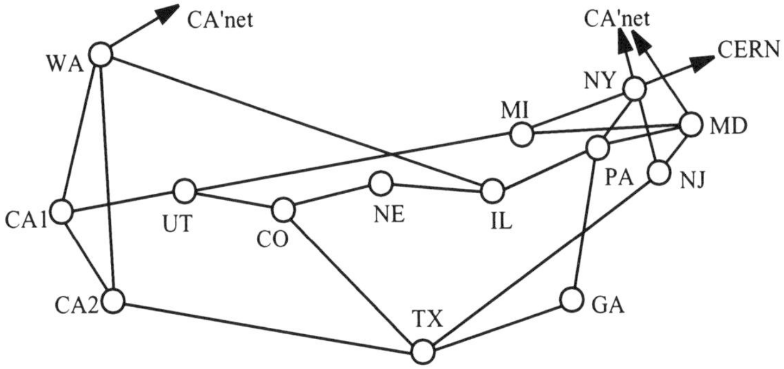

Figure 1.5. The NSFnet network.

Table 1.1. Traffic between cities in NSFnet (multiply by 1000 to get bytes per 15-min interval).

	WA	CA1	CA2	UT	CO	TX	NE	IL	PA	GA	MI	NY	NJ	MD
WA	0	268	117	27	196	8	53	249	34	18	311	96	44	191
CA1	719	0	610	301	586	261	398	1549	114	214	799	1031	552	775
CA2	109	475	0	466	85	363	86	856	100	46	516	62	139	215
UT	70	62	136	0	19	6	7	28	20	32	131	121	23	69
CO	1227	1590	190	34	0	40	107	622	240	179	721	1185	131	217
TX	18	165	34	55	34	0	26	26	8	38	60	48	15	69
NE	370	620	1023	44	220	79	0	1141	198	219	1540	933	236	1638
IL	149	2345	2103	85	282	26	970	0	439	330	900	711	202	889
PA	849	199	373	60	249	68	250	610	0	396	1106	1476	456	631
GA	18	419	102	37	223	94	49	570	68	0	363	261	126	143
MI	111	376	582	50	94	129	187	378	204	251	0	596	322	371
NY	312	1318	198	146	429	71	173	573	396	294	2116	0	279	659
NJ	393	553	186	75	84	8	44	243	1176	356	691	792	0	521
MD	819	2270	542	229	892	318	327	918	306	16	1296	1375	627	0

Table 1.2. Routes between cities in NSFnet.

	2	3	4	5	6	7	8	9	10	11	12	13	14
1	1-2	1-3	1-4	1-4-5	1-3-6	1-4-5-7	1-8	1-3-6-9	1-2-8-10	1-4-11	1-4-11-12	1-3-6-13	1-4-11-14
2		2-3	2-1-4	2-1-4-5	2-3-6	2-8-7	2-8	2-8-10-9	2-8-10	2-1-4-11	2-8-10-12	2-3-6-13	2-8-10-14
3			3-1-4	3-6-5	3-6	3-6-5-7	3-2-8	3-6-9	3-6-9-10	3-1-4-11	3-6-13-12	3-6-13	3-6-13-14
4				4-5	4-5-6	4-5-7	4-5-7-8	4-5-6-9	4-11-12-10	4-11	4-11-12	4-11-14-13	4-11-13
5					5-6	5-7	5-7-8	5-6-9	5-7-8-10	5-4-11	5-4-11-12	5-6-13	5-6-13-14
6						6-5-7	6-9-10-8	6-9	6-9-10	6-13-14-11	6-13-12	6-13	6-13-14
7							7-8	7-8-10-9	7-8-10	7-5-4-11	7-8-10-12	7-5-6-13	7-8-10-14
8								8-10-9	8-10	8-10-12-11	8-10-12	8-10-14-13	8-10-14
9									9-10	9-10-12-11	9-10-12	9-6-13	9-10-14
10										10-12-11	10-12	10-14-13	10-14
11											11-12	11-14-13	11-14
12												12-13	12-13-14
13													13-14

Random network $G(V, E)$ is generated by using the approach introduced in [10]. 100-200 nodes are distributed randomly over a rectangular coordinate grid. Each node is placed at a location with integer coordinates. A link between two nodes u and v is added by using the probability function $F(u, v) = \beta \exp(-d(u, v)/\gamma\delta)$, where $d(u, v)$ is the distance between u and v, δ is the maximum distance between any two nodes, and $0 < \beta, \gamma \leq 1$. Larger values of β produce graphs with higher link densities, in this case the network $G(V, E)$ consists of links. While small values of γ increases the density of short links relative to longer ones. In our simulations, γ and β both are set to 0.9-0.95. Graphs are generated and tested until a connected graph is found. In random networks, the traffic distribution matrices or communication connection requests are also randomly generated. The method of their generation may be dependent on the formulations of the problem discussed. To eliminate accidental factors, all simulations are done 50-100 times. We will take the mean values as the simulation results.

At **Step 6**, we will analyze the simulation results obtained in Step 5. The simulation analysis has two purposes which result in two parts. One is to see the effectiveness and efficiency of the proposed algorithms. Another is to justify the correctness of the established mathematical models by examining the relationship among various communication and network parameters.

3. Discussion

This chapter just provides some preliminary knowledge concerning WDM networks and combinatorial optimization, which will help readers to understand the problems that we will study and the methods that we will use. To know more about multiwavelength optical networks, readers may refer to surveys [7, 9]. To get more detailed and technical materials in this area, readers may refer to two monographs [2, 6]. For complete knowledge of combinatorial optimization, readers may refer to the classic monograph [1].

References

[1] M. R. Garey and D. S. Johnson, *Computers and Intractability: A Guide to the Theory of NP-Completeness*, W. H. Freeman, San Francisco, CA, 1979.

[2] P. E. Green, *Fiber Optic Network*, Englewood Cliffs, NJ: Prentice Hall, 1993.

[3] K.-C. Lee and V. O. K. Li, A wavelength-convertible optical network, *Journal of Lightwave Technology*, 11 (5/6) (1993), 962-970.

[4] B. Mukherjee, WDM-Based local lightwave networks Part I: single-hop system, *IEEE Network*, 6 (3) (1992), 12-26.

[5] B. Mukherjee, WDM-based local lightwave networks Part II: multihop systems, *IEEE Network*, 6 (4) (1992), 22-32.

[6] B. Mukherjee, *Optical communication networks*, New York: McGraw-Hill, 1997.

[7] B. Mukherjee, WDM optical communication networks: progress and challenges, *IEEE Journal on Selected Areas in Communications*, 18 (10) (2000), 1810-1824.

[8] B. Mukherjee, Light-trees: optical multicasting for improved performance in wavelength-routed networks, *IEEE Communications Magazine*, 37 (2) (1999), 67-73.

[9] K. M. Sivalingam and S. Subramaniam, *Optical WDM Networks: Principles and Practice*, ed., Boston, London: Kluwer Academic Publishers, 2000.

[10] B. M. Waxman, Routing of multipoint connections, *IEEE Journal on Selected Areas in Communications*, 6 (9) (1988), 1617-1622.

Chapter 2

ROUTING FOR LOAD BALANCE

The load balancing problem, as one of the most significant and extensively studied problems in computer communication networks, is addressed not only in wavelength routed optical networks but also in general circuiot/packet-switched networks. However, this problem is particularly important for the optimization of the usage of some WDM network resources, such as wavelengths and bandwidths.

In this chapter, the load of a link in a WDM network, called *link load*, is defined as the number of connections over the link. The *network load* is defined as the maximum link load in the system. We will focus on the *Load Balancing Problem* (LBP) in single-hop systems. The problem is how to route a set of connections properly such that the network load is minimized. The importance of this problem in WDM networks is not only for reducing the network congestion but also for saving wavelength resources. Since the number of wavelengths required in the system is at least equal to the network load, the goal of assigning wavelengths to connections (i.e., lightpaths) by using the minimal number of wavelengths can be achieved through solving the load balancing problem.

The rest of this chapter is organized as follows. Section 1 discusses the load balancing problem in ring networks. The first part presents Schrijver et al's result [8] for the case of unidirectional connections. The second part presents Wilfong and Winkler's result [9] for the case of bidirectional connections. Section 2 presents Banerjee and Mukherjee's approach [1] for the load balancing problem in arbitrary networks. All these work adopt the same methodology. It consists of three steps as follows: First, it formulates the problem as an *Integer Linear Programming* (ILP). Secondly, it solves a *Linear Programming* (LP) obtained by relaxing the integer constraints of the ILP. Thirdly, it produces an integral solution to ILP (the original problem) from the obtained fractional solution to the IP by some rounding techniques. Rounding process can be done either in a deterministic way as in Section 1.1 where the returned solution is proved to be a $\frac{3}{2}$-approximation solution and in Section 1.2

where the output solution turns out to be an optimal solution, or in a random way as in Section 2 where the output solution can be shown to be a good approximation solution in the probabilistic sense. Section 3 concludes the chapter.

1. Ring Networks

In this section we study the load balancing problem in ring networks. $G(V, E)$ is now a ring of n nodes labelled clockwise $\{v_1 = 1, v_2 = 2, \cdots, v_n = n\}$. For the simplicity of presentation, all arithmetic involving node labels will be done implicitly using module n operations. The problem is particularly called *undirected ring routing problem* for the case of bidirectional connections and *directed ring routing problem* for the case of unidirectional connections, where the network load is called ring load.

1.1 Bidirectional Connections

In this subsection we assume bidirectional connections. Thus the route for a connection is a undirected path in the ring. We denote the path between nodes v_s and v_t by $[s, t] = \{v_i \mid s \leq i \leq t\}$ which also denotes the closed integer interval between s and t. For the simplicity of presentation, we denote by $[i, j)$ and $(i, j]$ the half-closed integer intervals $\{i, i+1, \cdots, j-1\}$ and $\{i+1, i+2, \cdots, j\}$, respectively. The problem of undirected ring routing can be formulated as follows.

PROBLEM 2.1 *Undirected Ring Routing Problem*

Instance A positive integers n and nonnegative integers $\lambda_{i,j} \geq 0, 1 \leq i < j \leq n$.
Solution A routing of demands $\{\lambda_{i,j}\}$, i.e., 0-1 assignment to x_{ij} for i, j.
Objective Minimizing ring load $L = \max\{L_k = A_k + B_k | 1 \leq k \leq n\}$, where
$A_k = \sum\{\lambda_{i,j} \mid x_{ij} = 1 \text{ and } k \in [i, j-1]\}$, and
$B_k = \sum\{\lambda_{i,j} \mid x_{ij} = 0 \text{ and } k \notin [i, j-1]\}$.

In the above, $\lambda_{i,j}$ is called a traffic *demand* of the connection between nodes v_i and v_j. Setting $x_{ij} = 1$ means to route the traffic $\lambda_{i,j}$ around the ring in clockwise direction (that is, through node v_{i+1}), while setting $x_{ij} = 0$ means to route the traffic $\lambda_{i,j}$ around the ring in anticlockwise direction (that is, not through node v_{i+1}). The routing $\{x_{ij}\}$ induces link load L_k on link $[k, k+1]$ and the ring load L. In particular, A_k (B_k) is the sum of traffics routed in clockwise (anticlockwise) direction that go through link $[k, k+1]$.

Fig. 2.1 shows an instance of Problem 2.1 and its solution studied in [8]. In Fig. 2.1(a), there are four nonnegative demands between nodes in a ring of 8 nodes, $\lambda_{1,6} = \lambda_{7,8} = 1$ and $\lambda_{2,5} = \lambda_{3,4} = 2$. The solution as shown in Fig. 2.1(b) routes demands $\lambda_{1,6}$, $\lambda_{3,4}$ and $\lambda_{7,8}$ in clockwise direction while $\lambda_{2,5}$ in anticlockwise direction. It induces the minimum ring load of 3 that meets at links $[3, 4]$ and $[7, 8]$.

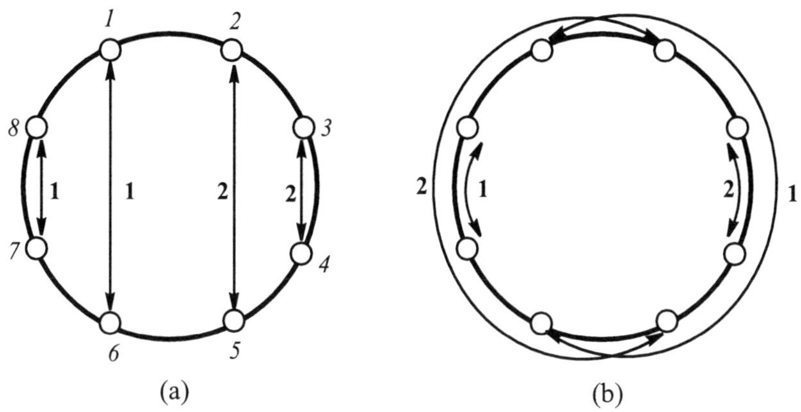

Figure 2.1. (a) An instance of the undirected ring routing problem, and (b) its solution.

For the case of $\lambda_{i,j} = \{0, 1\}$ for $1 \leq i, j \leq n$, the undirected ring routing problem can be solved in time polynomial of n (refer to the discussion in next subsection). In the following discussion, we will focus on the general case where $\lambda_{i,j} > 1$ for some i, j. The undirected ring routing problem turns out to be $\mathcal{NP}$-hard in the general case. A simple reduction is available from the *partition problem*, which is known to be $\mathcal{NP}$-complete as well. This problem is to decide whether a given set of positive integers can be partitioned into two subsets of equal sums of integers in the subsets. The following theorem [8] shows that undirected ring routing problem is $\mathcal{NP}$-hard.

THEOREM 2.1 *Problem 2.1 is $\mathcal{NP}$-hard.*

PROOF We consider the decision version of Problem 2.1. Given an instance of the problem and a bound $B > 0$, the problem is to decide if there is a routing method that produces the ring load $L \leq B$. We will show in the following that given a set of $a_1, a_2, \cdots, a_m$, it can be partitioned into two subsets of equal sum if and only if there is a routing method that makes ring load at most B.

Now we set traffic demands $\lambda_{i,m+2} = a_i$ for $1 \leq i \leq m$ and $\lambda_{m+1,m+2} = \lambda_{m+2,m+3} = \sum_i^m a_i/2$, and set all other demands equal to zero. In addition, set $n = m + 3$ and $B = \sum_i^m a_i$. See Fig. 2.2 for the demonstration of the reduction, where $m = 5$.

Suppose that there is a desired partition. Then we can construct a routing with ring load B as follows: route demands $\lambda_{m+1,m+2}$ and $\lambda_{m+2,m+3}$ in clockwise direction, and route all other demands in such a way that $L_{m+1} = L_{m+2} = B$.

Suppose that there is a desired routing with ring load B. It is easy to verify that demands $\lambda_{m+1,m+2}$ and $\lambda_{m+2,m+3}$ must be both routed in clockwise direction

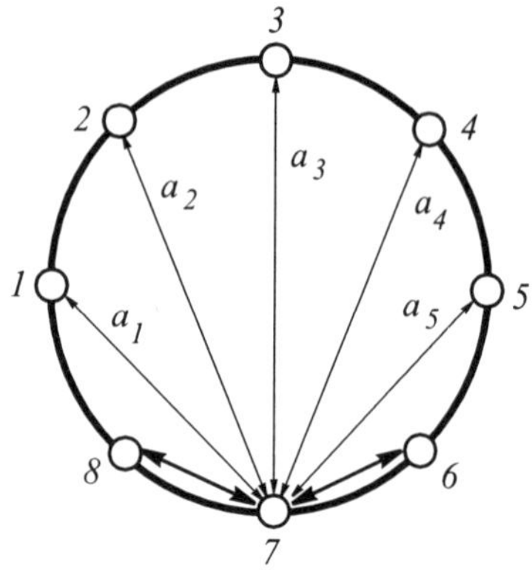

Figure 2.2. Reduction from the partition problem to Problem 2.1.

otherwise the ring load exceeds B. Moreover, it is clear that links $[m+1, m+2]$ and $[m+2, m+3]$ have equal link load B, i.e., $L_{m+1} = L_{m+2} = B$, otherwise one of the links must have a link load bigger than B. Now we can partition set $\{a_1, a_2, \cdots, a_m\}$ into two subsets of equal sums as follows: one subset consists of all demands a_i that are routed through $[m+1, m+2]$ while the other subset consists of all other demands. ■

In the following we will show that a theoretical guaranteed approximation solution to undirected ring routing problem can be obtained through solving its relaxed version. In the relaxed version, demands may be split, that is, traffics of a demand could be routed partly in clockwise direction and partly in anticlockwise direction.

PROBLEM 2.2 *Relaxed Undirected Ring Routing Problem*

Instance A positive integers n and nonnegative integers $\lambda_{i,j} \geq 0, 1 \leq i < j \leq n$.
Solution A routing of demands $\{\lambda_{i,j}\}$, i.e., 0-1 assignment to x_{ij} for i, j.
Objective Minimizing ring load $\overline{L} = \max\{\overline{L}_k = \overline{A}_k + \overline{B}_k | 1 \leq k \leq n\}$, where
$\overline{A}_k = \sum\{\lambda_{i,j} \,|\, \overline{x}_{ij} = 1 \text{ and } k \in [i, j-1]\}$, and
$\overline{B}_k = \sum\{\lambda_{i,j} \,|\, \overline{x}_{ij} = 0 \text{ and } k \notin [i, j-1]\}$.

Note that the ring load of an optimal solution to the relaxed version of Problem 2.2. provides a lower bound to the ring load of an optimal solution to the original version of Problem 2.1. These values in general are different. Consider a simple example shown in Fig. 2.3(a), where $n = 8$, $\lambda_{1,5} = \lambda_{3,7} = 1$ and all other demands are zero. Fig. 2.3(b) shows an optimal routing of Problem 2.1 that induces ring load 2 and Fig. 2.3(c) shows an optimal routing of Problem 2.2 that induces ring load 1, where each demand is split into two separate demands both carrying half of the traffic of the given demand. Notice that even if an optimal solution to Problem 2.2

has the same ring load as an optimal solution to Problem 2.1, it may be not be an optimal solution to Problem 2.1.

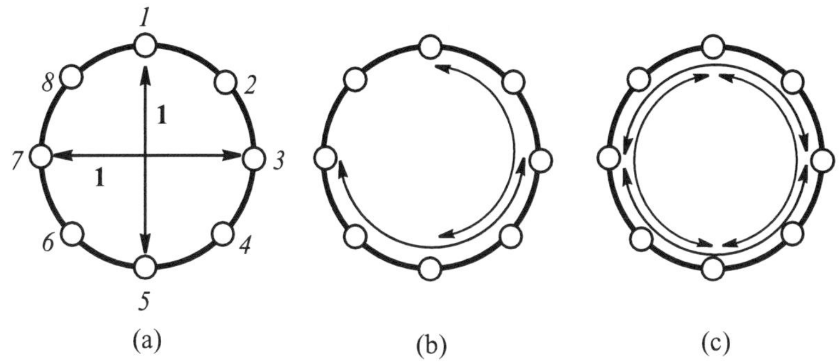

Figure 2.3. (a) An instance of undirected ring routing. (b) An optimal solution to the original version. (c) An optimal solution to the relaxed version.

Since Problem 2.2 can be formulated as a linear programming problem, it is solvable in polynomial time. In fact, we will see that an optimal solution can be obtained in a very fast greedy way, even if we require that it satisfies some additional properties. We will use the obtained optimal solution of Problem 2.2 to produce an approximate solution to Problem 2.1 whose ring load is not far away from the ring load of the optimal solution to Problem 2.1. To show how to achieve this, we introduce some useful concepts. Two demands $\lambda_{a,b}$ and $\lambda_{c,d}$ are said to *cross* each other if all the indices are distinct and exactly one of c and d lies in $[a, b]$ (as the case shown in Fig. 2.3(a) where $\lambda_{1,5}$ and $\lambda_{3,7}$ cross); Otherwise they are said to be *parallel.* In addition, a link which lies between two parallel demands is said to be *between* the demands. Finally, a routing $\{\overline{x}_{i,j}\}$ for Problem 2.2 is said to *split* a demand $\lambda_{a,b}$ if $0 < \overline{x}_{a,b} < 1$, and it is called a *fractional* routing.

The following lemma [8] characterizes the fractional routing of the relaxed version of Problem 2.2 that we need to build up an integral routing of the original version of Problem 2.2 (i.e., Problem 2.1).

LEMMA 2.1 *Let $\overline{x}_{opt}$ be an optimal routing for an instance of Problem 2.2 that induces link load $\overline{L}_k^{opt}$. Suppose that it is also minimal in the sense that no other routing has link load $\overline{L}_k \leq \overline{L}_k^{opt}$ for every k and $\overline{L}_i < \overline{L}_i^{opt}$ for some i. Then no link which lies between two parallel demands carries traffic from both demands.*

PROOF Assume otherwise, that there exists a pair of parallel demands $\lambda_{a,b}$ and $\lambda_{c,d}$ and link $[i, i+1]$ carries a quantity λ_1 of traffic from demands $\lambda_{a,b}$ and λ_2 from demand $\lambda_{c,d}$, where $\lambda_2 \geq \lambda_1$. Now rerouting a quantity λ_1 of traffic from each demand produces a new routing in which no traffic from demand $\lambda_{a,b}$ will go through link $[i, i+1]$ and traffic from demand $\lambda_{c,d}$ is split (one carries λ_1 while the other $\lambda_2 - \lambda_1$). It is easy to verify that no link has load increased and link load of $[i, i+1]$ is decreased by $2\lambda_1$. This contradicts the minimality of $\overline{x}_{opt}$. ■

Note that Lemma 2.1 does not hold for the original Problem 2.1. This can be seen from the example shown in Fig. 2.1, where links $[1,2]$ lies between two parallel demands $\lambda_{2,5}$ and $\lambda_{1,6}$, but it carries traffics from both demands.

Given a set of demands of connections, we now wish to find an optimal routing that satisfies the condition of Lemma 2.1. This can be done in an efficient way by putting each link in a tight *cut*. This concept is introduced from a more general version of Problem 2.2, where a capacity c_k of each link $[k, k+1]$ is appended to the instance of Problem 2.2. The problem now asks for a routing whose ring load $\overline{L}$ is minimal and link load $\overline{L}_k \leq c_k$ for each k. Each pair of links $[p, p+1]$ and $[q, q+1]$ with $p < q$ constitutes a *cut* of capacity $c_p + c_q$. Here, we may think of a cut as a chord connecting the midpoints of links $[p, p+1]$ and $[q, q+1]$; If a demand $\lambda_{i,j}$ crosses this chord, any routing will contribute load $\lambda_{i,j}$ to the cut. See Fig. 2.4.

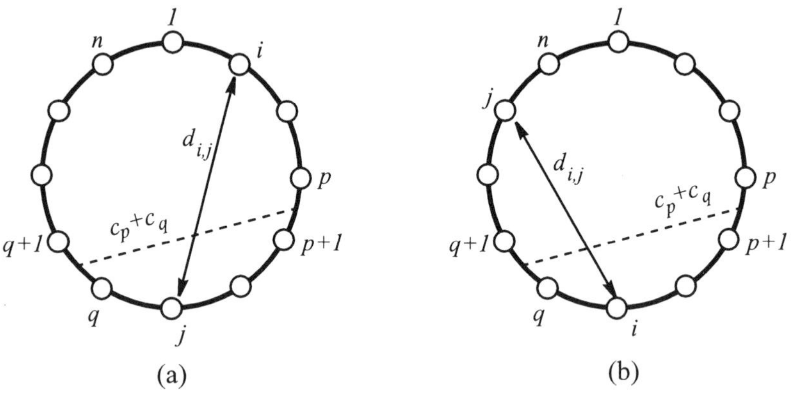

Figure 2.4. Two cases of a demand $\lambda_{i,j}$ crossing the cut.

Thus, if the instance of this general problem has a feasible routing, then the total traffic demand across the cut $\Lambda_{p,q} \equiv A'_{p,q} + B'_{p,q}$, where

$$\begin{aligned} A'_{p,q} &= \sum\{\lambda_{i,j} \,|\, i \leq p \text{ and } j \in (p,q]\}, \\ B'_{p,q} &= \sum\{\lambda_{i,j} \,|\, i \in (p,q] \text{ and } j > q\}. \end{aligned}$$

Clearly, $\Lambda_{p,q} \leq c_p + c_q$. In fact, the following lemma [8] claims that the converse is also true.

LEMMA 2.2 *Suppose that $\Lambda_{p,q} \leq c_p + c_q$ holds for each cut. Then there is a solution to Problem 2.2 satisfying the capacity constraint.*

PROOF Suppose, by contradiction argument, that there is an instance such that $\Lambda_{p,q} \leq c_p + c_q$ holds for each cut but no solution satisfies the capacity constraints. We consider a counterexample with minimal n and, subject to the minimality of n, having the least number of nonzero demands. In the following discussion, it will be useful to allow a degenerated cut $\{p, p\}$ having capacity $2c_p$ and demand $\Lambda_{p,p} = 0$. The cut constraint for these degenerated cuts is thus equivalent to nonnegative of the link capacities.

Now we take any nonzero demand, say $\lambda_{i,j} > 0$ and let $\{p, q\}$ minimize $\Delta = \Lambda_{p,q} - c_p - c_q$ subject to $i \leq p < q < j$; Thus $\{p, q\}$ is the tightest cut in the route for $\lambda_{i,j}$ in clockwise direction. Here a cut $\{p, q\}$ is said to be *tight* if $\Lambda_{p,q} = c_p + c_q$.

We then send $\min(\Lambda_{p,q}, \Delta/2)$ of the demand $\Lambda_{p,q}$ in clockwise direction and, if $\Lambda_{p,q} > \Delta/2$, send the remaining $\Lambda_{p,q} - \Delta/2$ in anticlockwise direction. When the capacities are decreased accordingly, we will have a new instance with one less nonzero demand $\lambda_{i,j} = 0$. If the new instance still satisfies the cut constraints, this will contradict the minimality of the number of nonzero demands.

Suppose that in the new instance some cut is violated. That cut must lie on the route for $\lambda_{i,j}$ in anticlockwise direction, since this demand has already been accounted for in cuts which it crosses, and cuts on the route in clockwise direction have sufficient slack by the way of choosing Δ. Then we have a cut $\{p', q'\}$ with $[p', q') \cap [i, j) = \emptyset$ such that

$$\Lambda_{p',q'} + 2(\lambda_{i,j} - \Delta/2) > c_{p'} + c_{q'},$$

where all quantities are computed in the original instance.

Call the cuts $\{p, q\}$ and $\{p', q'\}$ '*straight*' and the cuts $\{p, p'\}$ and $\{q, q'\}$ '*diagonal*'. Every demand must cross at least as many of the two diagonal cuts as the two straight cuts, where $\lambda_{i,j}$ crosses both diagonal cuts and neither straight cut. Hence,

$$\begin{aligned} \Lambda_{p,p'} + \Lambda_{q,q'} &\geq \Lambda_{p,q} + \Lambda_{p',q'} + 2\lambda_{i,j} \\ &> c_p + c_q - 2\frac{\Delta}{2} + c_{p'} + c_{q'} - 2(\lambda_{i,j} - \frac{\Delta}{2}) + 2\lambda_{i,j} \\ &= c_p + c_{p'} + c_1 + c_{q'} \end{aligned}$$

This means that one of the diagonal cuts must have violated the cut constraint. Note that nonviolation of degenerate cut $\{p, p\}$ ensures that the given routing of $\lambda_{i,j}$ is actually possible, that is, no link capacity will become negative afterward. ■

The following lemma [8] claims that a routing $\{\overline{x}_{i,j}\}$ which minimizes ring load $\overline{L}$ and satisfies the property stated in the conclusion of Lemma 2.2 can be found in an efficient way.

LEMMA 2.3 *There is an algorithm with running time $O(mn^2)$ for Problem 2.2 that can produce a solution minimizing $\overline{L}$ and satisfying the conclusion of Lemma 2.2, where m is the number of nonzero demands.*

PROOF This can be realized through putting each link in a tight cut as follows.

First, we compute all $\binom{n}{2}$ possible values $\Lambda_{p,q}$ for $1 \leq p < q \leq n$, and let the largest of these values be Δ. Since the ring with all capacities set to $\Delta/2$ satisfies the cut constraint, $\overline{L} = \Delta/2$. We now take the links in any order and lower their capacities as much as possible, that is, define capacities $\{c_i\}$ recursively by

$$c_p = \max\Big(\max q < p(\Lambda_{p,q} - c_q), \max q > p(\Lambda_{p,q} - \Delta/2)\Big).$$

Note that the obtained set $\{c_i'\}$ satisfies $c_i' \leq c_i$ for every i and $c_j' < c_j$ for some j, since the least such j would be part of a bad cut. Hence any feasible routing $\{\overline{x}_{i,j}\}$ satisfying these capacities is minimal solution of Problem 2.2 instance, and Lemma 2.2 applies. In particular, if $S = \{\{i,j\} \mid \lambda_{i,j}$ is split by $\{\overline{x}_{i,j}\}\}$, then every pair of chords in S crosses, this implies $|S| \leq n/2$.

Secondly, after reducing the capacities as above we can solve Problem 2.2 by routing each demand all clockwise or all anticlockwise until only mutually pairwise crossing demands remain. To see this, assume that there is still a parallel pair of unrouted demands, we choose a link between them and fix a tight cut containing that link. At most one of the two parallel demands crosses the cut; the other must be routed to miss the cut entirely.

Accordingly, an algorithm for solving Problem 2.2 can proceed as follows:

(1) Compute $\Lambda_{i,j}$ for $1 \leq i < j \leq n$ and $\overline{L} := \Delta/2$.

(2) Compute the minimal capacities $\{c_i\}$ as described above.

(3) While there are pairs of parallel demands, find the tightest cuts and route demands all clockwise or anticlockwise, and then reset capacities accordingly.

(4) When there is only one crossing cut, route as much as possible clockwise and the remainder anticlockwise.

The running time of this procedure is approximately of order mn^2, where m is the number of nonzero demands. ■

In any case, the algorithm proposed in Lemma 2.3 outputs a solution with at most only $n/2$ of the demands split. In the following we will show how to compute $\{\overline{x}_{i,j}\}$ and then '*unsplit*' the demands in S as carefully as possible in order to get a near-optimal routing for the original Problem 2.1.

From now on $\{\overline{x}_{i,j}\}$ will be fixed which is an optimal solution to Problem 2.2 with a set of split demands S specified as in the proof of Lemma 2.3. We will try to modify fractional routing $\{\overline{x}_{i,j}\}$ into an integral routing $\{x_{i,j}\}$ of Problem 2.1 in such a way that $L - \overline{L}$ is as small as possible.

If node k is not an endpoint of a split demand in S, then the difference between the loads on links $[i-1,i]$ and $[k,k+1]$ remains unchanged when $\{\overline{x}_{i,j}\}$ becomes $\{x_{i,j}\}$. Hence, node k can be removed from the original ring and these two links incident to k is now replaced by one single link in the new ring whose load becomes $\max(\overline{L}_{k-1}, \overline{L}_k)$. This operation can be repeated until every node in the ring is an endpoint of some split demand. As a result, $S = \{\{k, k+m\} \mid 1 \leq k \leq m\}$ and n is even.

Let us now define α_i and β_i to be the amount traffics of demand $\lambda_{i,i+m}$ routed clockwise and anticlockwise by $\{\overline{x}_{i,j}\}$, respectively. Then $\alpha_i, \beta_i > 0$ and $\alpha_i + \beta_i = \lambda_{i,i+m}$. If $\lambda_{i,i+m}$ is routed clockwise by $\{x_{i,j}\}$, then each link $[j, j+1]$ with $j \in [i, i+m)$ has its load increased by β_i which is the amount traffic routed anticlockwise by $\{\overline{x}_{i,j}\}$; At the same time, the rest of the links have their loads decremented by β_i. Similarly, if demand $\lambda_{i,i+m}$ is routed anticlockwise by $\{x_{i,j}\}$, then the load of each link in $[i, i+m]$ is decreased by α_i while the rest are increased by the same amount.

Therefore, if we set $\gamma_i = \beta_i$ when $x_{i,i+m} = 1$ and $\gamma_i = -\alpha_i$ otherwise, then we have

$$\Lambda_j = \overline{\Lambda}_j + \sum_{\substack{i\in[1,m]\\ j\in[i,i+m)}} \gamma_i - \sum_{\substack{i\in[1,m]\\ j\in[i+m,i)}} \gamma_i.$$

Notice that

$$L_j + L_{j+m} = \overline{L}_j + \overline{L}_{j+m}, \quad \text{for all } j.$$

Thus $L \leq 2\overline{L}$ for all choices of $\{x_{i,j}\}$. This implies that this method finds a solution to Problem 2.1 whose ring load is at most two times that of an optimal solution which is at least equal to $\overline{L}$. In fact, the following theorem [8] claims that we can do much better than that.

THEOREM 2.2 *Let $\{\overline{x}_{i,j}\}$ be an optimal solution with ring load $\overline{L}$ to Problem 2.2, and let δ be the maximum traffic of the demands split by $\{\overline{x}_{i,j}\}$. Then there is an polynomial-time algorithm for Problem 2.1 which produces a solution $\{x_{i,j}\}$ with ring load L such that $x_{i,j} = \overline{x}_{i,j}$ for all unsplit demands and $L - \overline{L} \leq \frac{3}{2}\Delta$.*

PROOF We set γ_i and $\{\overline{x}_{i,j}\}$ inductively so that

$$\sum_{i=1}^{k} \gamma_i \in [-\frac{\Delta}{2}, \frac{\Delta}{2}], \quad \text{for all } k,\ 1 \leq k \leq m.$$

This can be realized since, once $\gamma_1, \gamma_2, \cdots, \gamma_{k-1}$ are set and the partial sum $s = \sum_{i=1}^{k-1} \gamma_i$ lies in the required interval, the two possible values of $\sum_{i=1}^{k} \gamma_i$ lie on both sides of s and differ by only $\alpha_k + \beta_k \leq \Delta$. Now set

$$\Gamma_k := \sum_{i=1}^{k} \gamma_i - \sum_{i=k+1}^{m} \gamma_i = 2\sum_{i=1}^{k} \gamma_i - \sum_{i=1}^{m} \gamma_i \in [-\frac{3}{2}\Delta, \frac{3}{2}\Delta] \text{ and } \Gamma := \max_{1\leq k\leq m} |\Gamma_k|,$$

then we have

$$L - \overline{L} \leq \max_j (L_j - \overline{L}_j) = \Gamma \leq \frac{3}{2}\Delta.$$

The proof is then finished. ■

1.2 Unidirectional Connections

In this subsection we assume unidirectional connections. Thus the route for a connection is a directed path from the source to the destination. The formulation of the problem and its solution approach both are almost the same as in the case of bidirectional connections. Thus to make the presentation different, here we will only consider connections with uniform traffics and assume that all connections have traffic either one or zero. This restricted case is important because in some cases connections with nonuniform traffics can be split, but only at integral values. Thus this special case can be regarded as a multiplicity of connections with uniform traffics. We may consider $G(V, E)$ as a digraph consisting of two oppositely directed rings on the same set of n nodes labelled $\{v_1 = 1, v_2 = 2, \cdots, v_n = n\}$ in clockwise. Thus we define $[s, t]$ to be $\{v_i \mid s \leq i \leq t\}$ when $s < t$ and $[s, n] \cup [1, t]$ otherwise.

PROBLEM 2.3 *Directed Ring Routing Problem*

Instance Two positive integers n and m, and ordered pairs $(s_1, t_1), \cdots, (s_m, t_m)$, where $s_i \neq t_i$ and $1 \leq s_i, t_i \leq n$.
Solution A routing of connections from s_i to t_i, i.e., 0-1 assignment to x_i for i.
Objective Minimizing ring load $L = \max\{\max_k A_k, \max_k B_k\}$, where
$A_i \equiv |\{i \mid k \in [s_i, t_i - 1] \text{ and } x_i = 1\}|$, and
$B_i \equiv |\{i \mid k \in [t_i, s_i - 1] \text{ and } x_i = 0\}|$.

In the above, setting $x_i = 1$ means to route connection from s_i to t_i clockwise around the ring, and setting $x_i = 0$ indicates to route the connection anticlockwise around the ring. Any assignment of 0 and 1 to $\{x_i\}$ corresponds a routing of m connections. Hence A_k is the number of connections routed through (clockwise) link $k \to k+1$, which is its link load. Similarly, B_k is the number of connections routed through (anticlockwise) link $k+1 \to k$, which is its link load. The maximum of these two values is L that gives the ring load.

In the following we will show how to find an optimal solution to Problem 2.3 by applying the same approach as used for undirected ring routing problem. We first relax the integer constraints on variables x_i, and obtain the *relaxed* version of Problem 2.3. The optimal routing of the relaxed version can *split* a connection, we may consider this situation as that some fraction of data of the connection is sent in clockwise directed path while others is sent in anticlockwise directed path.

PROBLEM 2.4 *Relaxed Directed Ring Routing Problem*

Instance Two positive integers n and m, and ordered pairs $(s_1, t_1), \cdots, (s_m, t_m)$, where $s_i \neq t_i$ and $1 \leq s_i, t_i \leq n$.
Solution An assignment of a real value between 0 and 1 to $\overline{x}_i$ for $1 \leq i \leq m$.
Objective Minimizing ring load $\overline{L} = \max\{\max_k \overline{A}_k, \max_k \overline{B}_k\}$, where
$\overline{A}_i \equiv \sum\{\overline{x}_i \mid k \in [s_i, t_i - 1]\}$ and
$\overline{B}_i \equiv \sum\{(1 - \overline{x}_i) \mid k \in [t_i, s_i - 1]\}$.

We then can find an optimal solution to Problem 2.4, which can be formulated as a linear programming and be solved in polynomial-time. Let L_{opt} and $\overline{L}_{opt}$ denote the optimal ring load Problem 2.3 and Problem 2.4, respectively. Clearly, $\lceil \overline{L}_{opt} \rceil \leq L_{opt}$. However, instead of producing an optimal fractional routing of Problem 2.4, our strategy is to produce in polynomial-time a fractional routing $\{\overline{x}'_i\}$ that still satisfies $\overline{L}' \leq L_{opt}$ and have some additional properties. These properties enable us to modify $\{\overline{x}'_i\}$ into an integral routing, which can be proved to be an optimal routing of Problem 2.4. A fractional routing $\{\overline{x}_i\}$ is called *flush* if its sum of components $\sum_{i=1}^{m} \overline{x}_i$ is an integer. The following lemma [9] shows that a flush routing satisfying with upper bounded ring load can be efficiently produced.

LEMMA 2.4 *Given an instance of Problem 2.4, a flush routing $\{\overline{x}'_i\}$ with ring load $\overline{L}' \leq L_{opt}$ can be found in polynomial-time.*

PROOF It is clear that there is such a routing since any optimal routing of Problem 2.4 is a flush routing. For each possible value $F = 0, 1, \cdots, m$, we consider the flush enforced version of Problem 2.4 by adding the equality $\sum_{i=1}^{m} x_i = F$ to the constraints of Problem 2.4, which is still a linear programming. Hence in polynomial-time we can obtain an optimal routing $\{\overline{x}_i^F\}$ with ring load $\overline{L}_F$. We now choose a routing $\{\overline{x}'_i\}$ among these $m + 1$ flush routings with the minimal ring load $\overline{L}' = \min\{\overline{L}_F \mid F = 0, 1, \cdots, m\}$. Clearly, $\overline{L}' \leq L_{opt}$.

In fact, we can lower the upper bound $\overline{L}'$ by 1/2. Observe that when f takes any real number in $[0, m]$, the minimum ring load function $\overline{L}^f$ is concave. This is true, since if a routing $\overline{X}^f = \{\overline{x}_i^f\}$ yields ring load $\overline{L}_f$ and $\overline{Y}^g = \{\overline{y}_i^g\}$ yields $\overline{L}_g$, then for any λ between 0 and 1, $\lambda\overline{X}^f + (1-\lambda)\overline{Y}^g$ yields a ring load at most $\lambda\overline{L}_f + (1-\lambda)\overline{L}_g$. This means, $\overline{L}_{f+(1-\lambda)g} \leq \lambda\overline{L}_f + (1-\lambda)\overline{L}_g$.

Moreover, note that $\overline{L}_f$ is piecewise linear with slopes bounded by 1 in absolute value, since for any pair of two routings $\{\overline{x}'_i\}$ and $\{\overline{x}_i\}$ with $x'_i \geq x_i$ for each i, we have the following inequality

$$\overline{L}' \leq \overline{L} + \sum_{i=1}^{m} \overline{x}'_i - \sum_{i=1}^{m} \overline{x}_i.$$

Thus the obtained flush ring load $\overline{L}'$ cannot be more than $\overline{L}_{opt} + 1/2$, but may exceed $\lceil \overline{L}_{opt} \rceil$.

It follows that if $\overline{L}_f$ achieves the minimum at $f = r \in [0, 1, \cdots, m]$, then one of the values $\lfloor r \rfloor$ and $\lceil r \rceil$ must be the optimal solution to the minimization of function $\overline{L}_f$ under constraint $f \in \{0, 1, \cdots, m\}$. Therefore, if r is not an integer, we can just check those two values and take the less one. The proof is then finished. ■

For the simplicity of presentation, we now consider connections as directed chords in a circle representing the ring. Two connections (s_i, t_i) and (s_j, t_j) are said to be *parallel* if the intervals $[s_i, t_i]$ and $[t_j, s_j]$, or the intervals $[t_i, s_i]$ and $[s_j, t_j]$, intersect

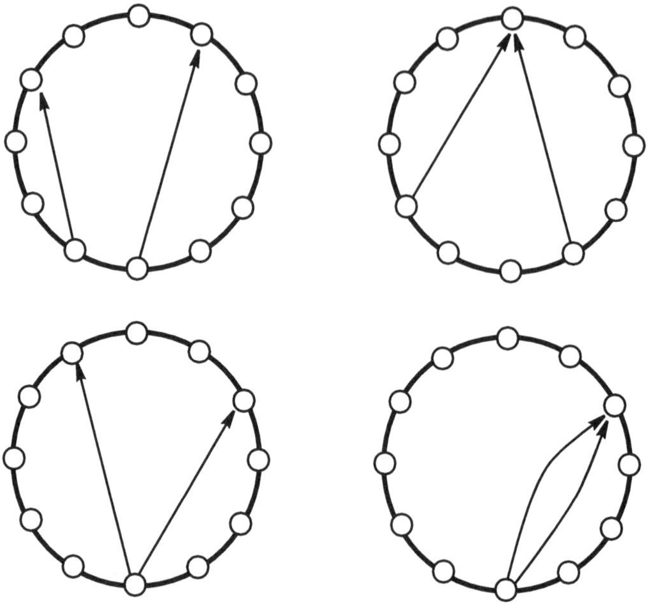

Figure 2.5. Parallel connection pairs.

at most at their endpoints. There are essentially four possible configurations of a pair of parallel connections. See Fig. 2.5, where it may be $s_i = s_j$ and $t_i = t_j$. In addition, we can consider a link as a chord. Thus a connection is said to be *parallel* to the link if the connection can be routed through that link. It is easy to see that any link partitions the connections into two groups: 1) those parallel to the link, and 2) those parallel to the reverse direction of the link.

In the following as in the preceding subsection, a fractional routing $\{\overline{x}_i'\}$ is said to *split* connection from s_i to t_i if $0 < \overline{x}_j' < 1$. The following lemma [9] shows that a flushing routing with more desired properties can be efficiently produced.

LEMMA 2.5 *Given an instance of Problem 2.4 and a flush routing $\{\overline{x}_i\}$ with ring load $\overline{L}$, a flush routing $\{\overline{x}_i'\}$ can be found in polynomial-time with ring load $\lceil \overline{L}' \rceil \le \overline{L}$, satisfying the property that no two parallel connections are both split.*

PROOF Clearly, we can assume that in any set of identical connections at most one is split. Thus suppose that there is a pair of unequal, parallel connections (s_i, t_i) and (s_j, t_j) with $0 < \overline{x}_i, \overline{x}_j < 1$. We will reroute these connections in such a way that one of them is no longer split, but their collective contribution to every link load is either maintained or reduced, and the routing sum $\sum_{i=1}^{m} \overline{x}_i$ remains unchanged.

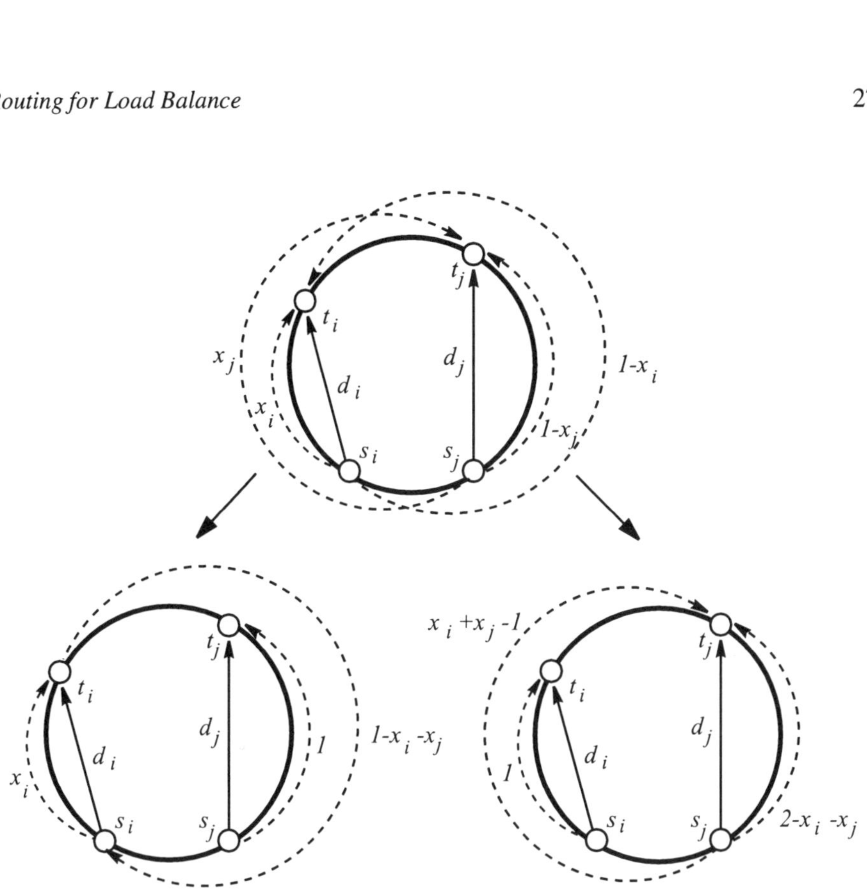

Figure 2.6. Untangling routings of split parallel connection pairs.

Since all connections are distinct, we may assume that the intervals $[s_i, t_i]$ and $[t_j, s_j]$ intersect in at most one node and do not cover the ring. We consider the two cases separately.

Case 1. $\overline{x}_i \leq 1 - \overline{x}_j$. Then we define a new routing $\{\overline{x}'_i\}$ by $\overline{x}'_i = \overline{x}_i + \overline{x}_j, \overline{x}'_j = 0$, and $\overline{x}'_i = \overline{x}_k$ for $k \notin \{i, j\}$. Thus links in $[s_i, t_i]$ and $[t_j, s_j]$ have the same load as before and all other links have the same or reduced loads.

Case 2. $\overline{x}_i > 1 - \overline{x}_j$. Then we define a new routing $\{\overline{x}'\}$ by $\overline{x}'_i = 1$, $\overline{x}'_j = \overline{x}_i + \overline{x}_j - 1$, and $\overline{x}'_i = \overline{x}_k$ for $k \notin \{i, j\}$. This will lead to the same result.

These two untangling routing methods of parallel connection pairs are illustrated in Fig. 2.6, where the connections have distinct endpoints. Since each untangling reduces by one (or two) the total number of split connections, at most m such procedures will produce the desired routing. ■

Now we are ready to prove the major result of this subsection obtained in [9].

THEOREM 2.3 *Problem 2.4 is polynomial-time solvable.*

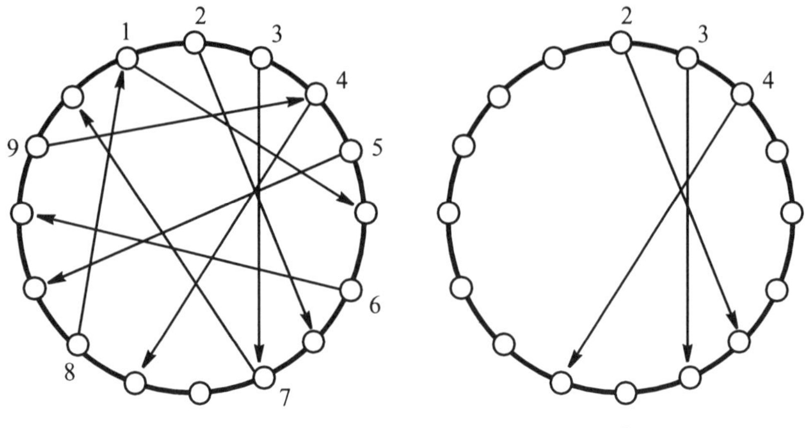

Figure 2.7. Parallel pairs of connections.

PROOF Due to Lemma 2.5 we can assume that a flush routing $\{\overline{x}'_i\}$ can be obtained in polynomial-time that has link loads $\overline{A}'_k$, $\overline{B}'_k$ and ring load $\overline{L}' \leq L_{opt}$, and satisfies the condition that no two parallel connections are both split. Since two nonparallel connections cannot share a source, the number of connections split by $\{\overline{x}'_i\}$ is at most n, the size of ring.

We assume, without loss of generality, that m given connections are labelled in such a way that the set of split connections is $S \equiv \{r_1, \cdots, r_q\}$. Moreover, since no two connections are parallel, we may order them clockwise simultaneously by source s_i and by destination t_i, as shown in Fig. 2.7. Hence for any clockwise link $k \to k+1$ there is an interval $[i_k, j_k] \subseteq \{1, \cdots, q\}$, interpreted if necessary "around the corner" modulo q which contains exactly the indices of the connections in S which are parallel to the link. For its anticlockwise counterpart $k+1 \to k$, the indices of the parallel links are just those in the complement of $[i_k, j_k]$, namely the interval $[j_k + 1, i_k - 1]$.

Note that unsplitting of the connections parallel to a link will affect the load on that link. Suppose that $\{x_i\}$ is the 0-1 routing with link loads A_k and B_k obtained from some unsplitting of the split requests of $\{\overline{x}'_i\}$. Then we have

$$A_k = \overline{A}'_k + \sum_{i \in [i_k, j_k]} (x_i - \overline{x}'_i),$$

$$B_k = \overline{B}'_k + \sum_{i \notin [i_k, j_k]} (\overline{x}'_i - x_i).$$

We now define an unsplitting $\{x_i\}$ recursively by setting

$$x_j = \begin{cases} 1, & \text{if } -\overline{x}'_j + \sum_{i=1}^{j-1}(x_i - \overline{x}'_i) < -1/2; \\ 0, & \text{otherwise.} \end{cases}$$

Then every partial sum $\sum_{i=1}^{j-1}(x_i - \overline{x}'_i)$ lies in the half-open real interval $[-1/2, 1/2)$. We study the following two cases separately.

Case 1. $i_k \leq j_k$. We have the following inequality as desired

$$A_k - A'_k = \sum_{i=1}^{j_k}(x_i - \overline{x}'_i) - \sum_{i=1}^{i_k-1}(x_i - \overline{x}'_i) < 1/2 - (-1/2) = 1.$$

Case 2. $i_k > j_k$. We have

$$A_k - A'_k = \sum_{i=1}^{q}(x_i - x'_i) + \sum_{i=1}^{j_k}(x_i - \overline{x}'_i) - \sum_{i=1}^{i_k-1}(x_i - \overline{x}'_i),$$

which lies in the half-open interval $[-3/2, 3/2)$. It follows from that $\{\overline{x}'_i\}$ is a flush routing and $\{x_i\}$ is an integral routing. Note that $\sum_{i=1}^{q}(x_i - \overline{x}'_i)$ is zero, since the value is an integer lying in interval $[-1/2, 1/2)$. Hence we have $A_k - \overline{A}'_k < 1$ as in Case 1.

In addition, using a symmetric argument for the anticlockwise links shows that $B_k - \overline{B}'_k < 1$ as well. Therefore, the ring load L induced by $\{x_i\}$ satisfies $L < \overline{L}' + 1$. In the end, due to Lemma 2.4 we can conclude that $L \leq \lceil \overline{L}' \rceil \leq L_{opt}$ since L is integral. And then the proof is finished. ■

2. General Networks

In this section, we focus on the load balancing problem in WDM networks with an arbitrary topology. It can be formulated as follows.

PROBLEM 2.5 *Load Balancing Problem*

Instance A network $G(V, E)$ and a set of m positive numbers $\{\lambda_{s,d} > 0\}$ representing the traffic of connection between source s and destination d.

Solution A routing of $\{\lambda_{s,d}\}$, that is a set of m paths in $G(V, E)$ between s and d representing the routes carrying the traffic of connection between s and d.

Objective Minimizing the network load.

The above problem is $\mathcal{NP}$-hard, since when restricted in ring network it is $\mathcal{NP}$-hard due to Theorem 2.1. We now formulate it as a problem of integer linear programming (ILP).

$$\begin{array}{lll} \text{Minimize} & L & (2.1) \\ \text{Subject to} & L \geq L_{i,j} \equiv \sum_{s,d} F_{ij}^{sd}, \quad \forall\, (i,j) \in E & (2.2) \\ & F_{ij}^{sd} = 0, \text{or } \lambda_{sd}, \quad \forall\, (i,j) \in E & (2.3) \\ & \sum_i F_{ij}^{sd} - \sum_k F_{jk}^{sd} = \begin{cases} -\lambda_{sd}, & \text{if } s = j, \\ \lambda_{sd}, & \text{if } d = j, \\ 0, & \text{otherwise.} \end{cases} \quad \forall\, \lambda_{sd} \text{ and } j. & (2.4) \end{array}$$

In the above, F_{ij}^{sd} denotes the traffic flowing from source s to destination d on link (v_i, v_j). Inequality (1.2) means that the network load L is the maximum of link loads $L_{i,j}$ over all links. Equality (1.3) means that traffic from source s to destination d is routed on link (v_i, v_j) if $F_{ij}^{sd} = \lambda_{sd}$, or not if $F_{ij}^{sd} = 0$. This also implies that the traffic between any source-destination pair is not allowed to split. Equality (1.4) means that at any intermediate node v_j in the path carrying traffic λ_{sd}, the traffic flowing into v_j equals the traffic flowing out v_j except at source node s and destination node d.

Problem 2.5 is a special case of the *Multicommodity Flow Problem* [4]. The problem studies how to ship several different commodities from their respective sources to their destinations in a given network with the total amount of flow going through a link limited by its capacity. In the formulation of Problem 2.5, the amount of each commodity we wish to ship from a source node to a destination node can be considered as the traffic of a connection between them, and the capacity is the traffic of connections that a link can support, which is known as *bandwidth*. An optimization version of multicommodity flow problem is *Concurrent Flow Problem* in which the goal is to find the maximum percentage ρ such that at least ρ percent of each demand of commodity can be shipped without violating the capacity constraint. When we focus on the concurrent flow problem with unit capacities, it is equivalent to the problem of finding a flow (disregarding capacities) that minimizes the maximum total flow, called the *congestion*, on any link.

It was proved that the concurrent flow problem with unit capacity is $\mathcal{NP}$-hard. However, when the integer constraint on flow is removed and the problem is reduced to a problem of linear programming (LP). Thus this nonintegral version of the problem can be solved in polynomial-time by using any linear programming method. Hence, the most popular way of solving the concurrent flow problem is to first solve nonintegral version of the problem and then reroute fractional flows by either using a deterministic method or a randomized technique.

Following this idea we can develop an algorithm [1] for solving Problem 2.5 as follows. First, we relax the requirement of the integral flows, that is, F_{ij}^{sd} is allowed to split and flow in different routes, and then solve the relaxed version of Problem 2.5. Secondly, note that in the obtained optimal solution to the relaxed version, some fractions of traffic λ_{sd} of connection between source s and destination d may flow through different paths connecting s and d. We find and use those paths as a set of candidate paths to route the traffic λ_{sd} of connection between s and d. In the end, we make a biased dice such that each face corresponds a candidate path and has a probability to face up, where the probability is based on the amount of traffic flowing over the path. And then we toss the dice to select the path, that faces up, over which to route traffic λ_{sd} of connection between s and d.

ALGORITHM 2.1 *Routing for Load Balance*

Step 1 *Solving the relaxed version of Problem 2.5*
  Obtain an optimal (fractional) routing by solving a LP,
  Set $f_{sd}(e)$ to be the flow of traffic λ_{sd} on edge $e \in E$.
Step 2 *Finding a set of paths that carry fractional flows of traffic* λ_{sd}
  for traffic λ_{sd} with $\lambda_{sd} > f_{sd}(e) > 0$ for some $e \in E$ **do**
    $P_{sd} := \emptyset$.
    $E' := \{e \in E \,|\, f_{sd}(e) > 0\}$.
    while $E' \neq \emptyset$ **do**
      $f_m := \min\{f_{sd}(e) \,|\, e \in E'\}$.
      find a shortest path between s and d in subgraph $G(V, E')$
        in terms of distance function $f_{sd}(e)$
      put the path into P_{sd} along with a weight f_m/λ_{sd}.
      $f_{sd}(e) := f_{sd}(e) - f_m$ for each $e \in E'$,
      $E' := \{e \in E \,|\, f_{sd}(e) > 0\}$.
    end-while
  end-for
Step 3 *Rerouting traffics of connections with fractional flows*
  for traffic λ_{sd} with $\lambda_{sd} > f_{sd}(e) > 0$ for some $e \in E$ **do**
    make a dice of $|P_{sd}|$ faces each of which corresponds a path in P_{sd},
    set the probability that a face shows up to the weight of the path.
    cast the dice,
    reroute traffic λ_{sd} of connection between s and d over the path
      whose face appears up.
  end-for

Let $\overline{L}_{opt}$ be the network load of an optimal solution to the relaxed version of Problem 2.5. Then it can be used as a lower bound on the network load of an optimal solution to Problem 2.5 and yields the estimation of the performance of Algorithm 2.1. The following theorem [6] claims that in the probabilistic sense that the network load L of output solution by Algorithm 2.1 is not very far away from $\overline{L}_{opt}$.

THEOREM 2.4 *Given* $0 < \epsilon < 1$. *If* $\overline{L}_{opt} \geq 2 \ln |E|$, *then the probability that* $L \leq \overline{L}_{opt} + \sqrt{3\overline{L}_{opt} \ln |E| / \epsilon}$ *is at least* $1 - \epsilon$.

A more sophisticated technique [4] uses a length function on the links to reflect congestion and iteratively reroute some traffics of connections from more congested paths to less congested paths. Based on such rerouting method, a randomized method can be developed to choose the flow paths. The following theorem [4] shows that this method has a better approximation performance.

THEOREM 2.5 *There is an algorithm running polynomial time in n that can produce a solution to Problem 2.5 whose network load* $L \leq \overline{L}_{opt} + O(\sqrt{\overline{L}_{opt} \log n})$.

3. Discussion

In this chapter we have discussed the load balancing problem in the physical topology of a given network, that is to find a path (consisting of some physical links) to carry given traffic between a source-destination node pair such that the maximum traffic flowing over a physical link is minimized. As we have presented, the general approach for solving this problem consists of three steps:

1) Formulate it as a problem of ILP;

2) Solve the integer relaxed version of ILP, which is a problem of LP;

3) Produce a solution to ILP by rounding the fractional solution to LP.

In the following chapters this approach will be used for some other problems as well.

In Chapter 4 we will study the logical topology design problem. It can be divided into some subproblems, one of them can be considered as the load balancing problem in the logical topology of a given network. A logical topology has the same vertex-set as the physical topology. There is a logical link between a node pair if and only if there is a lightpath between them. Routing the traffic of a connection between a node pair is to find multiple paths (consisting of some logical links) between them, each of them carries a fraction of the traffic. The problem studied there is how to route the traffics of connections between all node pairs such that the maximal traffic load on a logical link, that is called the network *congestion*, is minimized. Since the fractional traffic flow is allowed, this version of the load balancing problem becomes a problem of LP and can be solved in polynomial-time.

The load balancing problem also finds an important application in VLSI design, in which a collection of modules are separated by channels and connected by wires that are routed through the channels, for the purpose of regularity the channels have uniform width. Thus it is desirable to minimize the width in order to minimize the total area of the VLSI circuit.

References

[1] D. Banerjee and B. Mukherjee, A practical approach for routing and wavelength assignment in large wavelength-routed optical networks, *IEEE Journal on Selected Areas in Communications*, 14 (5) (1996), 903-908.

[2] T. Erlebach and K. Jansen, Scheduling of virtual connections in fast networks, *Proceedings of the 4th Workshop on Parallel Systems and Algorithms* (PASA), 1996, 13-32.

[3] V. Kumar and E. J. Schwabe, Improved access to optical bandwidth in trees, *Proceedings of the 8th Annual ACM-SIAM Symposium on Discrete Algorithms* (SODA), 1997, 437-444.

[4] P. Lkein, S. Plotkin, C. Stein, and É. Tardos, Faster approximation algorithms for the unit capacity concurrent flow problem with applications to routing and finding sparse cuts, *SIAM Journal on Computing*, 23 (3) (1994), 466-487.

[5] M. Mihail, C. Kaklamanis, and S. Rao, Efficient access to optical bandwidth, *Proceedings of the 36th Annual IEEE Symposium Foundations of Computer Science* (FOCS), 1995, 548-557.

[6] P. Raghavan and C. D. Thompson, Randomized rounding: A technique for probably good algorithms and algorithmic proofs, *Combinatorica*, 7 (4) (1987), 365-374.

[7] P. Raghavan and E. Upfal, Efficient routing in all-optical networks, *Proceedings of the 26th Annual ACM Symposium Theory of Computing* (STOC), 1994, 134-143.

[8] A. Schrijver, P. Seymour, and P. Winkler, The ring loading problem, *SIAM Journal on Discrete Mathematics*, 11 (1) (1998), 1-14.

[9] G. Wilfong and P. Winkler, Ring routing and wavelength translation, *Proceedings of the 9th Annual ACM-SIAM Symposium on Discrete Algorithms* (SODA), 1998, 333-341.

Chapter 3

WAVELENGTH ASSIGNMENT

Wavelengths are limited resource in WDM networks. State-of-the-art technology allows close to 300 wavelengths in the laboratory. However, less than 64 wavelengths can be used in real systems. Thus how to make wavelength assignment is crucial for taking full advantage of the potential of WDM networks.

In this chapter we will consider the *Routing and Wavelength Assignment Problem* (RWAP) in single-hop systems. The problem studies how to find a path and assign a wavelength for each of given lightpaths such that no two lightpaths sharing a link are assigned the same wavelength and the total number of wavelengths used is minimal. RWAP is usually decomposed into two separate subproblems, one is the load balancing problem, that has been discussed in Chapter 2, and the other is the *Wavelength Assignment Problem* (WAP) that is the focus of this chapter. WAP studies how to make an optimal wavelength assignment assuming that the routes of lightpaths are given.

The rest of this chapter is organized as follows. The first three sections discuss WAP. The key approach is to reduce the problem to coloring edges or vertices of graphs. The first part of Section 1 presents the results in undirected tree networks due to Erlebach et al [2, 3] and Raghavan and Upfal [14]. In this case the wavelength assignment is done through solving an edge-coloring problem. The second part of Section 1 presents the results in directed tree networks due to Erlebach et al [3]. In this case the wavelength assignment is done through properly coloring the edges of a bipartite graph. Section 2 presents result in rings due to Raghavan and Upfal [14] and Wilfong and Winkler [16]. In this case the wavelength assignment is done through coloring an interval graph. Section 3 introduces a standard approach used for solving the WAP in networks of arbitrary topologies. In this case the wavelength assignment is done through coloring vertices of a graph. Section 4 discusses RWAP and presents some results in ring networks due to Wilfong and Winkler [16] and Kumar [6, 7]. Section 5 concludes the chapter.

1. Tree Networks

In general the wavelength assignment problem can be considered as coloring paths, which represent the routes of connections, in a given graph, which represents the physical topology of a WDM networks. The problem can be formulated as follows.

PROBLEM 3.1 *Wavelength Assignment Problem*

Instance A graph $G(V, E)$ and a set P of paths in G between vertex pairs.
Solution An assignment of w wavelengths to paths in P such that two paths that share a link must be assigned two distinct wavelengths.
Objective Minimizing w, the number of wavelengths used.

In this section we consider Problem 3.1 in tree networks. The network load is particularly called *tree load.*

1.1 Bidirectional Connections

In this subsection we assume bidirectional connections. Thus the route of a connection is a (undirected) path in the tree. Let us first look at the instance shown in Fig. 3.1(a). Three paths in a tree of four vertices produce tree load 2. It is obvious that three wavelengths are necessary for these three paths since each of them shares a link with the other two paths. Moreover, three wavelengths are sufficient as well.

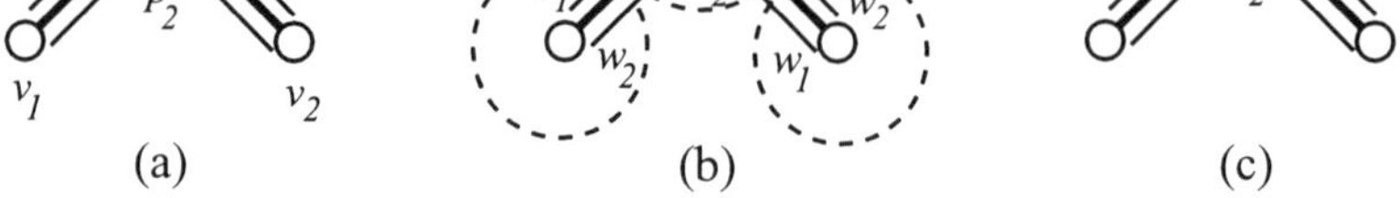

Figure 3.1. (a) An instance of Problem 3.1, (b) Problem 3.1 in a tree is decomposed into Problem 3.1 in stars, (c) Solutions to Problem 3.1 in stars can be merged into a solution to Problem 3.1 in a tree.

The following lemma [14] gives a lower bound on the number of wavelengths required for paths routed in a tree with load L.

LEMMA 3.1 *For any positive integer L, there is a tree and a set of paths routed in the tree yielding tree load at most L such that $3L/2$ wavelengths are both necessary and sufficient to assign these paths.*

PROOF Consider $L/2$ identical paths of p_i, for $i = 1, 2, 3$, that are routed in the tree as shown in Fig. 3.1(a) and yield tree load L. It is trivial that $3L/2$ wavelengths are sufficient for these $3L/2$ paths. Since each of $3L/2$ paths shares a link with all the others, one wavelength can only be used for one path. This means that $3L/2$ wavelengths are required for assigning all paths. ■

The following theorem [2] shows that the wavelength assignment problem is $\mathcal{NP}$-hard even in tree networks.

THEOREM 3.1 *The Problem 3.1 for bidirectional connections is $\mathcal{NP}$-hard in undirected tree networks.*

PROOF We consider the decision version of the Problem 3.1 in undirected tree networks. That is to determine whether, for given a set of undirected paths in a tree and k, k wavelengths can be assigned to the paths without causing wavelength conflict. We will construct a polynomial-time reduction from an instance of *edge-coloring problem* to an instance of the decision version of Problem 3.1. An instance I' of the edge-coloring consists of a graph $G(V', E')$ with maximal degree Δ. The problem is then to decide whether the edges of $G(V', E')$ can be colored with Δ colors such that edges are assigned different colors if they share an endpoint. This problem is known $\mathcal{NP}$-complete even for 3-regular graphs [5].

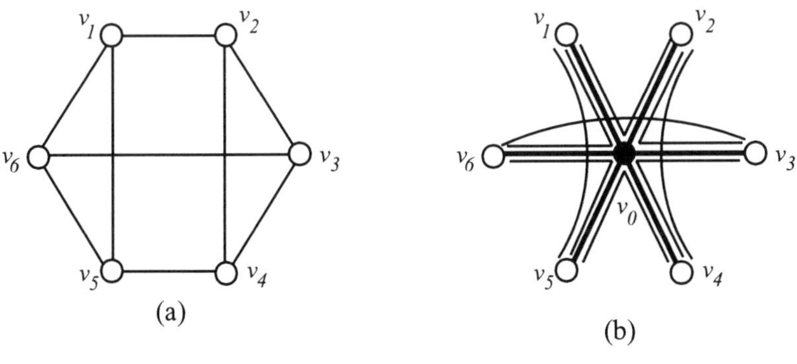

Figure 3.2. A polynomial-time reduction: (a) an instance I' of edge-coloring problem and (b) an instance I of wavelength assignment problem.

We now define an instance I of the decision version of Problem 3.1 in undirected tree networks. Graph $G(V, E)$ consists of vertex-set $V \equiv V' \cup \{v_0\}$, here $v_0 \notin V'$, and edge-set $E \equiv \{(v_0, v) \mid v \in V'\}$. $G(V, E)$ is a special case of tree graph with v_0 adjacent to every vertex in V'. Set the set of paths $P \equiv \{(u, v) \mid (u, v) \in E'\}$ and $k \equiv \Delta$. Fig. 3.2(a) shows a 3-regular graph of six vertices. Fig. 3.2(b) shows nine paths in a tree, each of which corresponds an edge in the graph of Fig. 3.2(a). Note that all paths pass through the center vertex v_0. It is not difficult to verify that I'

has a proper coloring of Δ colors if and only if I has a valid wavelength assignment of Δ wavelengths. Moreover, a proper coloring of edges in I' corresponds a valid wavelength assignment of paths in I. ■

The basic idea of the proposed approach for solving Problem 3.1 is to reduce the wavelength assignment problem on a given tree to the edge-coloring problem on a star as we do in the proof of Theorem 3.1. A *star* is a tree that at most one vertex (called *center*) in the tree has degree greater than two. Fig. 3.1(a) and Fig. 3.2(b) give two simple examples of stars. To realize this idea, we need to introduce some notations. For a given tree $G(V, E)$ and a set P of paths in G, denote by $P_v \subseteq P$ for any $v \in V$ the set of paths that passes v. In the example given in Fig. 3.1(a), we have

$$P_{v1} = \{p_1, p_2\}, P_{v_2} = \{p_2, p_3\}, P_{v3} = \{p_1, p_3\}, \text{ and } p_{v0} = \{p_1, p_2, p_3\}.$$

For each $v \in V$, let $w(P_v)$ denote the number of wavelengths that an optimal assignment for paths in P_v uses. In particular, denote by $w_{opt}(P)$ the number of wavelengths that an optimal assignment requires for all given paths. Observe that to assign wavelengths to paths in P_{v_1}, P_{v_2}, and P_{v_3} two wavelengths are both sufficient and necessary, while P_{v_0} requires three wavelengths. As we mentioned earlier, three wavelengths are sufficient for the whole problem. That is,

$$w(P_{v_1}) = w(P_{v_2}) = w(P_{v3}) = 2, w(P_{v0}) = 3 = w_{opt}(P).$$

The following lemma [2] shows that this is true for any set of paths in any tree.

LEMMA 3.2 *Suppose that P is a set of paths in a tree $G(V, E)$. Then $w_{opt}(P) = \max\{w(P_v) \mid v \in V\}$.*

PROOF Since $P_v \subseteq P$ for all $v \in V$, we have $w_{opt}(P) \geq \max\{w(P_v) \mid v \in V\}$. Thus it suffices to show that there is a method for assigning all paths in P by using $\max\{w(P_v) \mid v \in V\}$ wavelengths. Let $A(P_v)$ denote an optimal assignment for paths in P_v for any $v \in V$. Note that some path may appear in two subset of paths P_v and P_u for $v \neq u$, and it may be assigned different wavelengths by $A(P_u)$ and $A(P_v)$, respectively. See Fig. 3.1(b), where path $p_1 \in P_{v_1}$ is assigned wavelength w_1 by $A(P_{v_1})$ while it, as belonging to P_{v3}, is assigned wavelength w_2 by $A(P_{v3})$. We will show in the following how to combine $A(P_v)$ for all $v \in V$ into a method $A(P)$ for assigning all paths in P without using more wavelengths than the largest number of $w(P_v)$ over $v \in V$.

The process is based on a *merging* technique. We begin merging from a vertex v_0 with $w(P_{v0}) = \max\{w(P_v) \mid v \in V\}$. Initially, we set $A(P) := A(P_{v0})$ and mark vertex v_0 as the only vertex that has already been processed. We repeatedly merge $A(P)$ and an assignment $A(P_v)$ where v is a vertex that has not been processed yet but adjacent to a previously processed vertex. During the merging process no new wavelength has to be introduced. This assures that the merging process leads to an assignment that uses $w(P_{v0})$ wavelengths.

Observe that an unprocessed vertex v can only be adjacent to a previously processed vertex u, and all paths that contained in both P_v and $A(P)$ use edge (u, v). (Here we also use the notation $A(P)$ to denote the paths that $A(P)$ has assigned wavelengths.) We call them the *intersecting* paths. Let T_v denote the subtree of G that contains v and all vertices reachable from v without using edge (u, v). Note that the intersecting paths are the only paths in $A(P)$ that intersect subtree T_v. Moreover, the intersecting paths are the only paths in $A(P_v)$ that intersect subtree $G \setminus T_v$. Therefore, the intersecting paths are the only paths that we have to take care when merging $A(P)$ and $A(P_v)$.

Merging $A(P)$ and $A(P_v)$ can be done through permuting the wavelengths used by $A(P_v)$ in such a way that the intersecting paths in $A(P_v) \cap A(P)$ are assigned the same wavelengths while other paths are assigned accordingly. This can be accomplished, since $w(P_v)$ is the number of wavelengths that $A(P)$ uses and , $w(P_v) \leq w(P_{v_0})$. And then, the resulting assignment is set to be the new $A(P)$, which assigns all paths in old $A(P)$ and P_v. After that, we mark v as a processed vertex and repeat this process until all vertices in V become the processed vertices. The assignment $A(P)$ that we obtain in the end uses $w(P_{v_0})$ wavelengths to assign all paths in P. Hence, it is an optimal assignment for all paths in P in the original tree G. The proof is then finished. ■

The constructive proof of the above lemma suggests an algorithm for solving an instance of Problem 3.1 on tree networks through solving independently $|V|$ instances of Problem 3.1 on star networks and combining the obtained solutions into one by merging process. Unfortunately, the reduction used in the proof of Theorem 3.1 actually shows that Problem 3.1 is $\mathcal{NP}$-hard even in star networks. However, the following lemma [2] claims that a near-optimal assignment can be found in polynomial-time.

LEMMA 3.3 *Given a set of paths P, there is a polynomial-time algorithm A that can find an assignment of $w_A(P)$ wavelengths for P with $w_A(P) \leq 1.1 w_{opt}(P) + 0.8$, where $w_{opt}(P)$ is the number of wavelengths that an optimal wavelength assignment requires for P.*

PROOF We will transform the problem of assigning wavelengths to paths in a star network to the problem of coloring edges of a multigraph with self-loops. A *multigraph* is a graph in which there may be more than one edge between a pair of vertices. Given a star network that consists of a center vertex v_0 and k vertices adjacent to it, $v_1, v_2, \cdots, v_k$. We construct a multigraph $G_m(V', E')$ as follows: $V' = \{v_1, v_2, \cdots, v_k\}$. There is a self-loop $(v_i, v_i) \in E'$ if there is a path that contains v_i and v_0 is one of its endpoints, and there is an edge $(v_i, v_j) \in E'$ if there is a path that contains vertices v_i, v_j and passes vertex v_0. Fig. 3.3(a) shows nine paths in a star of $k = 5$. Fig. 3.3(b) shows corresponding nine edges in a multigraph. Observe that in Fig. 3.3(b) vertex v_5 has two self-loops and there are two edges between vertices v_1 and v_2. It is easy to verify that two paths in P share a common edge if and only if both of them contains a neighbor v_i of v_0, which is the case if and only if the corresponding edges in G_m share an endpoint v_i. Thus each proper edge-coloring of G_m corresponds to a feasible wavelength assignment for P in the star. In particular, an optimal wavelength assignment for paths in P corresponds to an optimal edge-coloring of G_m.

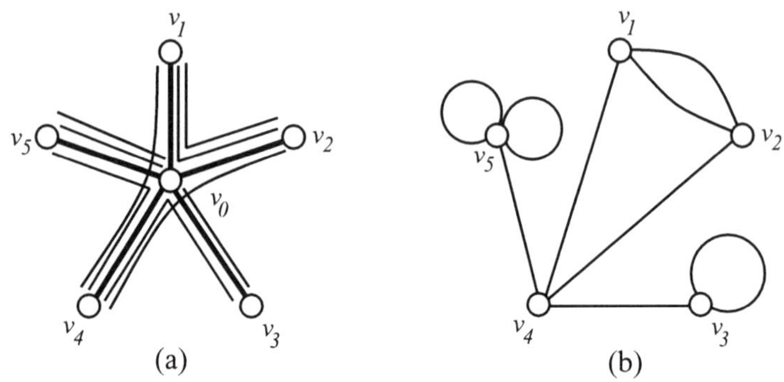

Figure 3.3. Transforming the wavelength assignment problem in a star to the edge-coloring problem of a multigraph.

Now using the algorithm proposed in [13], an edge-coloring for G_m can be found in polynomial-time that uses no more than $1.1\Delta(G_m) + 0.8$ colors, where $\Delta(G_m)$ is the maximum degree of vertices in G_m and it is a lower bound on the number of colors used by an optimal edge-coloring. However, this algorithm was originally designed for multigraphs without self-loops. To deal with self-loops in G_m, we can first ignore (or remove) the self-loops from G_m and apply this edge-coloring algorithm to the resulting multigraph (without self-loops). After that we can color the self-loops with the colors used by the algorithm (if possible) or with new colors (if necessary). Note that coloring self-loops in this greedy manner will not increase the ratio between the number of colors used by the proposed algorithm and that by an optimal algorithm. The proof is thus finished. ■

As the edge-coloring algorithm in [13] plays a major role in solving the WAP for bidirectional connections in tree networks, we now give an outline of this method.

Assume that a multigraph $G_m(V, E)$ is edge-colored with as a set of q colors, $c_1, c_2, \cdots, c_q$. Color c_k is a *missing color* of vertex $v \in V$ if none of the edges incident to v is colored c_k. Denote by $M(v)$ the set of all missing colors of v. Denote by $C(u, v)$ the set of colors assigned to the multiple edges joining vertices u and v. An edge colored c_k is called a *c_k-edge*. For two colors c_i and c_j, the spanning subgraph of G_m induced by all the edges colored c_i or c_j is called an *$c_i c_j$-subgraph*, and is denoted by $G_m[c_i, c_j]$. It is trivial that each connected component of $G_m[c_i, c_j]$ is either a path or a cycle, in which edges are colored alternately c_i or c_j. Such a path (a cycle) is called a c_ic_j-*alternating path* (a c_ic_j-*alternating cycle*) or simply a c_ic_j-*path* (a c_ic_j-*cycle*). It is obvious that interchanging the colors c_i and c_j of the edges in a c_ic_j-path or a c_ic_j-cycle yields another proper edge-coloring of G_m with the same set of colors. This interchanging process is called *recoloring* of a path or a cycle. If all the edges of G_m, except an edge $e = (u, v) \in E$, are colored with q colors, and if $c_i \in M(u)$ and $c_j \in M(v)$, then the c_ic_j-path between u and v, if any, is particularly called a c_ic_j-*critical path*, which is denoted by $p(c_i, c_j)$. By the definition, it is easy to see that the number of vertices in $p(c_i, c_j)$, denoted by $|p(c_i, c_j)|$, is odd. If there is no c_ic_j-critical path, then $|p(c_i, c_j)|$ is defined to be *infinite*.

The key operation of the edge-coloring algorithm is essentially an iterative process that colors the edges of G_m one by one. However, when it colors a uncolored edge, some previously colored edges may be recolored so that this edge can be colored with a previously used color. Thus this algorithm can be considered as a greedy method since it introduces a new color only when it is unable to color an edge by using one of currently used colors. Initially, q colors are available, where $q = 1.1\Delta(G_m) + 0.8$. The algorithm colors an edge (u, v) in five steps, each corresponding to the cases $|p(c_i, c_j)| = 3, 5, 7, 9$ or the case $|p(c_i, c_j)| \geq 11$. Each of the first four steps can increase $|p(c_i, c_j)|$ by two or more through recoloring, and it is then reduced to one of the latter cases. When we eventually have $|p(c_i, c_j)| \geq 11$, the last step makes u and v have a common missing color among missing color, which can be used to color (u, v). It can be proved that the algorithm colors one edge by repeating the recoloring of an alternating path or cycle at most a constant times, and the algorithm finishes in time $O(|E|(\Delta(G_m) + |V|))$.

Now from Lemma 3.2 and Lemma 3.3 we can deduce the following theorem [2].

THEOREM 3.2 *Given a set P of paths in a tree $G(V, E)$, there is a polynomial-time algorithm A that can find an assignment of $w_A(P)$ wavelengths with $w_A(P) \leq 1.1w_{opt}(P) + 0.8$, where $w_{opt}(P)$ is the number of wavelengths that an optimal wavelength assignment uses for P.*

PROOF We first find an assignment $A(P_v)$ for each $v \in V$ by using the method outlined in the proof of Lemma 3.3. And then we combine these $|V|$ assignments into one by using the merging process described in the proof of Lemma 3.2 for all paths in P. It is clear that the number of wavelengths used by this algorithm A is

$$\begin{aligned} w_A(P) &= \max\{w_A(P_v) \,|\, v \in V\} \\ &\leq \max\{1.1w_{opt}(P_v) + 0.8 \,|\, v \in V\} \\ &= 1.1w_{opt}(P) + 0.8 \end{aligned}$$

In addition, the proposed algorithm can make a wavelength assignment in time $O(|V||P|(|P| + |V|))$ since the number of edges in G_m is bounded above by the number of paths in P. The proof is thus finished. ■

Another wavelength assignment algorithm was described in [14] that produces a $\frac{3}{2}$-approximation solution to Problem 3.1 for bidirectional connections in tree networks. The basic idea of this method is the same as one described in Theorem 3.2. The difference is that it uses a different edge-coloring proposed in [1]. Note that when the optimal wavelength assignment uses more than two wavelengths the method presented here has a better performance, under the worst case analysis, than the method given in [14].

1.2 Unidirectional Connections

In this subsection we assume unidirectional connections. Thus the route of a connection is a directed path. Let us first look at an instance of the Problem 3.1 in this case shown in Fig. 3.4. Five directed paths in a tree of seven vertices, p_i for $i = 1, 2, \cdots, 5$, produce tree load 2. It can be easily verified that no three of them can share a wavelength. This means that three wavelengths are necessary for these five directed paths. In fact, three wavelengths are also sufficient, since p_1 and p_3 can share a wavelength, p_2 and p_4 can share another one while p_5 uses a different wavelength alone.

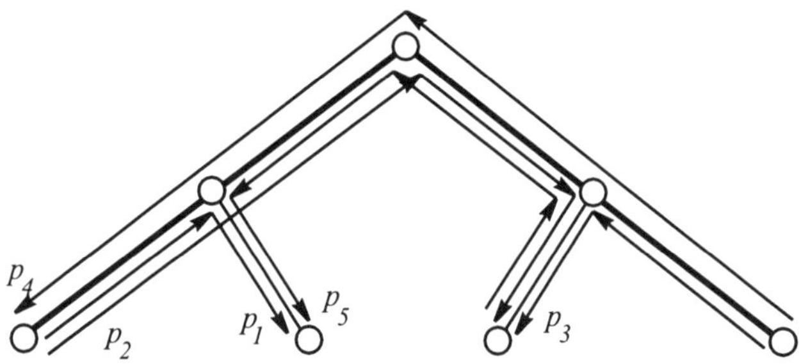

Figure 3.4. An instance of Problem 3.1 for unidirectional connections.

The following lemma [8] gives a lower bound on the number of wavelengths required for directed paths in a tree with maximal directed link load L.

LEMMA 3.4 *For any positive integer L, there is a tree and a set of directed paths in the tree yielding tree load at most L such that* $5L/4$ *wavelengths are both necessary and sufficient for these directed paths.*

PROOF Consider $L/2$ identical directed paths p_i, for $i = 1, 2, \cdots, 5$, in the tree shown in Fig. 3.4. They yield tree load L.

To show that $5L/4$ wavelengths $\{w_1, w_2, \cdots, w_{5L/4}\}$ are sufficient for these $5L/2$ directed paths, we assign $L/2$ wavelengths to each of the five sets consisting of $L/2$ identical directed paths (one wavelength for each directed path in the set) as follows.

$$\{\overbrace{p_1, p_1, \cdots, p_1}^{L/2}\} \leftarrow \{w_1, w_2, \cdots, w_{L/2}\}$$
$$\{\overbrace{p_2, p_2, \cdots, p_2}^{L/2}\} \leftarrow \{w_{L/2+1}, w_{L/2+2}, \cdots, w_L\}$$
$$\{\overbrace{p_3, p_3, \cdots, p_3}^{L/2}\} \leftarrow \{w_{L+1}, w_{L+2}, \cdots, w_{5L/4}\} \cup \{w_1, w_2, \cdots, w_{L/4}\}$$
$$\{\overbrace{p_4, p_4, \cdots, p_4}^{L/2}\} \leftarrow \{w_{L/4+1}, w_{L/4+2}, \cdots, w_{3L/4}\}$$
$$\{\overbrace{p_5, p_5, \cdots, p_5}^{L/2}\} \leftarrow \{w_{3L/4+1}, w_{3L/4+2}, \cdots, w_{5L/4}\}$$

It can be verified that under the above assignment no two paths that share the same directed link are assigned the same wavelength.

We now show that $5L/4$ wavelengths are also necessary for those $5L/2$ directed paths. Note that a wavelength can be used only once in each set of directed paths, and it cannot be used in more than two sets, since no three out of the five paths in $\{p_1, p_2, \cdots, p_5\}$ can share a wavelength. This implies that no wavelength can be used for three or more directed paths. As there are a total of $5L/2$ paths to be assigned, $5L/4$ wavelengths are necessary. ■

The following theorem [2] shows that the wavelength assignment problem is $\mathcal{NP}$-hard even in tree networks.

THEOREM 3.3 *The Problem 3.1 for unidirectional connections is $\mathcal{NP}$-hard in tree networks.*

PROOF We consider the decision version of Problem 3.1 for unidirectional connections in trees. The problem is to determine whether, given a set of directed paths in a tree and an integer $k > 0$, k wavelengths can be assigned to the directed paths without causing wavelength conflict. We will again construct a polynomial-time reduction again from edge-coloring to this problem. The basic idea of the transformation is the same as in the proof of Theorem 3.1, but the case of directed paths require a more involved construction. Let $G(V', E')$ be a 3-regular graph. It is $\mathcal{NP}$-complete to decide whether $G(V', E')$ can be properly edge-colored with 3 colors. We now show how to transform any 3-regular graph $G(V', E')$ into an instance of the Problem 3.1 for unidirectional connections in a tree $G(V, E)$.

The vertex-set V is set to contain all vertices in V' and a new vertex $v_0 \notin V'$, and nine additional vertices $v_1, v_2, v_3, v_{11}, v_{12}, v_{21}, v_{22}, v_{31}, v_{32}$ for each $v \in V'$. That is

$$V \equiv V' \cup \{v_0\} \cup \{v_1, v_2, v_3, v_{11}, v_{12}, v_{21}, v_{22}, v_{31}, v_{32} \,|\, v \in V'\}.$$

The edge-set E is set as follows: v_0 is the root of $G(V, E)$ and each $v \in V'$ is a child of v_0. For each $v \in V'$, v_1, v_2, v_3 are the children of v, and v_{i1}, v_{i2} are the children of v_i for $i = 1, 2, 3$. That is

$$E \equiv \{(v_0, v) \,|\, v \in V'\} \cup \{(v, v_i), (v_i, v_{ij}) \,|\, v \in V', i = 1, 2, 3, j = 1, 2\}.$$

The set P of directed paths is set to contain four directed paths associated with each edge $e = (u, v) \in E'$. $p_1(e) = (u_{i1}, v_{j2})$, $p_2(e) = (v_{j1}, u_{i2})$, and $p_3(3) = p_4(3) = (u_{i1}, u_{i2})$. The construction of tree network $G(V, E)$ and the set of directed paths is illustrated in Fig. 3.5. The vertices of graph $G(V', E')$ in Fig. 3.5(a) correspond to the children of the root of graph $G(V, E)$ in Fig. 3.5(b). The black vertices u and v in $G(V', E')$ correspond to the black children u and v of root v_0 in $G(V, E)$. The dashed edge between u and v in $G(V', E')$ corresponds to the four dashed directed paths indicated in $G(V, E)$. The subtrees rooted at the children of root v_0 of $G(V, E)$ are shown only for two black vertices u and v. It is easy to see that the construction can be done in polynomial-time.

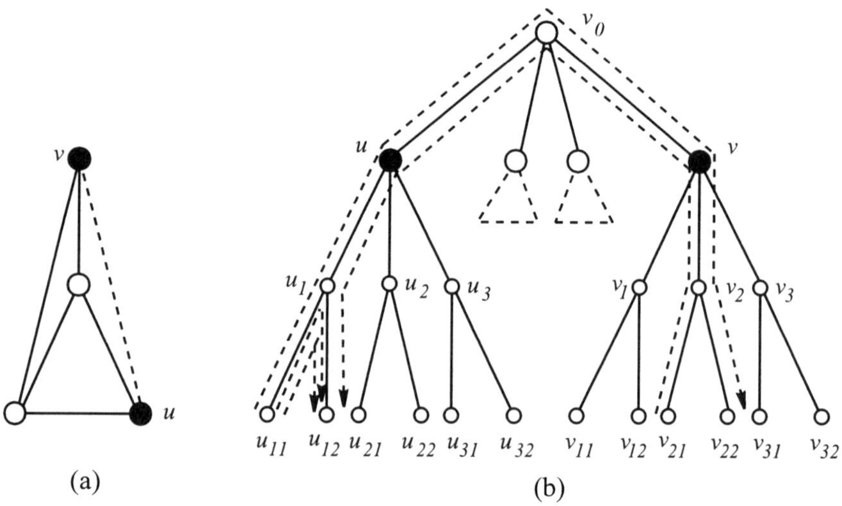

Figure 3.5. A polynomial-time reduction, (a) $G(V', E')$ and (b) $G(V, E)$ with a set of directed paths.

The indices of i and j are selected from $\{1, 2, 3\}$ in such a way that a different value of i (of j, respectively) is chosen for each edge incident to u (to v, respectively). Intuitively, directed paths $p_1(e)$ and $p_2(e)$ correspond to the single bidirectional connection that we use in the proof of Theorem 3.1, and the other two directed paths $p_3(e)$ and $p_4(e)$ are introduced to make sure that $p_1(e)$ and $p_2(e)$ are assigned the same wavelength.

We now show that the created set of directed paths in $G(V, E)$ has a valid assignment of three wavelengths if and only if the edges in $G(V', E')$ has a proper edge-coloring of three colors. Consider "if" part. For each edge $e \in E'$, the directed paths $p_1(e)$ and $p_2(e)$ are assigned the same wavelength (color) that edge e is colored. The directed paths $p_3(e)$ and $p_4(e)$ are assigned two different wavelengths (colors). Consider "only if" part. Assume that the created set of directed paths has a valid assignment of three wavelengths. Under this assignment, the directed paths $p_3(e)$ and $p_4(e)$ make $p_1(e)$ and $p_2(e)$ to be assigned the same wavelength. Notice that for edge $f \in E$ that is incident to $e \neq f$, no directed path $p_1(f)$ can be assigned this wavelength. As a result, if we color each edge $e \in E$ the color (wavelength) that directed path $p_1(e)$ is assigned, we obtain a proper edge-coloring of three colors. The theorem is thus proved. ■

In the following, we will outline the approximation algorithm proposed in [3] for the Problem 3.1 of unidirectional connections. For the simplicity of presentation, we assume that the given set of directed paths in P induces each directed link of the tree to has load exactly L. (If not, extra paths can be added to P.) We further assume that L is a multiple of 3, i.e., $L = 3l$ for some positive integer l. (The cases of $L = 3l + 1$ and $L = 3l + 2$ can be processed in similar ways.)

Initially the vertices of the given tree are processed in *Depth-First-Search* (DFS) order, starting at an arbitrary leaf vertex as a root. The algorithm begins with assigning the directed paths touching the root, this can easily be done by using L wavelengths. When the next vertex v (in the DFS order) is processed, all directed paths touching its parent or any other vertex with smaller DFS-number have already assigned wavelengths. And now the problem is reduced to how to extend the existing wavelength assignment to include all directed paths touching v. This problem can be reduced to the edge-coloring problem in a bipartite graph G_v as follows: Denote by v_0 the parent of v and by $v_1, v_2, \cdots, v_k$ the children of v. The bipartite graph G_v has left and right vertex-sets $\cup_{i=0}^{k}\{x_i, v_i\}$ and $\cup_{i=0}^{k}\{y_i, u_i\}$, respectively, and every edge in G_v is between a vertex in left vertex-set and a vertex in right vertex-set. An edge in G_v is associated with a directed path touching v in the following way (see Fig. 3.6):

1 There is an edge (x_i, y_j) in G_v if there is a directed path coming from vertex v_i and heading for vertex v_j.

2 There is an edge (x_i, u_i) in G_v if there is a directed path coming from vertex v_i and terminating at vertex v.

3 There is an edge (v_i, y_i) in G_v if there is a directed path starting at vertex v and heading for vertex v_i.

Note that in the bipartite graph G_v all vertices x_i and y_i have degree L, while the vertices v_i and u_i may have degrees less than L. In addition, there is no edge between vertices x_i and y_i and between vertices v_i and u_i for $i = 0, 1, \cdots, k$, respectively.

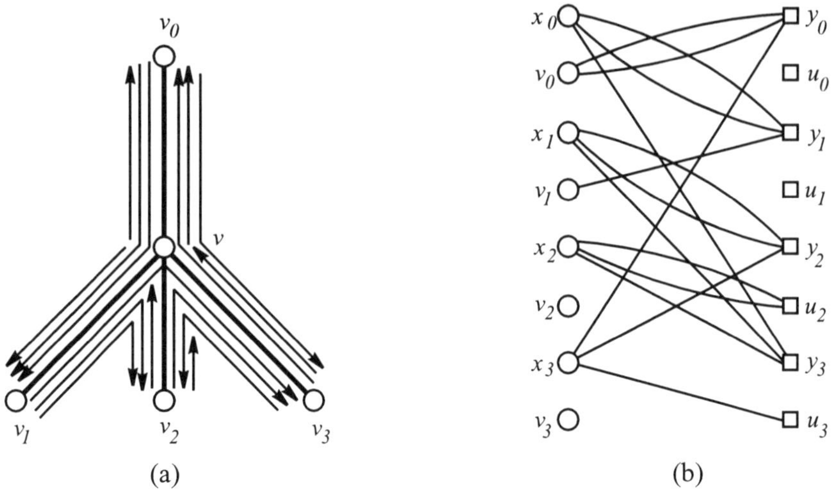

Figure 3.6. Constructing a bipartite multigraph: (a) directed paths touching vertex v and (b) bipartite graph G_v.

Under the above construction, it is easy to see that two directed paths touching v must be assigned different wavelengths if and only if the corresponding edges in G_v share a vertex. Hence, any proper edge-coloring of G_v can be transformed to a valid wavelength assignment for the directed paths touching v.

Now let us see how to color edges of G_v. In fact, the edges incident to x_0 and y_0 have already received a color (wavelength), since corresponding directed paths touch v_0 and have been colored (assigned) at some previous step. We call them *pre-colored* edges. In particular, the colors that appear on pre-colored edges of G_v are called *single* colors if they appear only once (in either x_0 or y_0 but not both), and *double* colors if they appear twice (in both x_0 and y_0). Thus our problem is how to color the uncolored edges in G_v without conflicting those pre-colored edges. In order to use minimal number of colors to color all directed paths, we adopt the same greedy strategy, as we used for coloring undirected paths in the preceding subsection, that we do not introduce a new color unless we have to. During the whole process, we will color in such a way that:

(i) The number of colors used for coloring directed paths that go through a two oppositely directed edges of a link is at most $4L/3 = 4l$.

(ii) The number of colors used for coloring all directed paths is at most $5L/3 = 5l$.

At the beginning, when we color the directed paths touching one leaf vertex (root), the above two requirements (i, ii) can be easily satisfied. We will show how a given coloring can be extended in polynomial time to include the directed paths touching an additional vertex while the requirements (i, ii) remain satisfied.

Denote by S the number of single colors, and by D the number of double colors. The requirement (i) ensures that $S + D \leq 4L/3$. Since we assume that every edge has load L, we get $S + 2D = 2L$. These two inequalities lead to $D \geq 2L/3$. In fact, we can assume that $D = 2L/3$. If $D > 2L/3$, we can simply "split" an appropriate number of double colors by recoloring one of the two pre-colored edges colored with the same double color a new color for the current stage of edge-coloring. A series of color exchanges can then rearrange the original colors on the pre-colored edges with the requirements (i, ii) satisfied.

Assume now that $S = D = 2L/3$, we show that G_v can be edge-colored by using at most $L/3$ new colors (that do not appear on the pre-colored edges), such that the number of colors used for edges incident to x_i or y_i (and v_i or u_i, respectively) is at most $4L/3$ for $i = 0, 1, \cdots, k$.

In order to make G_v to be L-regular, we add a dummy edge between v_j in the left vertex-set and u_i in the right vertex-set if there is an edge between (x_i, y_j). In Fig. 3.7(a) following from Fig. 3.6(b), the dummy edges are indicated with dashed lines. The resulting G_v is now an L-regular bipartite graph, where $L = 3$. Thus edges in G_v can be partitioned into $L = 3$ disjoint perfect matchings as shown in Fig. 3.7(b,c,d). By partitioning them into groups of 3 matchings in an appropriate way, we can obtain 3-regular subgraphs, each containing two single colors and four (not necessarily distinct) double colors. For our example in Fig. 3.7, partitioning process is unnecessary since $L = 3$. The uncolored edges of each such subgraph are then colored by using at most one new color and reusing some of the previously used colors. Hence, as required at most $D/2 = L/3$ new colors are used in the total. In addition, each 3-regular subgraph is edge-colored in such a way that by the number of colors used for edges incident to x_i or y_i (and v_i or u_i, respectively) is at most four, here $i > 0$. Our example is a simple case which does not need a new color, four previously used colors are sufficient. Fig. 3.8 gives an edge-coloring with four colors. The bipartite graph as shown in Fig. 3.7(a) is partitioned into four matchings as shown in Fig. 3.8(a, b, c, d), all edges in a matching is assigned the same color.

To realize the above idea, each of L matchings is classified according to the colors on its two pre-colored edges. A matching between two single colors is called a *SS-matching*. A matching between a single color and a double color is called a *ST-matching*. A matching between two different double colors are called a *TT-matching*. A matching between the same double color on both pre-colored edges is called a *PP-matching*. In the example as shown in Fig. 3.6(a), suppose that the directed path from v_3 to v_0 and one of the directed paths from v_0 to v_1 are preassigned the same wavelength (a double color), and the directed path from v to v_0 and the directed path from v_0 to v_3 are preassigned the same wavelength (another double color), while the directed path from v to v_0 and one of the directed path from v_0 to v_1 are preassigned different wavelengths (two single colors, respectively). Then the

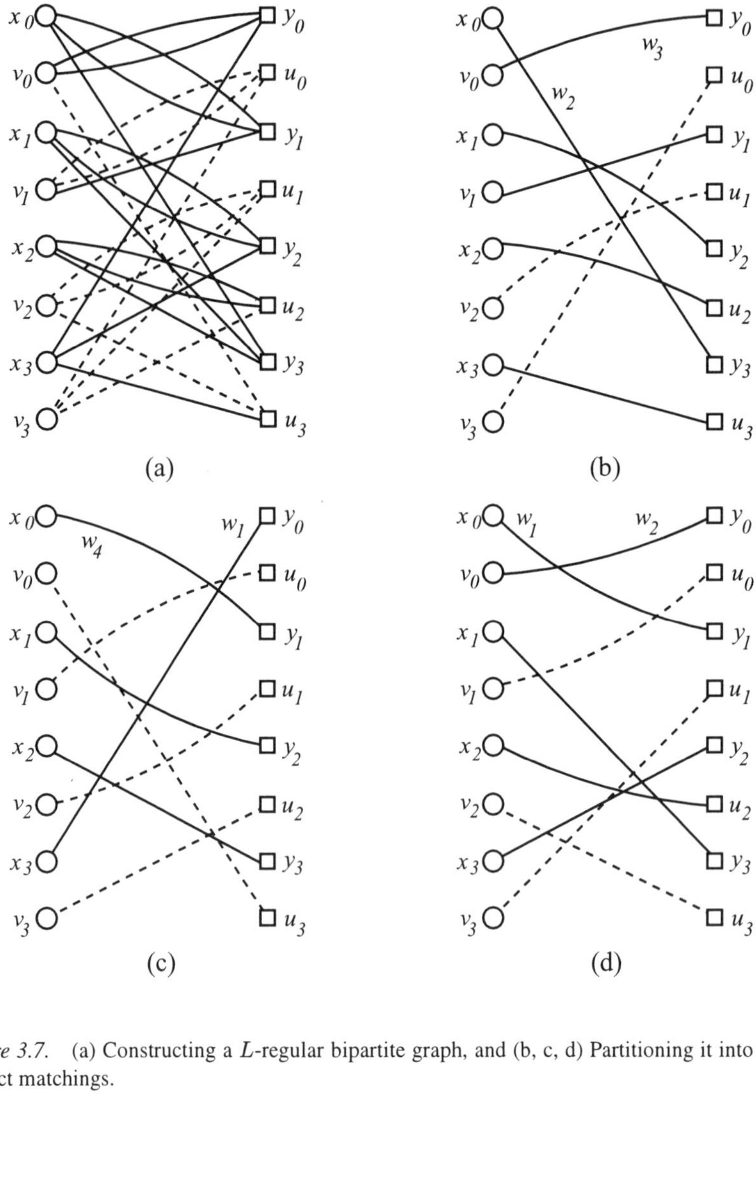

Figure 3.7. (a) Constructing a L-regular bipartite graph, and (b, c, d) Partitioning it into L disjoint perfect matchings.

matchings of Fig. 3.7(b,c) are two ST-matchings while the matching of Fig. 3.7(d) is a TT-matching.

Through a series of involved and sophisticated analysis on each of those matchings, Erlebach et al [3] proved the following theorem.

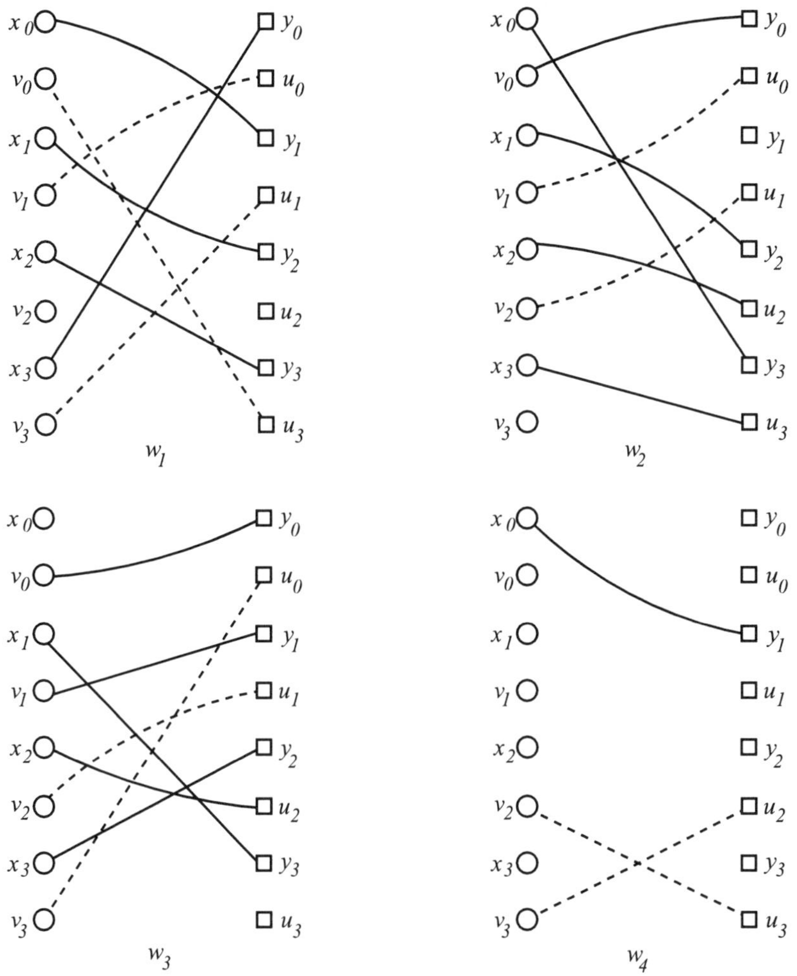

Figure 3.8. Coloring the bipartite graph with 4 colors.

THEOREM 3.4 *Given a set of directed paths in a tree with each directed link load at most L, there is a polynomial-time algorithm that can assign all the directed paths by using at most* $5L/3$ *wavelengths.*

2. Ring Networks

In this section we consider the WAP in ring networks. When the routes of connections are given in ring networks, as far as the wavelength assignment is concerned, there is no essential difference between bidirectional and unidirectional connections, since one instance of the problem for unidirectional connections can be treated as two independent instances of the problem for bidirectional connections. Accordingly in the following we will just consider the case of unidirectional connections.

The following lemma [16] gives a lower bound on the number of wavelengths required for paths routed in a tree with load L.

LEMMA 3.5 *For any positive integer $L > 0$, there is a set of directed paths in a ring with ring load L that requires at least $(2L-1)$ wavelength.*

PROOF Consider a ring of n vertices, where $n = 4(2L-1)$, and $(2L-1)$ directed paths from vertex $4i$ to vertex $(4i + 2(2L-1) - 1)$ routed all in clockwise direction for $0 \leq i < 2L - 1$. See Fig. 3.9(a), where $L = 2$ and $n = 12$. It can be verified that the ring load is L and these $(2L-1)$ directed paths are pairwise intersecting. Since none of them can share a wavelength with another, $(2L-1)$ wavelengths are required. ■

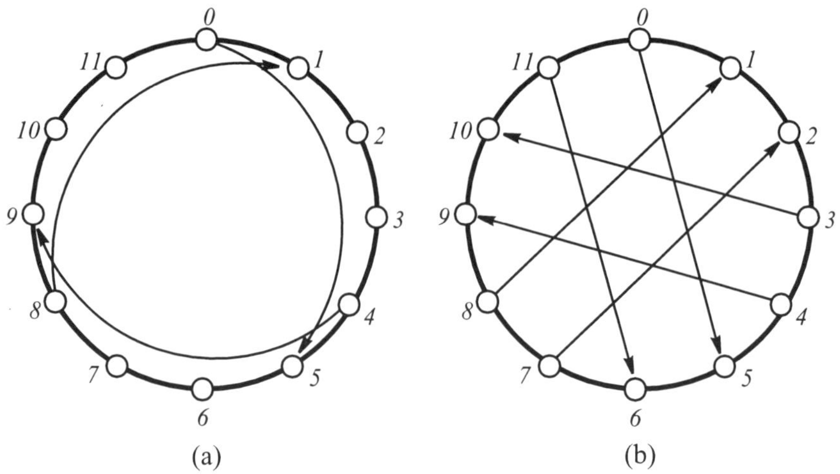

Figure 3.9. (a) An example of Lemma 3.5, and (b) an example of Theorem 3.5.

The following theorem [16] shows that the wavelength assignment problem is $\mathcal{NP}$-hard even in ring networks.

THEOREM 3.5 *The Problem 3.1 for unidirectional connections is $\mathcal{NP}$-hard in ring networks.*

PROOF It suffices to show that the problem of vertex-coloring circular-arc graphs, that was proved to be $\mathcal{NP}$-complete [4], can be reduced to the Problem 3.1 in polynomial-time. A graph is a *circular-arc graph* if its vertices can be represented by arcs of a circle such that there is an edge between two vertices in the graph if and only if the corresponding arcs intersect. Given an instance of the problem of vertex-coloring circular-arc graphs, each arc in the circle can be considered as a directed path in clockwise direction and the circle can be considered as a ring network. Thus each vertex (or equivalently arc) in the circular-arc graph is associated with a directed path in the ring. See Fig. 3.9(b). It is obvious that the circular-arc graph can be vertex-colored by using k colors if and only if k wavelengths is sufficient to assign all directed paths in the ring. ∎

The following theorem [16] describes a simple algorithm that produces a 2-approximation solution to Problem 3.1.

THEOREM 3.6 *Given a set of directed paths in a ring with ring load L, there is a polynomial time algorithm for the Problem 3.1 that assigns the directed paths by using at most $(2L-1)$ wavelengths.*

PROOF Without loss of generality, we assume that there exists at least one directed path starting from vertex v_0. Then we cut the ring at v_0 and obtain a line where we put v_0 at both ends of the line. Note that each of the directed paths that pass through v_0 was cut into two pieces. In Fig. 3.10 directed paths from v_3 to v_2 and from v_6 to v_4 are broken.

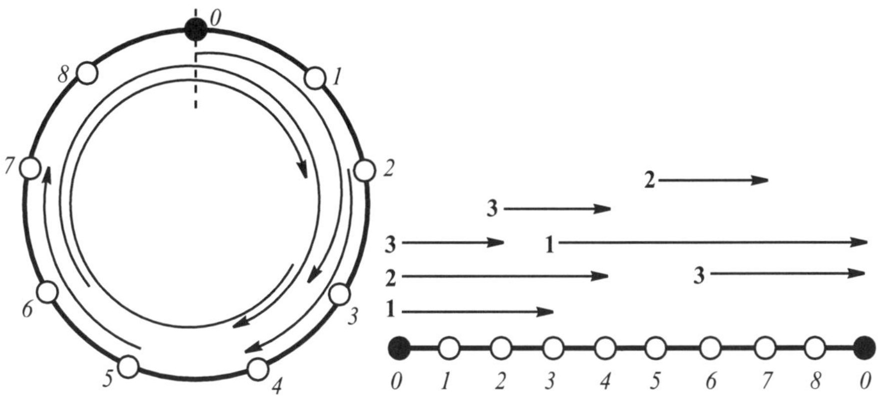

Figure 3.10. Breaking a ring into a line.

Now we first put all directed paths (or pieces of directed paths) in order from left to right by their source nodes (or the broken nodes). Then we allocate wavelengths to each of them in that order. The wavelength assignment is done again in a greedy way such that each directed path receives the least wavelength not already assigned to a directed path intersecting it. It is easy to see that L wavelengths are sufficient (refer to [15]). Since there are at most $(L-1)$ directed paths that were cut into two pieces, and there are at most $(L-1)$ directed half-paths that received wavelengths different from the wavelengths their counterparts received. See Fig. 3.10 for the illustration of wavelength assignment process, where two broken parts of directed paths from v_3 to v_2 are assigned two different wavelengths w_3 and w_1. Hence, we can simply choose an entirely new wavelength for each of these wavelength mismatched paths. In the total, at most $(2L-1)$ wavelengths are used. The proof is finished. ■

3. General Networks

In this section we study WAP in WDM networks of arbitrary topologies. We will introduce two approaches for solving this problem. Since they can be applied to both unidirectional and bidirectional connections, we will just consider the case of bidirectional connections.

3.1 Vertex-Coloring Approach

This approach is natural. It is different from the approach that we used for the case of tree networks in Section 1, where the WAP is transformed to edge-coloring problem. This most widely used approach transforms the problem of wavelength assignment into the problem of coloring vertices of graphs. The *Vertex Coloring Problem* is to, given graph $G'(V', E')$, use minimal number of colors to color all vertices in V' such that no two adjacent vertices are assigned the same color.

We first construct a graph $G'(V', E')$ such that V' is a set of given paths in P, i.e., $V' = P$, and there is an edge in E' between two nodes in V' in graph $G'(V', E')$ if the corresponding paths pass through a common physical link in $G(V, E)$. Clearly, the wavelength assignment problem in $G(V, E)$ is equal to the vertex coloring problem in $G'(V', E')$.

Since the vertex-coloring problem is $\mathcal{NP}$-complete (refer to [17]), it is very difficult to determine the minimum number of colors needed to coloring vertices of graph $G'(V', E')$. This number is usually called *chromatic number* of $G'(V', E')$ and denoted by $\chi(G')$. In fact, even worse than that, it was proved in [10] that the vertex-coloring problem has no approximation algorithm with a constant performance ratio unless $\mathcal{P} = \mathcal{NP}$. The following theorem gives an upper bound, in terms of the number of edges in G', on the number of wavelengths used by an optimal solution to wavelength assignment problem.

THEOREM 3.7 *Let $w_{opt}(G)$ be the number of wavelengths used by an optimal solution to Problem 3.1. Then $w_{opt}(G) \leq 1/2 + \sqrt{2|E'| + 1/4}$.*

PROOF Let C_{opt} be an optimal coloring of G' that uses $\chi(G')$ colors. The color set for each color used is the set of vertices that are assigned this color. Then G has at least one edge between any two color sets, otherwise we could have used the one color for both color sets. Thus, $|E'| \geq \chi(G')(\chi(G')-1)/2$. This, together with $\chi(G') = w_{opt}(G)$, implies the desired bound on $w_{opt}(G)$. ■

The following theorem [11] gives another upper bound in terms of the maximal degree of graph G'.

THEOREM 3.8 *Let $w_{opt}(G)$ be the number of wavelengths used by an optimal solution to Problem 3.1. Then $w_{opt}(G) \leq \Delta(G') + 1$.*

PROOF To prove the theorem, it suffices to find a way to color vertices in V' that uses at most $(\Delta(G')+1)$ colors. Consider the following greedy algorithm: Start from any fixed vertex order of G', $v'_1, v'_2, \cdots, v'_m$, color each vertex v'_i with the first available color c_j, which is the smallest positive integer j that has not been used to color any neighbor of v'_i among $v'_1, v'_2, \cdots, v'_{i-1}$. It is easy to see that this method will never use more than $(\Delta(G')+1)$ colors. ■

The simple greedy algorithm used in the proof of Theorem 3.8 is a widely used heuristic, which is called the *sequential coloring algorithm* [11]. Its performance is dependent on how the vertices are ordered. Denoted by $G'(v'_1, \cdots, v'_k)$ the subgraph of G' induced by vertices $v'_1, \cdots, v'_k$ for $1 \leq k \leq m$. Let $\deg_{G'(v'_1,\cdots,v'_k)}(v'_i)$ is the degree of vertex v'_i in $G'(v'_1, \cdots, v'_k)$. Then the following theorem [11] gives the performance of sequential coloring algorithms.

THEOREM 3.9 *There is a sequential coloring algorithm that can color all vertices in G' by using at most $\max_{1\leq i\leq m}\{1+\deg_{G'(v'_1,\cdots,v'_i)}(v'_i)\}$ colors.*

PROOF In the sequential coloring algorithm when we come to color the vertex v'_i, we do not need to consider any neighbor v'_j of v'_i with $j > i$. This means that it is sufficient to use $1 + \deg_{G'(v'_1,\cdots,v'_i)}(v'_i)$ colors. Taking the maximum of this value over i leads to the desired upper bound. ■

The order that minimizes $\max_{1\leq i\leq m}\{1+\deg_{G'(v'_1,\cdots,v'_i)}(v_i)\}$ can be found in the following way.

(1) Choose the vertex whose degree in G' is minimal as v'_m.

(2) Choose the vertex whose degree in $G' \setminus \{v'_m, \cdots, v'_{i-1}\}$ is minimal as v'_i for $i = m-1, \cdots, 1$.

It can be verified that the obtained order satisfies

$$\deg_{G'(v'_1,\cdots,v'_i)}(v'_i) = \min_{1\leq j\leq i} \deg_{G'(v'_1,\cdots,v'_i}(v'_j), \ 1 \leq i \leq m.$$

This is called the *smallest-last vertex ordering*.

On the other hand, intuitively, if graph G' has only a few vertices of very large degree, then coloring these nodes early will avoid the need of using a very large set of colors. This observation leads to the following theorem [11].

THEOREM 3.10 *There is a sequential coloring algorithm that can color all vertices in G' by using at most $\max_{1\leq i\leq m}\min\{1, 1+\deg_{G'}(v'_i)\}$ colors.*

PROOF Consider the order of vertices, $\deg(v'_i) \geq \deg(v'_{i+1})$. It is easy to see that the sequential coloring algorithm using this order has the desired bound. ■

3.2 Integer Linear Programming Approach

This approach is to transform the problem of wavelength assignment into the problem of ILP. To do so, we introduce the following notations and variables.

- i: subscripts to number the connections, here $1 \leq i \leq m$;
- (u, v): to indicate physical link between nodes u and v in graph $G(V, E)$;
- q: superscripts to number the wavelengths, here $1 \leq q \leq \omega$;
- w_i^q: wavelength assignment variables, $w_i^q = 1$ if the i-th connection is assigned the q-th wavelength, otherwise $w_i^q = 0$;
- $w_i^q(u, v)$: routing variables, $w^{k,q}(i, j) = 1$ if the i-th connection that uses link (u, v) is assigned wavelength w_q, otherwise $w_i^q(u, v) = 0$.

Now Problem 3.1 can be easily formulated by the ILP approach as the following:

$$\text{Minimize} \quad \sum_{q=1}^{\omega}\sum_{i=1}^{m} w_i^q \tag{3.1}$$

$$\text{Subject to} \quad \sum_{q=1}^{\omega} w_i^q = 1, \quad \text{for all } i, \tag{3.2}$$

$$w_i^q(u, v) = w_i^q, \quad \text{for all } i, q, (u, v), \tag{3.3}$$

$$\sum_i^m w_i^q(u, v) \leq 1, \quad \text{for all } q, (u, v), \tag{3.4}$$

$$w_i^q, w_i^q(u, v) \in \{0, 1\}, \quad \text{for all } i, q, (u, v), \tag{3.5}$$

The equality (3.2) ensures that exactly one wavelength is assigned to each connection. The equality (3.3) ensures that the i-th connection is assigned the q-th wavelength if and only if the same wavelength is used on every link in the path that the connection is routed. The equality (3.4) ensures that if more than one connection use link (u, v) then at most one of them can be assigned a specified wavelength. The objective function (3.1) is, in fact, a constant m (the number of paths in P). Thus the problem of above ILP (3.1-5) is just to find a feasible solution satisfying constraints (3.2-4). In other words, this corresponds to the problem of deciding if the given m connections can be assigned by using ω wavelengths. Therefore, to solve the Problem 3.1 we can solve the problem of ILP (3.1-5) with at most $\log_2 m$ different inputs ω by using binary search technique, because the problem of ILP (3.1-5) with $\omega = m$ has a trivial solution (m connections can be assigned by using m wavelengths).

Now what we need to do is to find an optimal solution to each of the $\log_2 m$ instances of ILP (3.1-5), this can be done by using the branch-and-bound algorithm. However, this method demands exponential time in terms of instance input in worst case since the problem of ILP in general is $\mathcal{NP}$-hard. One way to find a good approximate solution to the problem of ILP is to, first, get a problem of LP by relaxing the integer constraint of ILP; Secondly, find an optimal solution to LP (this can be done in polynomial-time); Finally, round the obtained optimal (fractional) solution to get a (provably) good integer solution to the problem of ILP. Rounding can be done either deterministically or randomly based on the information contained in the optimal fractional solution. This standard technique was used in Section 2 of Chapter 2 for solving the load balancing problem in general networks (i.e., Problem 2.5).

4. Routing for Wavelength Assignment

In this section, we consider the *Routing and Wavelength Assignment Problem* (RWAP), which is also called the *Wavelength Routing Problem.* RWAP is different from Problem 3.1, the wavelength assignment problem, which assumes that the routes of connections are given. Here, only the source-destination node pairs of connections are given, we are allowed to find and choose a proper route for each of the given connections in order to assign them by using minimal number of wavelengths. The RWAP in general can be formulated as follows, where we do not specify connections to be bidirectional or unidirectional.

PROBLEM 3.2 *Routing and Wavelength Assignment Problem*

Instance A graph $G(V, E)$ and a set of pairs of vertices $s_i, t_i \in V$, $i = 1, \cdots, m$.
Solution A path $p(s_i, t_i)$ between s_i and t_i in G for each i with an assigned wavelength such that paths $p(s_i, t_i)$ and $p(s_j, t_j)$ are assigned different wavelengths if they include a common link for $i \neq j$.
Objective Minimizing the number of wavelengths required for m paths.

Although we have proved in Section 2 that the wavelength assignment problem in ring networks is $\mathcal{NP}$-hard, this does not necessarily imply that the routing and wavelength assignment problem in ring networks is also $\mathcal{NP}$-hard, because in the latter we have the additional freedom to choose one of two possible routes for each of given connections. Nevertheless, the following theorem [16] gives a positive answer.

THEOREM 3.11 *The Problem 3.2 for unidirectional connections is $\mathcal{NP}$-hard in ring networks.*

PROOF We consider the decision version of the Problem 3.2 for unidirectional connections. The problem is to decide whether, given the source-destination pairs of m connections and a positive integer $k > 0$, the connections can be routed in such a way that they can be assigned by using at most k wavelengths. In the following, as in the proof of Theorem 3.5 we will convert an instance of the problem of determining the chromatic number of a circular-arc graph to the decision version of Problem 3.2 in such a way that the latter has a desired wavelength assignment of given connections if and only if the former has a proper coloring of the circular-arc graph.

Consider an instance of chromatic number of circular-arc graph consisting of the collection arcs $(s_1, t_1), \cdots, (s_m, t_m)$ and a positive integer $k < m$, where the i-th arc is from s_i to t_i. Note that s_i and t_i can be assumed to be nonnegative integers. Let $n \leq 2m$ be the number of distinct s_i and t_i. Without loss of generality, we can further assume that s_i and t_i are nonnegative integers less than n. From such an instance, an instance of the decision version of the Problem 3.2 in a ring can be constructed as follows. The ring has $(k+1)n$ vertices, the bound is k, and m (long) connections $((k+1)s_1, (k+1)t_1), \cdots, ((k+1)s_m, (k+1)t_m)$ on ring R. We also introduce k (short) connections $(j, j-1)$ for every vertex j in the ring, $0 \leq j < (k+1)n$.

Suppose that the given circular-arc graph is k-colorable. Then the long connections can be routed clockwise and assigned wavelengths accordingly, while the short connections can be routed anticlockwise and assigned wavelengths arbitrarily subject to each set of k identical short connections being assigned all k wavelengths.

Now suppose, on the other hand, that the circular-arc graph is not k-colorable, but the constructed instance of Problem 3.2 can still be assigned with k wavelengths. Then at least one of the long connections, say (s_1, t_1) must be routed anticlockwise. Hence for each link between $(k+1)s_1 - i$ and $(k+1)s_i - i - 1$, $0 \leq i < k+1$, at least one short connection must be routed in the clockwise direction around the ring. This adds $k+1$ to the load of any other clockwise link, which contradicts the assumption that k wavelengths are sufficient for the instance of Problem 3.2. The proof is then finished. ∎

4.1 Optimization through Load Balancing

Since the network load, in terms of number of connections, is a lower bound of the minimal number of wavelengths required, one may expect to use less number of wavelengths by reducing the network load. Fig. 3.11(a) gives a routing of four bidirectional connections in a ring of four vertices, $(0,1), (1,2), (2,3)$ and $(3,0)$. It yields ring load one, thus one wavelength is sufficient for all four connections. Fig. 3.11(b) gives another routing that yields ring load three, and 4 wavelengths are required for the same four connections.

The following theorem [14, 16] shows that a simple method, which deals with routing and wavelength assignment separately, can produce a good approximate solution for the Problem 3.2 in ring networks.

THEOREM 3.12 *There is a 2-approximation algorithm for the Problem 3.2 in the ring networks.*

PROOF Given a set of connections from source s_i to destination t_i for $i = 1, 2, \cdots, m$, we can first find a directed path $p(s_i, t_i)$, by running the algorithm in Section 2 of Chapter 2, such that they induce the minimum ring load, denote it by L_{opt}. And then

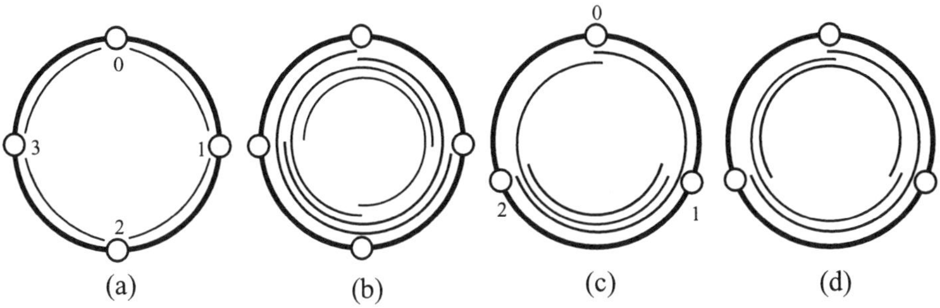

Figure 3.11. (a-b) The number of wavelengths required is reduced when the ring load is reduced; (c-d) The number of wavelengths required remains unchanged when the ring load is reduced.

by applying the algorithm described in Section 2, we can assign directed paths of connections in $\{p(s_i, t_i) \mid i = 1, 2, \cdots, m\}$ by using at most $(2L_{opt} - 1)$ wavelengths. Since an optimal algorithm for the Problem 3.2 must cause at least ring load L_{opt} (no matter how it routes the connections) and use at least L_{opt} wavelengths. Therefore, this two-phrase algorithm outputs a 2-approximate solution to the Problem 3.2 in ring networks. ■

The methodology used in the proof of Theorem 3.12 can be extended to general networks, that is to solve Problem 3.2 through solving two separate problems, namely the load balancing problem (Problem 2.5) and the wavelength assignment problem (Problem 3.1). However, it is very difficult (if not impossible) to design an approximation algorithm, based on this two-phrase approach, with guaranteed worst case performance. The reason is that it is very difficult to establish the relationship between the network load and the minimal number of wavelengths required. Fig. 3.11(c, d) gives a simple example showing that lowering the network load may not necessarily reducing the number of wavelengths required. In Fig. 3.11(c) three connections are routed in the worst way that makes network load three, so three wavelengths are required. In Fig. 3.11(d) the same three connections are routed in a better way so that the network load is reduced from three to two. However, three wavelengths are still needed.

Besides the above widely used approach, we can consider another option that is to route each of the given connections in its shortest path in terms of the number of links. This simple routing method does not directly aim at minimizing either the network load or the number of wavelengths used. Nevertheless it results in that the routes of given connections have few chance to overlap one another, so that one wavelength is able to be assigned to more connections. The following fact partially supports this idea. For ring networks, it can be proved that the network load induced by this simple routing method is at most two times that of the optimal routing method. This yields a 4-approximation solution to the Problem 3.2. Although compared with the result of Theorem 3.12 this approach is not good, it is simple and easy to be extended to networks of arbitrary topologies.

4.2 Integrated Approach

Instead of dealing with routing and wavelength assignment separately, we can tackle RWAP in an integrated way. This can be accomplished by applying the integer linear programming approach as we have done for WAP in Section 2 of Chapter 3. To do so, we need to add into the Problem of ILP (3.1-5) the following constraints to specify the path that the i-th connection uses for each i.

$$\begin{cases} \sum_u w_i^q(u,v) = w_i^q, & \text{if } v \text{ is the destination;} \\ \sum_u w_i^q(v,u) = w_i^q, & \text{if } v \text{ is the source;} \\ \sum_u w_i^q(u,v) = \sum_u w_i^q(v,u), & \text{Otherwise.} \end{cases}$$

Note that in the above $w_i^q(u,v)$ become variables since the route of the i-th connection needs to be determined for each i. The above equations ensure that if the i-th connection is assigned the q-th wavelength for some q, then every link on its route must be assigned the q-th wavelength as well.

In the following we outline the approach proposed by Kumar [7] for the Problem 3.2 in ring networks. This is a randomized algorithm that achieves an approximation ratio about 1.5. This is a better result than the deterministic method described in Theorem 3.12.

The algorithm first routes some (not all) of connections in an optimal way. And then it applies the integer linear programming approach to route the rest of connections and assign wavelengths to all connections. Finally, it rounds the fractional solution to an integer solution by randomized rounding.

The key technique of the algorithm is so called *parallel routing*. This is a similar technique that was used in Section 1 of Chapter 2 to solve the load balancing problem in ring networks. For the simplicity of presentation, we can consider a ring network as a circle, nodes in the ring as points on the circle, and paths in the ring as arcs on the circle. As a result the instance I of the Problem 3.2 in ring networks is now a collection of connections (node pairs) and arcs (all unassigned wavelengths, but some may have been routed). A collection C of arcs is said to be *obtainable* from I if C can be obtained from I by routing the connections in I. Denote by $O(I)$ the collection of all such sets of arcs obtainable from I. A (feasible) solution for instance I is some $C \in O(I)$ with wavelengths properly assigned to arcs in C.

A *conflicting pair* of arcs is a pair of arcs such that every point on the circle is contained in at least one of them and there is some point on both of them. A *parallel routing* is a collection of arcs that does not include any conflicting pairs. The following lemma [7] shows the importance of parallel routing.

LEMMA 3.6 *For any given instance of the Problem 3.2 in ring networks, there is an optimal solution whose arcs make a parallel routing.*

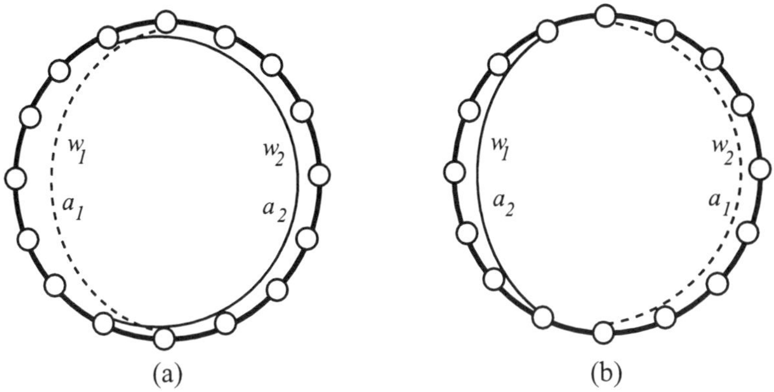

Figure 3.12. (a) A conflicting pair of arcs a_1 with wavelength w_1 and a_2 with wavelength w_2, (b) A conflicting pair disappeared by rerouting of connections and reassignment of wavelengths.

PROOF Suppose that an optimal solution contains a conflicting pair a_1 and a_2 whose corresponding connections are assigned wavelengths w_1 and w_2, respectively. See Fig. 3.12. Now we route these two connections in different way (there are two ways to route any connection in a ring network), and shift the wavelengths that they used. It is easy to see that this will not cause wavelength conflict. Moreover, this modification reduces the number of conflicting pairs in the solution without introducing any new wavelength. Thus repeating this process will produce a desired optimal solution. ■

The above lemma implies that we can find an optimal solution among parallel routings. The following lemma [7] gives some properties of parallel routings which can help us to locate an optimal solution.

LEMMA 3.7 *Let C_p be a parallel routing. Then for every link e there is another link e' such that no arc of C_p contains both e and e'.*

PROOF Let S_e denote the set of arcs in C_p that contains link e of the ring. Let a (a') be the arc in S_e whose clockwise (anticlockwise) endpoint is farthest from e. Note that S_e can not contain the whole circle, since otherwise a and a' would contain the whole circle and overlap each other over e. Thus a and a' constitute a conflicting pair, this contradicts that S_p is a parallel routing. Now there exists some link e' that is not included in either a or a' since a and a' do not contain the whole circle. Therefore, there does not exist an arc containing both e and e'. ■

Each of two links e and e' in the above lemma is called a *complement* of the other. The removal of a complement pair e and e' would partition the ring into two pieces, which we call a *complementary bisection* of the ring and denote it by $CB(e, e')$. The following lemma [7] gives some properties of complementary bisection which can help us to locate an optimal solution.

LEMMA 3.8 *If both the endpoints of an arc lie in one of two halves of some complementary bisection $CB(e, e')$, then the arc is contained entirely in that half.*

PROOF If not so, the arc will contain both e and e'. This contradicts Lemma 3.7. ■

Based on Lemma 3.7-8 a randomized algorithm [7] is designed as follows. First select a link e randomly. Assume first that a link e' could be found that is the complement of e in some optimal solution C_{opt} whose arcs constitute a parallel routing. According to Lemma 3.6, there exists a such an optimal solution. Then we consider the following two cases of connections and route them in two different ways:

Case 1. The source and destination are both in one of two halves of $C_{opt}(e, e')$. The connection is routed in that half, because in C_{opt}, it must be routed in that way according to Lemma 3.8. Let I' be the resulting instance after all connections belonging to this case are routed. Clearly, C_{opt} is still obtainable from I'.

Case 2. The source and destination are in different halves of $C_{opt}(e, e')$. All connections in both cases are routed and assigned wavelengths by using the integer linear programming approach.

Since we do not know C_{opt}, we cannot find the desired e'. To get this problem around, we just repeat the above process $(n - 1)$ times by trying all possible links of e' and take the best from the obtained solutions. It is obvious that this produces a solution no worse than the one we would assume that we know e'. Kumar [7] obtained the performance of this algorithm through a series of arguments.

THEOREM 3.13 *Given any instance I of the Problem 3.2 in ring networks, the number of wavelengths required by the randomized algorithm is no more than $(1.5 + 1/2e + o(1))Opt(I) + O(\sqrt{Opt(I)\ln|V|})$, where $Opt(I)$ is the number of wavelengths required for instance I by an optimal algorithm.*

Recently Kumar [6] has improved the above result by proposing another randomized approximation algorithm which has an asymptotic performance ratio of $(1 + 1/e)$.

5. Discussion

In this chapter we have studied the routing and wavelength assignment problem, in particular, the wavelength assignment problem. Some best results obtained so far are summarized at Table 3.1. By "αL being necessary" we mean that assigning some set of connections with the maximal link load L requires at least αL wavelengths.

In this chapter we have assumed implicitly that there is only one fiber on each link between two nodes. Li and Sinha [9] studied the WAP (Problem 3.1) in multifiber networks. Under the assumption that each link has k fibers and any wavelength channel on the i-th fiber can be switched to the same wavelength channel on the j-th fiber with $1 \leq i, j \leq k$. They proved that at most $(\frac{k+1}{k^2}L - 1)$ wavelengths

Table 3.1. The number of wavelengths required for connections in tree and ring networks.

Networks	Unidirectional connections	Bidirectional connections
Tree	$\frac{5}{3}$-approximation available [8]	$\frac{3}{2}$-approximation available [14]
	$\frac{5}{4}L$ wavelengths are necessary [8]	$\frac{3}{2}L$ wavelengths are necessary [14]
Ring	2-approximation available [14, 16]	
	$(2L-1)$ wavelength are necessary [16]	

are required on each fiber to support any set of connections with network load L. Note that if a k-fiber ring is just simply decomposed into k parallel rings with approximately L/k network load on each of them, $(2L/k - 1)$ wavelengths are sufficient for supporting any set of connections with network load L according to Theorem 3.6. This means that a proper wavelength assignment to connections in multifiber networks can reduce the number of wavelengths required for multiple single-fiber networks.

In the formulations of WAP (Problem 3.1) and RWAP (Problem 3.2), we assume that the connections are given at one time. This can be considered as a static traffic model. Mokhtar and Azizoglu [12] studied the RWAP under dynamic traffic model. In this case the connections are requested one after another. When a connection request arrives, we need to find a route and assign it a wavelength for the connection. If no path or no wavelength can be found available for the connection, it has to be blocked. The objective is to minimize the blocking probability. They investigate five adaptive routing and wavelength assignment algorithms by considering different sorting mechanisms of the wavelength set, where if no path is found after searching exhaustively the wavelength set, the connection request is blocked. *Packing algorithm* attempts to route the connection on the most utilized wavelength first, i.e., wavelengths are searched in descending order of utilization, in order to maximize the utilization of available wavelengths. *Spread algorithm* attempts to route the connection on the least utilized wavelength first, i.e., wavelengths are searched in ascending order of utilization, in order to achieve a near-uniform distribution of the load over the wavelength set. *Exhaustive algorithm* searches all of the wavelengths for the shortest available path and the shortest path among them is selected. *Random algorithm* searches the wavelength set in a random order with a uniform distribution over the set of all permutations. *Fixed algorithm* searches the wavelength set according to a fixed order *a priori*. Their study shows that the spread algorithm is the most efficient, closely followed by the random algorithm and the fixed algorithm, and then the pack algorithm, in the end the exhaustive algorithm. But in terms of blocking performance the order is reversed.

References

[1] C. Berge, *The Theory of Graphs and its Applications*, John Wiley, 1962.

[2] T. Erlebach and K. Jansen, Scheduling of virtual connections in fast networks, *Proceedings of the 4th Workshop on Parallel Systems and Algorithms* (PASA), 1996, 13-32.

[3] T. Erlebach, K. Jansen, C. Kaklamanis, and P. Persiano, An optimal greedy algorithm in directed tree networks, *DIMACS Series in Discrete Mathematics and Theoretical Computer Science*, 40 (1998), 117-129.

[4] M. R. Garey, D. S. Johnson, G. L. Miller, and C. H. Papadimitriou, The complexity of coloring circular arcs and chords, *SIAM Journal on Algebraic Discrete Methods*, 1 (2) (1980), 216-227.

[5] I. Holyer, The NP-completeness of edge-coloring, *SIAM Journal on Computing*, 10 (4) (1981), 718-720.

[6] V. Kumar, An approximation algorithm for circular arc coloring, *Algorithmica*, 30 (2001), 406-417.

[7] V. Kumar, Approximating circular arc coloring and bandwidth allocation in all-optical ring networks, *Lecture Notes in Computer Science*, 1444 (1998), 147-158.

[8] V. Kumar and E. J. Schwabe, Improved access to optical bandwidth in trees, *Proceedings of the 8th Annual ACM-SIAM Symposium on Discrete Algorithms* (SODA), 1997, 437-444.

[9] G.-Z. Li and R. Sinha, On the wavelength assignment problem in multifiber WDM star and ring networks, *IEEE/ACM Transactions on Networking*, 9 (1) (2001), 60-68.

[10] C. Lund and M. Yannakakis, On the hardness of approximating minimization problems, *Journal of the ACM*, 41 (5) (1994), 960-981.

[11] D. W. Matula, G. Marble, and J. D. Isaacson, Graph coloring algorithms, in *Graph Theory and Computing*, R. C. Read, Ed., New York and London: Academic, 1972, 109-122.

[12] A. Mokhtar and M. Azizoglu, Adaptive wavelength routing in all-optical networks, *IEEE/ACM Transactions on Networking*, 6 (2) (1998), 197-206.

[13] T. Nishizeki and K. Kashiwagi, On the 1.1 edge-coloring of multigraphs, *SIAM Journal on Discrete Mathematics*, 3 (3) (1990), 391-410.

[14] P. Raghavan and E. Upfal, Efficient routing in all-optical networks, *Proceedings of the 26th Annual ACM Symposium Theory of Computing* (STOC), 1994, 134-143.

[15] M. Slusarek, A coloring algorithm for interval graphs, *Lecture Notes in Computer Science*, 379 (1989), 471-480.

[16] G. Wilfong and P. Winkler, Ring routing and wavelength translation, *Proceedings of the 9th Annual ACM-SIAM Symposium on Discrete Algorithms* (SODA), 1998, 333-341.

[17] M. Yannakakis and F. Gavril, Edge dominating sets in graphs, *SIAM Journal on Applied Mathematics*, 38 (1980), 364-372.

Chapter 4

DESIGN OF LOGICAL TOPOLOGIES

As we addressed in the previous chapter, in single-hop systems of WDM networks a connection is set up by selecting a path of physical links and assigning a (reserved) wavelength to these links. Because of limitations on the number of available wavelengths on the fibers, and hardware constraints at the nodes, it is impossible to set up a lightpath for each of connections between every pair of source-destination nodes. The set of lightpaths that are set up on the physical topology constitutes a logical (also called virtual) topology. One of the major advantages of WDM networks is that it is able to reconfigure its logical topology to adapt to changes of traffic patterns and network conditions.

In this chapter we will study the *Logical Topology Design Problem* (LTDP). It aims at improving some network performances, including minimum network congestion and maximum network throughput, under given network conditions, such as traffic between node pairs and the number of available wavelengths.

The rest of this chapter is organized as follows. Section 1 presents work [4, 7] due to Sivarajan et al on the LTDP in WDM networks without wavelength converters aiming at minimization of the network congestion. Section 2 presents Banerjee and Mukherjee's work [1, 5] on the LTDP in WDM networks with wavelength converters aiming at maximization of the network throughput. Although these works are presented in two sections (as they were studied separately in the literature) and their goals appear to be quite different, we will show that these two variants of LTDP, in fact, are closely related to each other. Both of them can be formulated as mixed integer linear program and solved either by rounding the fractional solutions to its linear program relaxation as in Section 1, or by some heuristics as in Section 2. Moreover, the maximization of the network throughput can be achieved through minimization of the network congestion. Section 3 concludes the chapter addressing some related issues under this topic.

1. Minimization of Network Congestion

In this section we will focus on the problem of constructing logical topologies over a wavelength-routed optical network with no wavelength converters. We will present a general liner formulation which takes into account most of important parameters of WDM networks, including traffic matrix, the maximal number of hops a lightpath is permitted to take, multiple logical links, and symmetry restrictions on the logical topologies. The objective is to minimize the congestion of the logical topologies.

1.1 Network Model and Problem Description

A *physical topology* is a graph representing the physical interconnection of the wavelength routing nodes by means of fiber optical cables. We assume that an edge in the physical topology represents a pair of fibers, one in each direction. For the simplicity of the presentation, each edge is indicated by a single edge with arrow at both ends. Fig. 4.1(a) shows a physical topology of a six-node network with seven edges in [4]. The numbers on the edges represent the physical distances between end nodes, which could also be used as relative propagation delays between them.

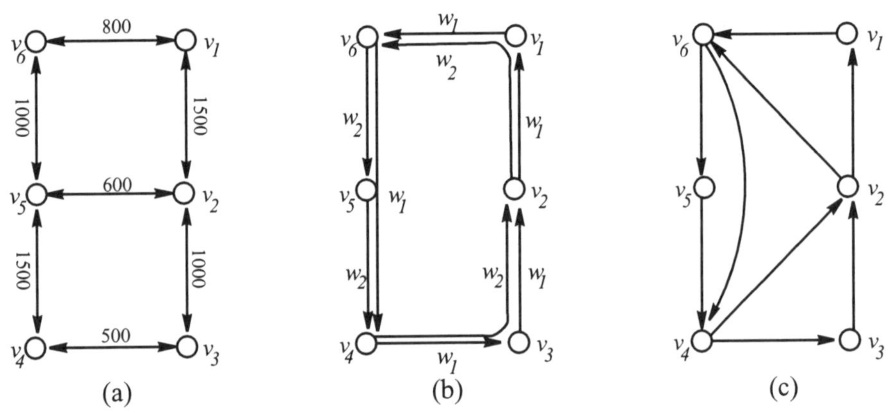

Figure 4.1. (a) Physical topology, (b) Configuration of routing nodes, (c) Logical topology.

Fig. 4.1(b) shows a possible configuration of the wavelength routing nodes of Fig. 4.1(a). There is an directed path from vertex v_2 to vertex v_6. This means that the data from routing node v_2 to routing node v_6 traverse the optical network in the optical domain only, carrying wavelength w_2. In this case there is no electronic conversion in the intermediated routing node v_1. Notice that another option to send the data from node v_2 to node v_6 is to use two directed paths carrying wavelength w_1, from vertex v_2 to vertex v_1 first and then from vertex v_2 to vertex v_1. It is important that the data could be sent from node v_2 to node v_6 on wavelength w_2 through node v_1 via physical fiber connecting v_2 and v_1, and simultaneously another data could be sent from node v_2 to node v_1 on wavelength w_1 via the same physical fiber between v_2 and v_1.

A *logical topology* is a directed graph that is obtained after the lightpaths are set up by properly configuring the wavelength routing nodes. Fig. 4.1(c) shows the logical topology of the configuration of Fig. 4.1(b). Each arc in the logical topology corresponds a lightpath in the physical topology, which is called a *logical link*.

Given a physical topology, there are different ways to set up lightpaths, which result in different logical topologies. A trivial logical topology is a complete digraph, that is, a lightpath is set up from each node to each other node. Thus there will be $|V| \times |V|$ lightpaths in total. However, this is generally impossible due to the following two reasons.

1 The number of wavelengths available is limited, this imposes a restriction on the number of lightpaths that can be supported. Study [6] shows that on average in an 128-node network of degree four, in total about 640 full-duplex lightpaths that use 32 wavelengths can be set up. For each node only approximately 12 full-duplex lightpaths can be set up, this is much less than the 127 lightpaths required to be connected with all other 127 nodes.

2 Each node can serve as the source and destination nodes for only a limited number of lightpaths. This is determined by the amount of optical hardware that a given WDM network has at the node and also by the amount of information that the node can handle.

Another trivial logical topology is the physical topology. In this case, as most of node pairs are not directly connected via lightpaths, they must use many lightpaths through intermediate nodes to communicate, that is, the number of physical links in the shortest path between them in terms of hops. Optical-electronic conversion will occur at every intermediate node. Thus it will result in longer delay and heavier congestion. In general, to design a good logical topology we need to take into account the following factors.

(1) Traffics of connections between nodes. The *traffic matrix* is a $|V| \times |V|$-matrix. Each entry gives the average traffic from one node to the other in the physical topology. It may be expressed as arriving packets per second, or a quantized bandwidth requirement, or some other suitable units. The matrix provides in numerical terms the nature of how the total network traffic is distributed between different source-destination node pairs, that is, the pattern of the network traffic.

(2) Symmetry of logical topologies. A logical topology is said to be *symmetric* if whenever there is a logical link from node v to node u, there is a logical link from node u to node v, and the corresponding lightpaths pass through the same set of intermediate wavelength routing nodes; Otherwise it is called *asymmetric*. Since

the logical topology can reflect the traffic intensities between the various nodes, a symmetric logical topology will result in the physical topology load balanced.

(3) Degrees of vertices in logical topologies. In a logical topology, the *in-degree* of a vertex is equal to the number of arcs terminating on it and the *out-degree* of a vertex is equal to the number of arcs originating from it. A Δ-*regular* logical topology is one where all the vertices have the same in-degree Δ and out-degree Δ. We simply say the vertex degree when the in-degree and out-degree of a vertex is assumed to be equal (as usually they are). The degree of a vertex in a logical topology is essentially determined by the degree of the node in the physical topology, but it is also restricted by the size of wavelength router equipped at the node.

(4) Multiplicity of logical topologies. A logical topology is said to have *multiplicity* κ if the maximal number of arcs between any node pair is κ.

(5) Hop lengths of logical links. The *hop length* of a logical link is the number of physical links whose corresponding lightpath traverses. For example, in Fig. 4.1(c) logical links (v_6, v_4), (v_4, v_2) and (v_2, v) all have hop lengths two while other links have hop lengths one. Observe that although there is a fiber connection between node v_2 and node v_3 in the physical topology, for unidirectional connection from v_2 to v_3 the data would have to go through three lightpaths (or equivalently use three logical links), from v_2 to v_6, and then from v_6 to v_2, in the end from v_4 to v_3. A logical topology is said to be *hop limited* if there is a restriction on the maximal number of hops a lightpath is allowed to take. The hop length can be used as a measure of the number of nodes encountered while setting up a logical link. If the hop length of a logical link is large, then there would be degradation of the optical signal by attenuation and crosstalk at the intermediate nodes. Therefore, it is important to keep the hop lengths of the logical links small.

(6) Wavelengths available. A logical topology is said to be *wavelength limited* if there is a restriction on the maximal number of wavelengths that can be used when setting up the lightpaths in the physical topology. Wavelengths are a scarce resource of WDM networks, so they have to be used in an optimal way.

(7) Congestion of logical topologies. The *congestion* of a logical link is defined as the aggregate traffic flowing through the corresponding lightpath. The *congestion* of a logical topology is defined as the maximum congestion over all logical links in the logical topology.

To construct a logical topology we need to establish some logical links, allocate wavelengths to them and route the traffics of connections over the logical links. An informal description of the *Logical Topology Design Problem* (LTDP) is to construct a logical topology such that the resources of the network subject to the given constraints are used optimally. In the following we will study the LTDP whose objective is to minimize the congestion. The reason for choosing the congestion (among other parameters of the network) as the target of our optimization is that the electronic processing (switching speed) requirement is proportional to the congestion. If the switching speeds at the nodes are limited, then minimizing congestion would make the overall speed of the network faster.

PROBLEM 4.1 *Logical Topology Design Problem*

Instance A physical topology $G(V, E)$, the traffic between node pairs, the number of wavelengths available on the fibers, the maximal number of hops that a logical link is allowed to take, the number of transmitters and receivers at the nodes, symmetry/asymmetry requirement.

Solution A logical topology satisfying the given requirements.

Objective Minimizing the congestion of the logical topology.

1.2 Formulation of Combined Optimization

In this subsection we will give a general linear formulation of Problem 4.1 of TLDP. It produce a combined optimization problem through integrating setting up lightpaths between pairs of nodes with routing the traffics of connections between nodes. We will introduce and use the following notations to facilitate the mathematical formulation of the problem.

$\lambda_{s,d}$: traffic from source s to destination d in $G(V, E)$;
(u, v): physical link between nodes u and v in $G(V, E)$;
(i, j): the logical link from nodes i to j in the logical topology;
k: k-th multiple logical link in the logical topology, here $k \leq \kappa$, that is a given bound;
w_q: the q-th wavelength, here $q \leq \omega$, that is a given bound;
t_i, r_i: number of transmitters and receivers at node i, here $t_i \leq \tau, r_i \leq \gamma$, that are two given bounds;
$h_{i,j}$: maximal number of hops that (i, j) is allowed to take, here $h_{i,j} \leq H$, that is a given bound;
$e_{u,v}$: indicator of a physical link between nodes u and v, i.e., if $e_{u,v} = 1$ then there is a fiber link $(u, v) \in E$, otherwise $e_{u,v} = 0$;
$l^k(i, j)$: logical link variables, $l^k(i, j) = 1$ if there exists a k-th multiple logical link (i, j), otherwise $l^k(i, j) = 0$;
$w^{k,q}(i, j)$: wavelength assignment variables, $w^{k,q}(i, j) = 1$ if the k-th logical link (i, j) is assigned wavelength w_q, otherwise $w^{k,q}(i, j) = 0$.
$w_{u,v}^{k;q}(i, j)$: routing variables, $w_{u,v}^{k;q}(i, j) = 1$ if the k-th logical link (i, j) with w_q is routed through (u, v), otherwise $w_{u,v}^{k;q}(i, j) = 0$);
$\lambda_{s,d}^k(i, j)$: flowing variables on the k-th logical link (i, j) that carry the fraction of traffic $\lambda_{s,d}$ from s to d;
$\lambda^k(i, j)$: loading variables on the k-th logical link (i, j), that is, $\lambda^k(i, j) = \sum_{s,d} \lambda_{s,d}^k(i, j)$;
L_{max}: network congestion, that is, $L_{max} = \max\{\lambda^k(i, j) | (i, j), k\}$.

Now in terms of the above notations and variables we can easily specify the constraints of Problem 4.1.

The *logical link degree constraint* can be formulated as follows.

$$\begin{cases} \sum_{k=1}^{\kappa} \sum_j l^k(i,j) \le t_i, & \text{for all } i, \\ \sum_{k=1}^{\kappa} \sum_j l^k(j,i) \le r_i, & \text{for all } i, \\ l^k(i,j) \in \{0,1\}. \end{cases} \tag{4.1}$$

The above constraint ensures that the number of logical links originating (out-degree) and terminating (in-degree) at node i is less than or equal to the number of transmitters and receivers at the node. When $\kappa = 1$, no multiple logical link is allowed.

The *unique wavelength constraint* can be formulated as follows.

$$\begin{cases} \sum_{q=1}^{\omega} w^{k,q}(i,j) = l^k(i,j), & \text{for all } (i,j) \text{ and } k, \\ w_{u,v}^{k,q}(i,j) \le w^{k,q}(i,j), & \text{for all } (i,j), (u,v), k \text{ and } q. \end{cases} \tag{4.2}$$

The equality in the above constraint means that if logical link $l^k(i,j)$ exists then only one wavelength (among ω available wavelengths) is assigned to it. The inequality in above constraint ensures that only those $w_{u,v}^{k,q}(i,j)$ could be nonzero whose corresponding $w^{k,q}(i,j)$ variables are nonzero. If wavelength w_q is chosen for the k-th logical link (i,j). Then $w^{k,q}(i,j) = 1$. This means, for all other wavelengths $w_{q'}$ with $q' \neq q$, $w^{k,q'}(i,j) = 0$. Thus the inequality would force $w_{u,v}^{k,q'}(i,j) = 0$, for all (u,v) and $q' \neq q$.

The *wavelength conflict constraint* can be formulated as follows.

$$\sum_k \sum_{i,j} w_{u,v}^{k,q}(i,j) \le 1, \quad \text{for all } (u,v) \text{ and } q. \tag{4.3}$$

In the above inequality we sum up all possible logical links (i,j) traversing a physical link (u,v) and for a wavelength w_q, by which we ensure that there is no wavelength conflict at physical link (u,v). That is, no two logical links traversing through the physical link (u,v) will be assigned the same wavelength.

The *wavelength conservation constraint* can be formulated as follows.

$$\sum_{q=1}^{\omega} \sum_u w_{u,v}^{k,q}(i,j) e_{u,v} - \sum_{q=1}^{\omega} \sum_u w_{v,u}^{k,q}(i,j) e_{v,u}$$

$$= \begin{cases} l^k(i,j), & \text{if } v = j, \\ -l^k(i,j), & \text{if } v = i, \\ 0, & \text{if } v \neq i \text{ and } v \neq j. \end{cases} \quad \text{for all } (i,j), k \text{ and } v \tag{4.4}$$

The above equation ensures that a wavelength is conserved at every node for a logical link $l^k(i,j)$. This is analogous to the flow conservation equations in multicommodity flow problems. Refer to equations (2.4). If logical link $l^k(i,j)$ uses wavelength w_q, then this constraint guarantees that there is a path in the physical topology from node i to node j assigned wavelength w_q.

The *traffic routing constraint* can be formulated as follows.

$$\begin{cases} \lambda_{s,d}^k(i,j) \le l^k(i,j)\lambda_{s,d}, & \text{for all } (i,j), (s,d) \text{ and } k, \\ \lambda^k(i,j) = \sum_{s,d} \lambda_{s,d}^k(i,j), & \text{for all } (i,j) \text{ and } k, \\ \lambda^k(i,j) \le L_{max}, & \text{for all } (i,j) \text{ and } k. \end{cases} \tag{4.5}$$

The first inequality of the above constraint ensures that $\lambda_{s,d}^k(i,j)$ can have a nonzero value if there exists the k-th logical link (i,j) (i.e., $l^k(i,j) \neq 0$), and the traffic on it flowing from node s towards node d is upper bounded by the total traffic $\lambda_{s,d}$ from s to d. The other two inequalities ensure that the load on any logical link is no greater than the maximum load L_{max}, which is to be minimized.

The *flow conservation constraint* can be formulated as follows.

$$\sum_k \sum_j \lambda_{s,d}^k(i,j) - \sum_k \sum_j \lambda_{s,d}^k(j,i)$$

$$= \begin{cases} \lambda_{s,d}, & \text{if } s = i, \\ -\lambda_{s,d}, & \text{if } d = i, \\ 0, & \text{if } s \neq i \text{ and } d \neq j. \end{cases} \quad \text{for all } (s,d) \tag{4.6}$$

The above equality ensures flow conservation for the traffic from node s to node d at each node i. It is worthwhile to point out that the traffic from node s to node d may be bifurcated, that is, some fractions of traffic may flow through different lightpaths from s to d. This is unlike the case of the load balancing problem (Problem 2.5) addressed in Chapter 2.

The *hop bound constraint* can be formulated as follows.

$$\sum_{u,v} w_{u,v}^{k,q}(i,j) \le h_{i,j}, \quad \text{for all } (i,j), k \text{ and } q.$$

The left of the above inequality sums up all the physical links (u,v). It ensures that the number of hops in the k-th logical link is bounded by $h_{i,j}$.

The *symmetry constraint* can be formulated as follows.

$$w_{u,v}^{k,q}(i,j) - w_{v,u}^{k,q}(j,i) = 0, \quad \text{for all } q, (i,j), k, \text{ and } (u,v). \tag{4.7}$$

The above inequality ensures that the number of logical links from node i to node j is equal to the number of logical links from node j to node i. Moreover, the routing and wavelength assignment for the lightpaths associated with logical links between node i and to node j traverse the same set of physical links and are assigned the same wavelength.

In addition, if multiple fibers in the physical topology are allowed, say ρ multiple fibers on link (u, v), then we need to modify the variable $w_{u,v}^{k;q}(i,j)$ to $w_{u,v}^{k,q,p}(i,j)$, where $p \in \{1, 2, \cdots, \rho\}$, and then modify the related constraints accordingly.

In the end, the logical topology design problem (Problem 4.1) can be formulated as the following optimization problem.

$$
\begin{array}{ll}
\text{Minimize} & L_{max} \\
\text{Subject to} & \text{Constraints } (4.1 - 4.7); \\
& l^k(i,j), w^{k,q}(i,j), w_{u,v}^{k;q}(i,j) \in \{0,1\}, \lambda_{s,d}^k(i,j) \geq 0, \\
& \text{for all } (i,j), (u,v) \in V \times V, 1 \leq k \leq \kappa, 1 \leq q \leq \omega.
\end{array}
$$

This is a problem of *Mixed Integer Linear Program* (MILP) [2], since variables $l^k(i,j), w^{k,q}(i,j), w_{u,v}^{k;q}(i,j)$ should be integers (in fact, either 0 or 1) while $\lambda_{s,d}^k(i,j)$ are not. This problem has $O(\kappa\omega|E||V|^2)$ constraints and the same magnitude of variables. Since the problem of ILP in general is $\mathcal{NP}$-hard, an optimal solution to this problem can only be found for moderate sized networks, for example, using the cutting plane or the branch-and-bound methods. For larger networks, we can only expect to find approximate solutions by using some heuristics.

1.3 Solution Approach

In this subsection we outline a heuristic proposed in [4] for LTDP applying integer linear programming approach. The basic idea is to decompose Problem 4.1 into several subproblems. Each of them is solved one after another separately.

ALGORITHM 4.1 *Designing a Logical Topology*

Step 0 Relax the integer constrains on variables $l^k(i,j)$, $w^{k,q}(i,j)$, and $w_{u,v}^{k;q}(i,j)$, that is, use the constraint, $0 \leq l^k(i,j), w^{k,q}(i,j), w_{u,v}^{k;q}(i,j) \leq 1$, to replace the constraint, $l^k(i,j), w^{k,q}(i,j), w_{u,v}^{k;q}(i,j) \in \{0,1\}$, and obtain a linear programming (IP), and then find an optimal solution to IP.

Step 1 *Produce a logical topology*
Round $l^k(i,j)$ to 1 or 0 as follows. List fractional $l^k(i,j)$ in descending order. Round each successive value of $l^k(i,j)$ to 1 if the degree constraints are not violated, and to 0 otherwise.

Step 2 *Assign wavelengths to logical links*
Round $w^{k,q}(i,j)$ as follows. If $l^k(i,j) = 0$ then $w^{k,q}(i,j) = 0$ for all q. If $l^k(i,j) = 1$ then $\max\{w^{k,q}(i,j) \mid q\}$ is set to 1 and the rest to 0. (A tie may be broken by choosing the largest index q.)

Step 3 *Find paths for logical links*
Round and $w_{u,v}^{k;q}(i,j)$. Suppose that set $l^k(i,j) = 1$ and $w^{k,q}(i,j) = 1$. Among the possible set of physical paths from node i to node j with w_q, pick a path as follows. Set $w_{i,u}^{k,q}(i,j) = \max\{w_{i,v}^{k,q}(i,j) \mid v\}$.
(i) If $u = j$ then stop. Else set $w_{u,y}^{k;q}(i,j) = \max\{w_{u,z}^{k;q}(i,j) \mid z\}$.

(ii) If $y = j$ then stop. Else assign y to u and then go back to (i).
Set $w_{u,v}^{k,q}(i,j) = 1$ if (u,v) is in the picked path, $w_{u,v}^{k,q}(i,j) = 0$ otherwise.

Step 4 *Reassign wavelengths to the logical links*
Assign wavelengths to lightpaths by proposed approach in Chapter 3.

In Step 0, the integer constraints on some variables are relaxed so that an LP is obtained whose optimal solution can be found efficiently.

In Step 1, the simplest way to round fractional solution is used. It sets the variables whose values are closer to one to one, and the variables whose values are closer to zero to zero, while maintaining feasibility. At the end of this step, integral $t^k(i,j)$ are obtained so that a logical topology can be produced. However, the routes and wavelength assignment for logical links have not determined. These will be done in Step 2.

In Step 3, the conservation of wavelength equations guarantees that if $x(i,j) = 1$ and $w^{k,q}(i,j)$ then there is at least one path from node i to node j with wavelength assignment w_q.

In Step 4, after obtaining a logical topology and the lightpaths of logical links with assigned wavelengths, there may exist two lightpaths sharing a common physical link that are assigned the same wavelength. Thus wavelengths need to be reassigned to make current wavelength assignment conflict free. A simple method is adopted here. It completely ignores the obtained assignment and makes a new wavelength assignment. Another option is just to reassign conflicted wavelengths.

By using the above approach, the optimal solution may not be found, since the decomposition is merely approximate not exact. In practice, a even worse case may occur that the solution returned is not feasible for the original problem. Hence some of the constraints (4.1-7) may not be put into the MILP of Problem 4.1. One of the constraints which is often relaxed is that the maximal number of wavelengths available on fibers. The relaxed problem will become easy to be solved. Although its solutions may not satisfy all the constraints (4.1-7), they can provide some lower bounds on the optimal solution to the original problem. This methodology will be used in the next subsection.

1.4 Solution Analysis

In this subsection, we will demonstrate by some simple examples in [4] how the congestions of logical topologies are affected by various parameters of a network, such as the traffic matrix, number of wavelengths available on the fibers, number of transmitters and receivers at nodes, the hop lengths of the logical links, the multiplicity restrictions on the physical topology, and symmetry restrictions.

We first investigate the relationship between the congestion L_{max} and equality constraints on the number of transmitters $t_{i,j}$ and receivers $r_{i,j}$, here no multiplicity is allowed, i.e., $\kappa = 1$. Consider a simple network where there is a link between every pair of four nodes, and a traffic matrix is given in Table 4.1a.

Table 4.1a. Traffic matrix that produces logical topologies in Fig. 4.2.

Nodes	v_1	v_2	v_3	v_4
v_1	0	0	1	0
v_2	1	0	0	1
v_3	0	1	0	1
v_4	1	1	0	0

Table 4.1b. Traffic matrix that produces logical topologies in Fig. 4.4.

Nodes	v_1	v_2	v_3	v_4
v_1	0	1	1	0
v_2	1	0	0	1
v_3	1	0	0	1/3
v_4	0	1	1/3	0

Table 4.2. Traffic matrix that produces logical topologies in Fig. 4.3.

Nodes	v_1	v_2	v_3	v_4	v_5	v_6
v_1	0	1	0	0	0	1
v_2	1	0	1	1	0	0
v_3	0	1	0	1	1	0
v_4	0	1	1	0	1	0
v_5	0	0	1	1	0	1
v_6	1	0	0	0	1	0

Table 4.3. Traffic matrix that produces logical topologies in Fig. 4.5.

Nodes	v_1	v_2	v_3	v_4	v_5	v_6
v_1	0.000	0.537	0.524	0.710	0.803	0.974
v_2	0.391	0.000	0.203	0.234	0.141	0.831
v_3	0.060	0.453	0.000	0.645	0.204	0.106
v_4	0.508	0.660	0.494	0.000	0.426	0.682
v_5	0.480	0.174	0.522	0.879	0.000	0.241
v_6	0.950	0.406	0.175	0.656	0.193	0.000

Fig. 4.2(a) shows the logical topology obtained by solving the MILP of Problem 4.1 without the equality constraint on the degrees of vertices. Its congestion is $L_{max} = 1$. Fig. 4.2(b) shows the logical topology obtained by solving the MILP of Problem 4.1 with equality constraint $t_{i,j} = r_{i,j} = 2$, it has congestion $L_{max} = 1.33$. Fig. 4.2(c) shows the logical topology obtained by solving the MILP of Problem 4.1 with symmetry constraint, whose congestion is $L_{max} = 1.5$. Consider another simple network with physical topology as in Fig. 4.1(1), and a traffic matrix is given in Table 4.2.

Fig. 4.3(a) shows the logical topology obtained by solving the MILP of Problem 4.1 with the symmetry constraint but without equality constraint on the degrees of vertices. It has congestion $L_{max} = 1$. Fig. 4.3(b) shows the logical topology obtained by solving the MILP of Problem 4.1 with both the symmetry constraint and equality constraint $t_{i,j} = r_{i,j} = 3$. Its congestion is $L_{max} = 1.33$.

These examples suggest that the equality and symmetry constraints may at times increase the congestion. The reason for this surprising discovery is that due to the regular nature of the logical topology the traffic from a source node to a destination node may traverse more logical links than necessary, which tends to increase the congestion.

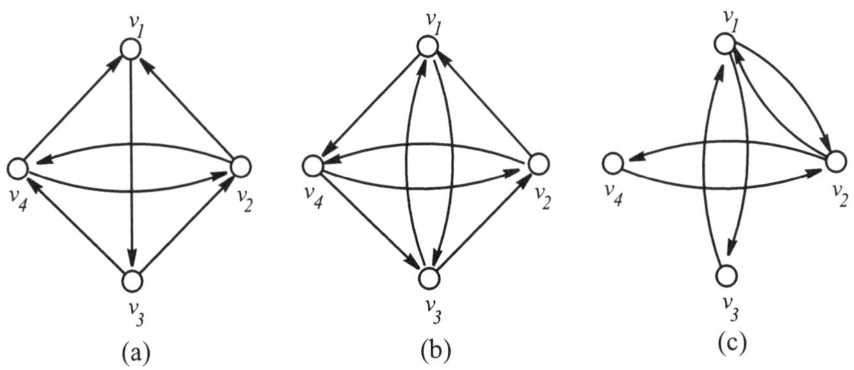

Figure 4.2. (a) The asymmetric logical topology with unequal vertex-degreess has $L_{max} = 1$. (b) The asymmetric logical topology with the same vertex-degrees has $L_{max} = 1.33$. (c) The symmetric logical topology with unequal vertex-degrees has $L_{max} = 1.5$.

Figure 4.3. (a) The symmetric logical topology with unequal vertex-degrees has $L_{max} = 1$. (b) The symmetric logical topology with the same vertex-degrees has $L_{max} = 1.33$.

Next we investigate the relationship between the congestion L_{max} and the multiplicity constraint κ. Here the equality constraint on the degrees of vertices is set to $t_{i,j} = r_{i,j} = \tau = \gamma$. Consider the same network of four nodes again but with a different traffic matrix as in Table 4.1b.

Fig. 4.4(a) shows the logical topology obtained by solving the MILP of Problem 4.1 with no multiplicity, i.e., $\kappa = 1$. It has congestion $L_{max} = 1$. Fig. 4.4(b) shows the logical topology obtained by solving the MILP of Problem 4.1 with multiplicity $\kappa \leq 2$. It has congestion $L_{max} = 1.33$.

This example suggests that by permitting multiple links the congestion may be reduced.

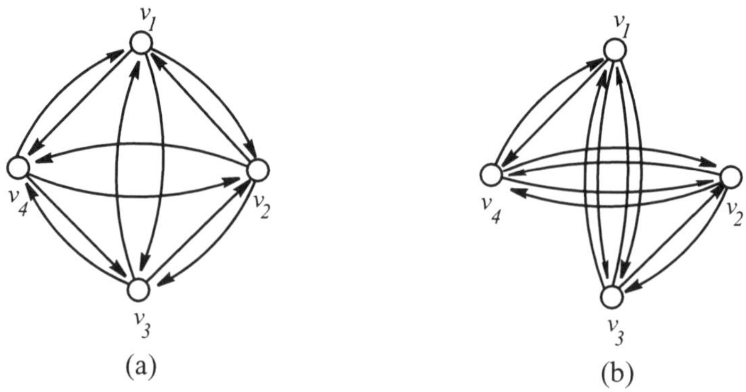

Figure 4.4. (a) The symmetric logical topology with $\tau = \gamma = 3$ and no multiple links has $L_{max} = 1$. (b) The symmetric logical topology with $\tau = \gamma = 3$ and multiple links has $L_{max} = 1.33$.

We now investigate the relationship between the congestion L_{max} and the number of wavelengths available on the fibers ω and hop-bound H. Consider the same network of six nodes again but with a different traffic matrix given in Table 4.3, whose entry is generated randomly from a uniform distribution in $(0, 1)$.

Fig. 4.5(a) shows the logical topology obtained by solving the MILP of Problem 4.1 with $\omega = 1$, $\tau = \gamma = 1$ and $H = 1$. Note that in this situation each node is equipped with one transmitter and one receiver. Thus there can only be two possible solutions for the logical topologies, one is the clockwise ring and the other is the anticlockwise ring. The former has congestion $L_{max} = 9.36$ and the latter as shown in Fig. 4.5(a) has the minimum congestion $L_{max} = 7.36$. Fig. 4.5(b) shows the logical topology obtained by solving the MILP of Problem 4.1 with $\omega = 1$ and $\tau = \gamma = 1$ but $H = 2$. We see that increasing the hop-bound H the congestion L_{max} is reduced from 7.36 to 7.077. In fact, solving the MILP of Problem 4.1 without constraints on the number of wavelengths ω and hop-bound H outputs a logical topology whose congestion is still 7.077. This implies that the congestion cannot be decreased further by increasing the values of ω and H. Fig. 4.5(c) shows the logical topology obtained by solving Problem 2 with $\omega = 1$, $\tau = \gamma = 2$, and $H = 1$. The congestion L_{max} is now reduced from 7.36 to 2.340.

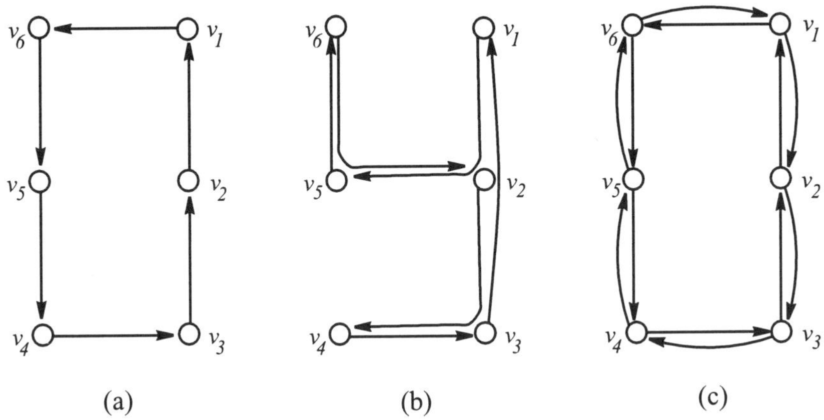

Figure 4.5. (a) The logical topology with one wavelength, equal vertex-degree one and hop-bound one has $L_{max} = 7.360$. (b) The logical topology with one wavelength, equal vertex-degree one and hop-bound two has $L_{max} = 7.077$. (c) The logical topology with one wavelength, equal vertex-degree two and hop-bound one has $L_{max} = 2.340$.

Fig. 4.6(a) shows the logical topology obtained by solving the MILP of Problem 4.1 with $\omega = 1$, $\tau = \gamma = 2$, and $H = 2$. The congestion L_{max} is now further reduced from 2.340 to 2.210. The reason behind this phenomenon is that two logical links of hop two, one from v_3 to v_1 and the other from v_4 to v_2, reduce the congestion of logical links from v_3 to v_2 and from v_4 to v_5 in Fig. 4.5(c), respectively. Fig. 4.6(b) shows the logical topology obtained by solving the MILP of Problem 4.1 with $\omega = 2$, $\tau = \gamma = 2$, and $H = 2$. The congestion L_{max} is now further reduced from 2.210 to 2.042. The reason behind this phenomenon is that more routes were produced when one more wavelength was introduced, so that the congestion could be reduced. In fact, the congestion cannot be further reduced by increasing the number of wavelengths and the number of hop-bound.

In the end, we study how the logical topology is affected by the traffic between node pairs $\lambda_{s,d}$. In the above examples, we notice that if there is a heavy traffic between some source-destination pair, then there is a logical link between them. For instance, in Fig. 4.2(a) the traffic from v_1 to v_3 is one and the traffic from v_3 to v_1 is zero, so there is a logical link from v_1 to v_3 and no logical link from v_3 to v_1. In Fig. 4.4(b) the traffics from v_1 to v_3 and from v_3 to v_1 are both one, the traffic from v_2 to v_4 and from v_4 to v_2 are both one, so there are two multiple logical links between v_1 and v_3, v_2 and v_4, respectively. This is exactly what we expect. That is, if there is a heavy traffic from node v_i to node v_j, then the objective function would tend to

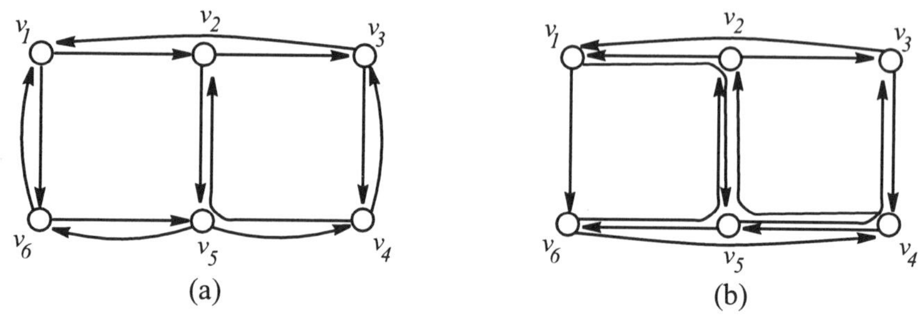

Figure 4.6. Logical topology with equal node-degrees two and hop-bounds two, (a) when only one wavelength is available and (b) when two wavelengths are available.

yield an arc (v_i, v_j) in the logical topology. If this does not happen, then the traffic from node v_i to v_j have to go through many logical links before it is delivered to its destination v_j, which makes the congestion to be increased. In general, when nodes have small degrees, it is more important to set up lightpaths for node pairs with fewer hops rather than those with heavier traffic. When nodes have large degrees, it is more helpful to create logical links for node pairs with heavier traffic.

2. Minimization of Average Packet Hop Distance

In this section, we turn to the LTDP in WDM networks employ wavelength routers/switches equipped with wavelength converters at the routing nodes studied in [1]. In this case, the wavelength continuity requirement is no longer assumed, that means, a lightpath does not need to be on the same wavelength across all the fibers that it traverses. As a result, the wavelength assignment of the LTDP will become considerably simple.

2.1 Network Model and Problem Description

Besides those parameters of WDM networks considered in last section, we will consider some other parameters in this subsection.

The *lightpath length bound* is a given parameter such that the length of the lightpath from node u to node v is no more than this bound times the length of the shortest path from node u to node v.

The *channel capacity* is a given parameter that is normally expressed in bits/s (but it can be converted to units of packets/s provided that the mean packet length is known). Each channel is allowed to carry at most this value of traffic from one node to another. When the traffic of a connection from a node to another is greater than the value, multiple lightpaths between them should be set up.

The *channel load* is a given parameter that prevents the queueing delay on a lightpath from getting unbounded by avoiding excessive link congestion. Here the queueing delay will not be incorporated explicitly in the formulation of the problem, since they are negligible as long as the value is chosen in a proper way [5]. Moreover, compared to propagation delays for a large network such as NSFnet, the queueing delays are negligibly small except under extremely heavy load. Although the queuing delays may be large in current (congested) backbone networks, the use of high-capacity WDM links along with silicon routers in the switching nodes will considerably alleviated this problem in future network.

The objective of the LTDP we study in this section is to minimize the average packet hop distance. The *average packet hop distance* is defined in [1] as the number of lightpaths that a packet has to traverse on average, and it is a function of the logical topology. The reason for choosing it as the target of our optimization is as follows. Under balanced load of all channels, we have

$$\text{Network throughput} \leq \frac{\text{Channel−capacity} \times \text{Number of channels}}{\text{Average packet hop distance}}. \qquad (4.8)$$

Therefore, minimizing the average packet hop distance is equivalent to maximizing the network throughput.

PROBLEM 4.2 *Logical Topology Design Problem*

Instance A physical topology $G(V, E)$, the traffic matrix, the maximum propagation delay that a lightpath is allowed to suffer, the maximum load that a lightpath is allowed to carry, the number of transmitters and receivers at the nodes, and the maximum number of wavelengths available on fibers.

Solution A logical topology.

Objective Minimizing the average packet hop distance.

2.2 Formulation of Combined Optimization

Most of the notations introduced in last subsection are kept in this subsection. Besides of those the following notations are introduced and will be used to specify the constraints.

- $d_{u,v}$: fiber distance from node u to node v. It is used as a measure of propagation delay. Here $d_{u,v} = d_{v,u}$ and $d_{u,v} = \infty$ if $e_{u,v} = 0$.
- $p_{u,v}$: shortest path delay from nodes u to v, that is the sum of propagation delays. It can be computed from $d_{u,v}$ by any shortest path algorithm.
- β: lightpath length bound β on the delay over a lightpath.
- α: capacity of each channel.
- $l(i, j)$: variables denoting the number of logical links from vertices v_i to v_j.
- $P_{u,v}(i, j)$: variables denoting the number of lightpaths from vertices v_i to v_j that are routed through fiber link $(u, v) \in E$.
- L_{max}: maximum load that a channel is allowed to carry.

Most of the constraints defined in last subsection are similar to the constraints in this subsection (except those three constraints concerning wavelengths). The key difference between them is that some parameters given as bounds previously are now considered as variables that need to be optimized.

The *logical link degree constraint* can be formulated as follows.

$$\sum_j l(i,j) \leq t_i, \quad \forall\, i; \qquad \text{and} \qquad \sum_i l(i,j) \leq r_j, \quad \forall\, j. \tag{4.9}$$

The above inequalities ensure that the number of lightpaths originating from a node is at most the number of transmitters at the node. Similarly the number of lightpaths terminating at a node is at most the number of receivers at the node. When $l(i,j) > 1$, it means that there is more than one lightpath from the source to the destination nodes. These lightpaths may follow the same route or different routes through the network.

The *unique wavelength constraint* is unnecessary, since it is assumed that each routing node in the network is equipped a converter that can provide full wavelength conversion.

The *lightpath routing constraint* can be formulated as follows.

$$\sum_u P_{u,k}(i,j) = \sum_u P_{k,v}(i,j), \qquad \text{if } k \neq i,j; \tag{4.10}$$

$$\sum_v P_{i,v}(i,j) = l(i,j), \ \sum_u P_{u,j}(i,j) = l(i,j); \tag{4.11}$$

$$\sum_{i,j} P_{u,v}(i,j) \leq \omega \times e_{u,v}, \tag{4.12}$$

The above equations (4.10-11) are multicommodity-flow-based equations governing the routing of lightpaths from a source node to a destination node. The inequality (4.12) ensures that the number of lightpaths traversing a fiber link does ont exceed the maximum number of available wavelengths ω.

The *traffic routing constraint* can be formulated as follows.

$$\sum_j \lambda_{s,d}(s,j) = \lambda_{s,d}, \quad \sum_i \lambda_{s,d}(i,d) = \lambda_{s,d}; \tag{4.13}$$

$$\sum_i \lambda_{s,d}(i,k) = \sum_j \lambda_{s,d}(k,j), \quad \text{if } k \neq s,d; \tag{4.14}$$

$$\sum_{s,d} \lambda_{s,d}(i,j) \leq \alpha L_{max} l(i,j). \tag{4.15}$$

$$\lambda_{s,d}(i,j) \leq \lambda_{s,d} \times l(i,j), \tag{4.16}$$

The equalities (4.13-14) are the multicommodity-flow equations governing the traffic flowing through the logical topology. The inequality (4.15) ensures that the traffic over logical link (i, j) does not exceed the capacity of lightpaths from vertex v_i to vertex v_j, while inequality (4.16) guarantees that traffic can only flow through an existing lightpath. It needs to be emphasized again that the traffic from node s to node d may be bifurcated, that is, some fractions of traffic may flow through different lightpaths from s to d. This is unlike the case of the load balancing problem (Problem 2.5) addressed in Section 2.

The *lightpath length constraint* can be formulated as follows.

$$\sum_{u,v} P_{u,v}(i,j)d_{u,v} \leq \beta p_{i,j}\, l(i,j). \tag{4.17}$$

The above inequality ensures that the average length of $l(i, j)$ lightpaths between vertex v_i and vertex v_j is at most β times the length of the shortest paths between them (there are $l(i, j)$ shortest paths in total). This prevents long lightpaths, that is, a shorter route is more desired than a longer one. Here the delay bound β is given as a uniform bound on the average delay between every node pairs v_i and v_j. This can be replaced by possibly different bounds on the average delay for different vertex pairs. That is, β could be replaced by $\beta_{i,j}$.

The *logical topology constraint* can be formulated as follows.

$$e_{u,v} = 1 \Rightarrow l(u,v) \geq 1 \text{ and } P_{u,v}(u,v) \geq 1. \tag{4.18}$$

The above constraint ensures that the physical topology is embedded into the logical topology. That is, there is a lightpath from a node to another if there is a physical link between them. This constraint guarantees that the tightest delay constraints on the packets are satisfied; Moreover, the lightpaths corresponding to the physical topology may also be used to route network control messages efficiently so that the network management could be simplified.

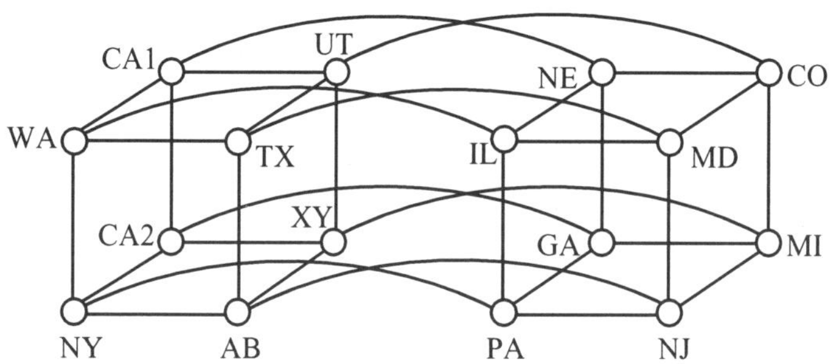

Figure 4.7. Logical topology of NSFnet backbone.

The *regular topology constraint* requires that the logical topology has some regularity. Hypercubes or shuffles as logical topologies have many advantages. They are well understood. Therefore, the routing is simple and the analysis of the bounds and averages is easy. In the case when the physical topology has fewer nodes than the chosen logical topology, some fictitious nodes have to be added into it. Fig. 4.7 shows a logical topology of NSFnet backbone as in Fig. 1. It is a 4-cube consisting of 16 nodes, where nodes AB and XY do not exist in NSFnet. Notice that most of nodes have four neighbors but some, such as node UT, have fewer neighbors.

The objective function minimizes the average packet hop distance in the network, and it can be formulated as follows.

$$\text{Minimize } H_{avg} \equiv \frac{\sum_{i,j}\sum_{s,d}\lambda_{s,d}(i,j)}{\sum_{s,d}\lambda_{s,d}}. \tag{4.19}$$

The objective function is linear, since the numerator is a linear sum of variables $\lambda_{s,d}(i,j)$ and the denominator is a constant for a given traffic matrix.

In the end, Problem 4.2 of LTDP can be formulated as the following optimization problem, which is similar to the formulation of Problem 4.1 presented in the preceding section.

$$\begin{array}{ll} \text{Minimize} & \sum_{i,j}\sum_{s,d}\lambda_{s,d}(i,j) \\ \text{Subject to} & \text{Constraints } (4.9-4.18); \\ & l(i,j) \text{ and } P_{u,v}(i,j) \text{ are natural numbers}, \lambda_{s,d}(i,j) \geq 0, \\ & \text{for all } (i,j),(u,v) \in V \times V, \end{array}$$

Again it is a problem of *Mixed Integer Linear Program* (MILP), since variables $l(i,j), P_{u,v}(i,j)$ should be integers while $\lambda_{s,d}(i,j)$ are not. This problem has $O(|E||V|^2)$ constraints and the same magnitude of variables.

2.3 Solution Approach

In the preceding subsection, we have formulated Problem 4.2 as a problem of MILP just as we have done for Problem 4.1 in the preceding section. It is clear that the methodology and Algorithm 4.1 designed for Problem 4.1 can be modified to solve Problem 4.2.

In fact, these two problems can be considered as a primal-dual pair. Problem 4.1 aims at minimizing the network congestion, while Problem 4.2 aims at minimizing the average packet hop distance H_{avg} and setting the maximal loading (i.e., the congestion) as a constraint. From the traffic routing constraint (4.5) of Problem 4.1, we can obtain the following inequality.

$$\sum_k\sum_{i,j}\sum_{s,d}\lambda^k_{s,d}(i,j) \leq \kappa|V|^2 L_{max}. \tag{4.20}$$

The left side of the above inequality is exactly the objective function of the MILP of the problem 4.2, i.e., $\sum_{i,j}\sum_{s,d}\lambda_{s,d}(i,j)$. Therefore, the average packet hop distance H_{avg} can be minimized through minimizing the network congestion L_{max}.

The above analysis suggests a simple greedy heuristic for Problem 4.2. The basic idea is to establish lightpaths between the source-destination node pairs with the heaviest traffics, subject to constraints on the number of transceivers at the two end nodes, and other constraints. This idea is supported by the results of solution analysis in preceding section.

ALGORITHM 4.2 *Designing a Logical Topology*

Step 1 *Traffic-Dependent Construction*
    $\Lambda := \{\lambda_{s,d} \,|\, s, d \in V\}$.
    while $\Lambda \neq \emptyset$ **do begin**
        $\lambda_{i,j} := \max\{\lambda_{s,d} \,|\, \lambda_{s,d} \in \Lambda\}$.
        if $t_i \geq \tau$ or $r_j \geq \gamma$ **then**
          $\Lambda := \Lambda \setminus \{\lambda_{s,d}\}$.
        else $m := \{\tau - t_i, \gamma - r_j, \lceil \lambda_{i,j}/(\alpha L_{max}) \rceil\}$
            find $m' \leq m$ feasible lightpaths from vertex v_i to vertex v_j
            such that they can be assigned with available wavelengths
            create m' logical links from vertex v_i to vertex v_j.
            $t_i := \tau - m'$;
            $r_j := \gamma - m'$;
            $\Lambda := \Lambda \setminus \{\lambda_{i,j}\}$.
    end-while
Step 1 *Traffic-Independent Construction*
    $U := \{(i, j) \,|\, t_i < \tau$ and $r_j < \gamma\}$,
    while $U \neq \emptyset$ **do begin**
        choose $(i, j) \in U$ with minimum t_i and r_j.
        if a feasible lightpath from vertex v_i to vertex v_j could be found
          such that an available wavelength can be assigned to it,
        then create a logical link from vertex v_i to vertex v_j,
            $t_i := t_i - 1$,
            $r_j := r_j - 1$.
        else $U := U \setminus \{(i, j)\}$.
    end-while
    Output the obtained logical topology

In Step 1, as many as possible logical links are created between node pairs with the heaviest traffic. In order to find feasible lightpaths such that available wavelengths can be assigned to them, one can use *pruning* technique [5]. It is based on tracking a limited number of alternate shortest paths between source-destination paries, such that the selected routes are within a constant factor of the shortest path distance between the given source-destination pair. Then the lightpath between them will be chosen among these alternate paths.

In Step 2, when there are still some free transceivers between some source-destination pairs after Step 1 finishes, we can create some logical links between them so that there will be more freedom for routing the traffic. In this case, logical links are created between those whose transceivers are least loaded so that the transceivers at nodes in the network are loaded as evenly as possible without considering the traffic between node pairs.

In fact, a logical topology can be constructed totally regardless of the traffic between source-destination pairs [5] as follows. First, places logical links between all one-hop neighbors in the physical topology, then between all two-hop neighbors provided that there are no logical links yet between them and the vertex-degree constraints are not violated, and so on. Since lightpaths consists of as few physical links as possible, they have few chances to overlap with each other and share a common physical link. This traffic totally independent heuristic may be used when the number of wavelengths is very limited or the traffic is uniformly distributed among node pairs.

When the logical topology is constructed, routing traffic between source-destination pairs can be done through solving the multicommodity flow problem. In Section 2 of Chapter 2 we have showed how the load balancing problem is formulated as a variant of the multicommodity flow problem and solved. Here our routing problem can be considered as the load balancing problem in the obtained logical topology (not the physical topology). Since fractional flows are allowed, the routing problem in this case becomes a linear program, so it can be solved in polynomial-time.

2.4 Solution Analysis

In this subsection, we will mainly investigate, through studying a simple example in [7], the relationship between the average packet hop distance H_{avg} (along with the congestion) and the delay constraint L_{max}.

Consider again the same network of six node as shown in Fig. 4.1(a) with the traffic matrix as in Table 4.3. The equality constraint on the degrees of vertices is enforced, $t_{i,j} = r_{i,j} = \tau = \gamma = 1$, but other constraints are free. Note that in this situation each node is equipped with one transmitter and one receiver. Thus the logical topology has a ring structure and there are fifteen possible rings with different permutations of six nodes and two directions. The optimal logical topologies for different delay bounds β are obtained by solving the MILP of problem 4.2. Fig. 4.8(a) shows the logical topology for $\beta \geq 2.8$, which means weak or no delay constraint. It has average packet hop distance $H_{avg} = 2.768$ and network congestion $L_{max} = 7.077$. Observe that the links in this logical topology bear little resemblance to the links in the physical topology. It includes four logical links (lightpaths) having two physical links (hops). Fig. 4.8(b) shows the logical topology for $2.5 \leq \beta < 2.8$. The average packet hop distance H_{avg} is increased to 2.806 and network congestion L_{max} is increased to 7.185. Fig. 4.5(a) also shows the logical topology for $1.933 \leq \beta < 2.5$. The average packet hop distance H_{avg} is further increased to 2.890 and network congestion L_{max} is further increased to 7.337. Observe that all links in this logical topology are also links in the physical topology. In addition, for the case of $\beta < 1.933$, there is no feasible logical topology.

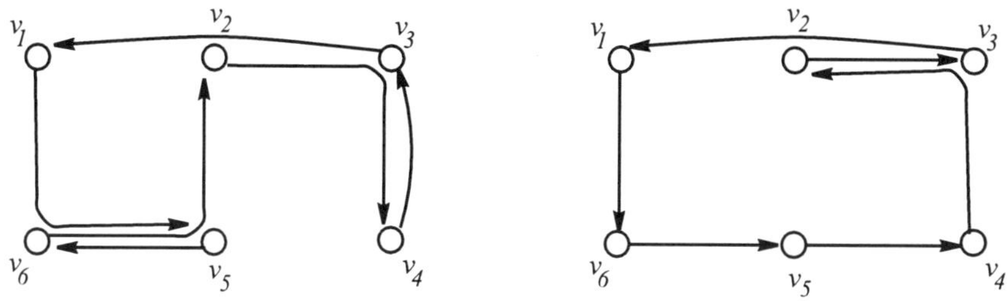

Figure 4.8. (a) The logical topology for $\beta \geq 2.8$ has $H_{avg} = 2.768$ and $L_{max} = 7.077$. (b) The logical topology for $2.5 \leq \beta \leq 2.8$ has $H_{avg} = 2.806$ and $L_{max} = 7.185$.

From these examples we can see that the average packet hop distance H_{avg} becomes shorter as the delay bound β turns larger (i.e., the delay constraint is more relaxed). At the first sight, this should not be the case. Since when β is larger, the lightpaths would have longer lengths and more physical hops. This can be observed by comparing the logical topology for $\beta \geq 2.8$ in Fig. 4.8(a) with the logical topology for $1.933 \leq \beta < 2.5$ in Fig. 4.5(a)). However, the reason behind this unexpected result is that because the delay constraint is not strict, given a node pair there will be more possible routes (physical paths) that can be chosen as a lightpath between them. Minimization of the network congestion favors those lightpaths that carry more traffic between node pairs, so that heavy traffic traverse few hops of lightpaths while light traffic travel many hops of lightpaths. This makes the average packet hop distance shorter.

From these examples we can also see that the network congestion L_{max} has the same behavior as the average packet hop distance H_{avg} due to the same reason. This result well justifies our analysis at the beginning of this section, that the minimization of the average packet hop distance H_{avg} can be achieved through the minimization of the network congestion L_{max}. At the same time it proves the effectiveness of Algorithm 4.2.

To investigate the relationship between the average packet hop distance H_{avg} and the number of transceivers per node, the simulation study [1] is done on the LTDP (Problem 4.2) in NSFnet, where the equal vertex-degree constraint is assumed, $t_i = r_j = \tau = \gamma$, and $\tau = \gamma$ varies from 4 to 8 while the number of wavelengths available on fibres ω takes values between 1 and 8. The experimental results show that the average packet hop distance H_{avg} decreases with a balanced increase in the number of transceivers and wavelengths in the network. Increasing the number of transceivers with ω fixed marginally improves the quality of the solutions, that is, shortens the average packet hop distance H_{avg}. As a contrast, increasing the number of wavelengths with $\tau = \gamma$ fixed considerably improves the quality of the solutions.

The relationships between the average packet hop distance and other parameters of the network or the constraints on the logical topology can be obtained through studying the relationship between the network congestion in preceding section.

3. Discussion

In this chapter, we have studied two versions of LTDP, Problem 4.1 discussed in Section 1 is to minimize the network congestion, and Problem 4.2 addressed in Section 2 is to minimize the average packet hop distance. In fact, they are closely related although they were studied separately in the literature.

In our presentation and formulation of LTDP, a static traffic model is adopted. When the traffic between some of node pairs changes or the physical topology changes, which may be due to failure of network components or addition/upgrading of network components, the current logical topology must be reconstructed to adapt to these changes. An interesting problem is how to reconfigure the network nodes and redesign a logical topology from the existing one under the change of traffic matrix.

When designing an algorithm for reconfiguring the logical topology, two issues must be taken into account. One is how to minimize the number of switch retunings that equals the number of disrupted lightpaths. The other is how to minimize the changes required to obtain a new logical topology form the current logical topology. In theory, given a small change in the traffic matrix, we would expect that there is little change in the logical topology, in terms of lightpaths routing and wavelength assignment. This means that it is desirable to minimize the changes in the number of *Wavelength Routers/Switches* (WRS) configurations needed to adapt from the existing logical topology to the updated logical topology. More numerically, it would be preferable if a large number of the variables, such as $l^k(i,j)$ or $l(i,j)$, $w_{u,v}^{k,q}(i,j)$ or $P_{u,v}(i,j)$, remain unchanged in the two solutions, without sacrificing the quality of the solution in terms of the network congestion L_{max} or the average packet hop distance H_{avg}. This concern can be set as either an objective function or a constraint. But in either way the formulation will not be linear any more.

Usually the network resources, such as the number of wavelengths available on fibers and the number of transceivers at nodes, are taken as constraints in the formulation of the logical topology design problem, as they were treated in Section 1 and Section 2. However, when studying the logical topology design problem, we should also consider the resource budgeting issue. It is clear that a network with a very large number of transceivers at a node, but very few wavelengths on a fiber and there are few fibers between node pairs, most of transceivers at nodes can not be used because lightpaths could not be established between them due to wavelength constraints. Similarly, in a network with few transceivers at nodes but a large number of available fibers and wavelengths on the fibers, most of wavelengths will be wasted because of limited transceivers.

The mismatch in transceiver utilization versus wavelength utilization yields a direct impact on the cost of the network. In general, the number of wavelengths supported in the network determines the cost of the switching equipment. A WRS with δ input ports and δ output ports that supports ω wavelengths requires $2\omega\delta^2$ wavelength insensitive optical switches for nonblocking operation. Adding fibers between two nodes will increase δ (i.e., the size of WRS), and putting more wavelengths into use will increase the number of cross-point switching elements required. Thus in either case the cost of the switching equipment would increase. The number of transceivers at nodes also determines the cost of the terminating equipment. The concerns about the utilization of network resources gives birth to the resource budgeting problem that is how to balance the network resources in order to maximize the utilizations of both the transceivers and the wavelengths in the network. Study in [1] shows that the transceiver utilization decreases as the number of wavelengths is reduced and/or the number of transceivers is increased, and the wavelength utilization decreases when the number of wavelengths is increased and/or the number of transceivers is reduced. Or more formally, in order to get a network with balanced loading across all wavelengths, the number of transceivers at a node should be approximately

$$t_i = r_i \approx \frac{\text{Number of fibers } \times \text{ Number of wavelengths}}{\text{Number of nodes } \times \text{ Average length of lightpaths}}.$$

For other issues and techniques which are related to the logical topology design problem, readers may refer to a nice survey on this topic by Dutta and Roukas [3].

References

[1] D. Banerjee and B. Mukherjee, Wavelength-routed optical networks: linear formulation, resource budgeting tradeoffs, and a reconfiguration study, *IEEE/ACM Transactions on Networking*, 8 (5) (2000), 598-607.

[2] D. Bienstock and O. Gunluk, Computational experience with a difficult mixed-integer multicommodity flow problem, *Mathematical Programming*, Series A, 68 (2) (1995), 213-237.

[3] R. Dutta and G. N. Rouskas, Design of logical topologies for wavelength routed networks, in *Optical WDM Networks: Principles and Practice*, edited by K. M. Sivalingam and S. Subramaniam, Kluwer Academic Publishers, (2000), 79-102.

[4] R. M. Krishnaswamy and K. N. Sivarajan, Design of logical topologies: a linear formulation for wavelength-routed optical networks with no wavelength changers, *IEEE/ACM Transactions on Networking*, 9 (2) (2001), 186-198.

[5] B. Mukherjee, D. Banerjee, S. Ramamurthy, and A. Mukherjee, Some principles for designing a wide-area optical network, *IEEE/ACM Transactions on Networking*, 4 (5) (1996), 684-696.

[6] R. Ramaswami and K. N. Sivarajan, Routing and wavelength assignment in all-optical networks, *IEEE/ACM Transactions on Networking*, 3 (5) (1995), 489-500.

[7] R. Ramaswami and K. N. Sivarajan, Design of logical topologies for wavelength-routed optical networks, *IEEE Journal of Selected Areas on Communications*, 40 (6) (1996), 840-851.

II

ADVANCED STUDY OF WDM NETWORKS

Chapter 5

PLACEMENT OF WAVELENGTH CONVERTERS

The number of wavelengths available in a network is always limited due to the restriction of hardware structure of optical routers/switches. An important goal of the design of WDM networks is to use less wavelengths to serve more communication needs. There are two basic approaches to achieving the goal. The first one is to find the proper routing and wavelength assignment methods, that is the routing and wavelength assignment problem (RWAP) discussed in Chapter 3. In this chapter we will focus on the second approach, that is to use of wavelength converters. Existing study has shown that the more converters installed in a network, the less number of wavelengths is needed, given the same network load (the maximal number of channels over a link). In fact, by using enough number of wavelength converters at the network nodes, the number of wavelengths required can be made equal to the network load, this is most ideal situation that we can expect since the number of wavelengths required in a system is no less than the network load. A simple example of achieving this feature, called the *Load-Wavelength Assignability* (LWA), is to equip every node in the network with a wavelength converter. However, it is too expensive to do so, because this will not only increase the cost of network hardware but also the complexity of routing and wavelength assignment.

In this chapter we will study the *Wavelength Converter Placement Problem* (WCPP) in WDM networks. The problem is how to place the minimal number of wavelength converters at some nodes in the network such that the number of wavelengths required can be made equal to the network load (that is to achieve LWA), or within a small constant times of the network load (that is to achieve relaxed LWA).

The rest of this chapter is organized as follows. Section 1 discusses the optimal placement of wavelength converters achieving LWA. Section 2 discuss the optimal placement of wavelength converters achieving the relaxed LWA. The basic ideas and theoretical analysis in these works [5, 6, 9] are originally from Wilfong and Winkler's work [15]. Section 3 concludes this chapter.

1. Placement for Load-Wavelength Assignability

There are basically two types of architectures of WDM network systems: single-hop systems and multihop systems. In single-hop systems, each pair of communication nodes has a logical channel configured and a single wavelength is used through out the route of the channel. There is no wavelength conversion in the intermediate nodes in the route of a channel. In multihop systems, the channel of a pair of nodes can consist of several path segments, each of them may use a different wavelength. Wavelength conversion is needed at a node where the input and output fibers of the channel use different wavelengths. A node capable of wavelength conversion must be equipped with a wavelength converter. In a network with sparse wavelength conversion, only a fraction of nodes can be equipped with converters because of the high cost of wavelength converters. We assume that each converter has *full conversion capability* [13]. That is, a converter can realize all possible permutations of input and output wavelengths.

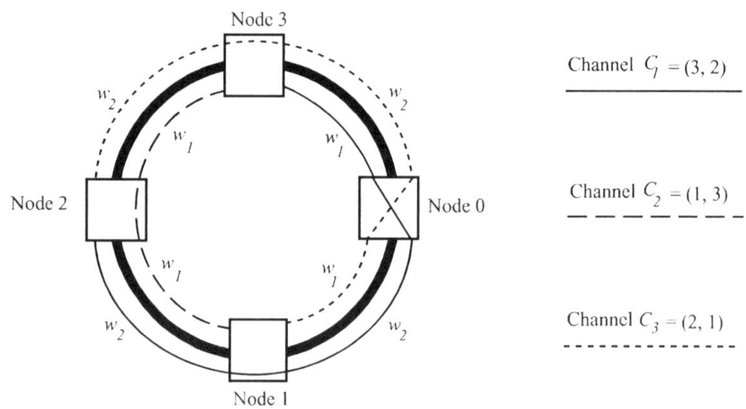

Figure 5.1. Wavelength conversion in a WDM network.

Fig. 5.1 illustrates wavelength conversion in a WDM network. In the network of Fig. 5.1, there are two wavelengths, w_1 and w_2, on each link, and three channels C_1, C_2, and C_3. C_2 between node 1 and node 2 uses wavelengths w_1 throughout its route. As C_1 between node 3 and node 2 shares links (3,0) and (0,1) with C_3 which is between node 2 and node 1, and C_2 shares links (1,2) and (2,3) with C_1 and C_3, respectively. C_1 and C_3 must use two different wavelengths on their routes, where wavelength conversion is required at node 0.

A set of nodes $S \subseteq V$, is said to achieve *Load-Wavelength Assignability* (LWA) if, by equipping each node in S with a wavelength converter, the number of wavelengths required for all channels can be made equal to the network load. The optimal converter placement for LWA is formulated as follows.

PROBLEM 5.1 *Converter Placement for Load-Wavelength Assignability*

Instance A graph $G(V, E)$.
Solution A subset set $S \subseteq V$ that guarantees LWA.
Objective Minimizing the cardinality of set S, i.e., $|S|$.

1.1 Optimization Formulation and Theory

The basic approach for Problem 5.1 is based on a simple but powerful technique introduced in [15]. It decomposes a general topology network into edge-disjoint simple subgraphs, such as paths or spiders. A *spider* is a tree that consists of several paths, called *legs*, with one end of each of these paths incident to a common vertex, called *body*. Clearly, paths and stars are two special cases of spiders.

The decomposition of a graph is done by a *splitting operation* described as the follows. Given a graph $G(V, E)$ and a subset $S \subseteq V$, a new graph $G_S(V, E)$ is generated through splitting each vertex $s \in S$ into $\delta(s)$ vertices, where $\delta(s)$ is the degree of s in the original graph $G(V, E)$. Each edge (s, t) in $G(V, E)$ becomes edge (s', t) in $G_S(V, E)$, where s' is a newly generated vertex from splitting s in $G(V, E)$. Fig. 5.2 illustrates the splitting operation on S consisting of three vertices in black.

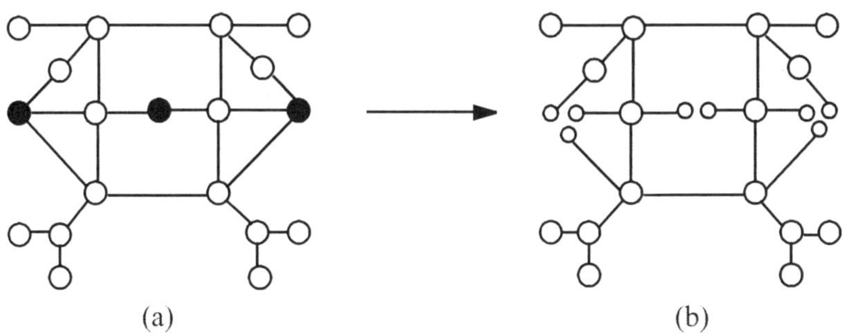

Figure 5.2. (a) The original graph $G(V, E)$. (b) The obtained graph $G_S(V, E)$ through the splitting operation on vertices in S.

By studying the obtained graph $G_S(V, E)$, we can divide the Problem 5.1 in networks with arbitrary topologies into the same problem but in networks with some simple topologies, such as paths and spiders. The Problem 5.1 in these simple graphs becomes easier to be solved. In the following we will discuss two cases separately, bidirectional channels and unidirectional channels. The first one is simple while the second one turns out to be complicated.

1.1.1 Bidirectional Channels

In this subsection, we assume bidirectional channels and LWA means LWA for bidirectional channels. The following lemma characterizes the property of sets that guarantee LWA.

LEMMA 5.1 *Given a graph $G(V, E)$, a subset $S \subseteq V$ guarantees LWA if and only if every connected component of $G_S(V, E)$ is a path.*

PROOF "If": Given any set of routed channels with the maximal link load L in $G(V, E)$, we examine them in $G_S(V, E)$. These channels which go through vertices in S are broken into several pieces, and each connected component is a path in $G_S(V, E)$. Wavelengths can be assigned to the channels which go through the entire or part of a path in the following way. See Fig. 5.3. The method is the same as the one used in the proof of Theorem 3.6.

Step 1. Order these channels from left to right according to the starting points in the path. Ties are broken arbitrarily.

Step 2. Assign each of these channels the wavelength which is *free* and has the least value. A wavelength becomes free once the channel that receives the channel terminates.

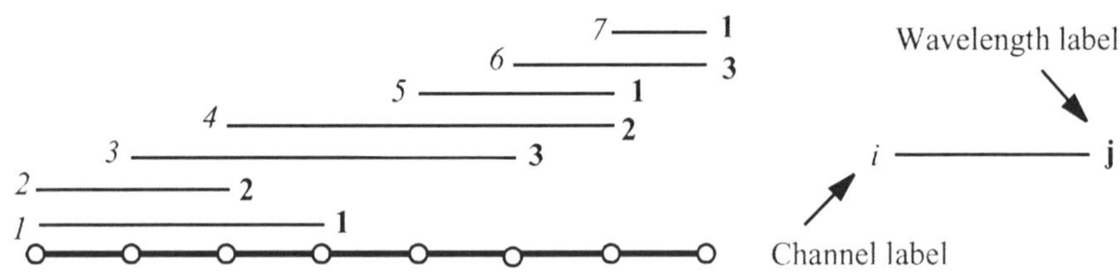

Figure 5.3. Wavelength assignment for channels in a path.

Since the maximal link load is L, it is obvious that Step 2 uses at most L wavelengths to assign wavelengths to channels without causing wavelength conflict. Moreover, the wavelength assignment in each connected component of $G_S(V, E)$ is independently, and there is no wavelength conflict between the channels in different components. This is because each vertex in S has full wavelength conversion capability.

"Only if": We prove it by contradiction. Assume that one of the connected components in $G_S(V, E)$ is not a path. There are only following two possibilities of this component. We will analyze them separately at below.

Case 1. At least one vertex in the component has degree greater than two. In this case, there exists a vertex adjacent to (at least) three other vertices in the same

Figure 5.4. (a) Three channels on a star. (b) Three channels on a ring.

component. Consider three channels in Fig. 5.4(a). Apparently, three wavelengths are needed even though the maximal link load is two, because each of three channels shares a link with the other two. This contradicts the load-wavelength assignability.

Case 2. The component is a ring. Consider three channels in Fig. 5.4(b). Again, three wavelengths are needed, but the maximal link load is two, because each of these three channels shares a link with the other two. This results in the same contradiction. The proof is then finished. ■

THEOREM 5.1 *The Problem 5.1 for bidirectional channels can be solved in time of* $O(|E| + |V|)$.

PROOF According to Lemma 5.1, the solution to the WCPP for LWA is to, given a graph $G(V, E)$, locate a set S that makes every component of $G_S(V, E)$ a path, and equip every vertex in S with a converter. Thus we can find such a set S by identifying all the vertices whose degrees are greater than two in $G(V, E)$, or breaking the ring if $G(V, E)$ is a ring. This can be done simply by searching original graph $G(V, E)$. If $G(V, E)$ is a ring, then let S include a vertex in V; Otherwise let S include all vertices whose degrees are greater than two. It is not difficult to see that S, constructed in such a way, is the minimal to make every component of $G_S(V, E)$ a path, and it can be obtained in linear time of $O(|E| + |V|)$. Hence, S is the optimal solution to the WCPP for LWA. ■

1.1.2 Unidirectional Channels

In this subsection, we assume unidirectional channels and LWA means LWA for unidirectional channels. Different from bidirectional channels, the wavelength conflict rule for unidirectional channels requires that channels over the same link and in the same direction are assigned different wavelengths. As we will see that Problem 5.1 becomes more complicated for unidirectional channels. The following lemma characterizes the property of sets that guarantee LWA.

LEMMA 5.2 *Given a graph $G(V, E)$, a subset $S \subseteq V$ guarantees LWA if and only if each connected component in $G_S(V, E)$ is a spider.*

PROOF "If": Given any set of routed channels with the maximal link load L in $G(V, E)$, we examine them in $G_S(V, E)$. For a connected component in $G_S(V, E)$, which is a spider, we first consider the channels crossing the spider body from one leg to another. Then, we consider the channels on a leg. See Fig. 5.5(a), where the spider body is marked in black.

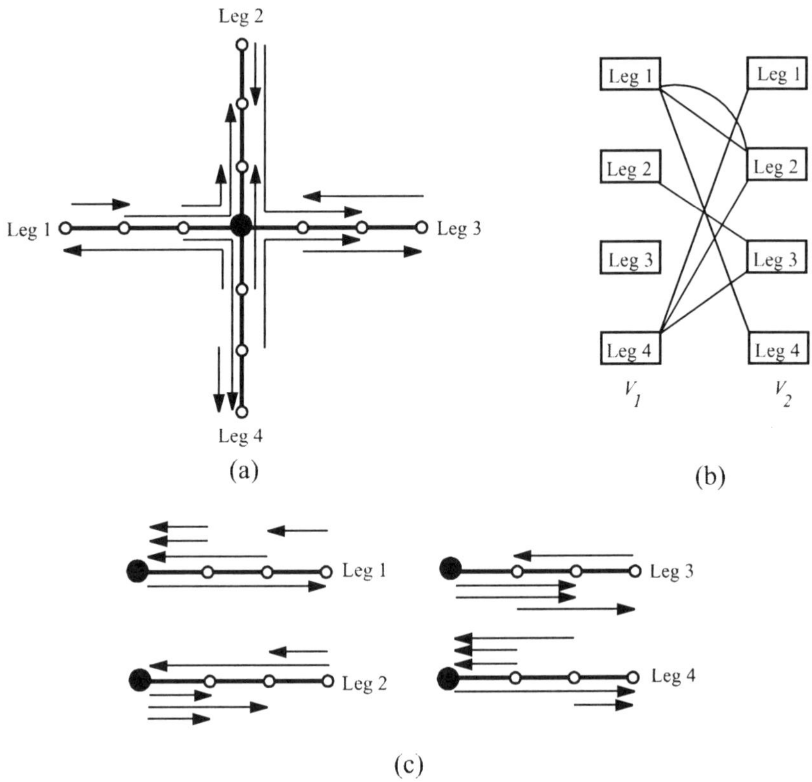

Figure 5.5. (a) Wavelength assignment in a spider. (b) Assigning wavelengths to channels crossing the body. (c) Assigning wavelengths to channels within legs.

Case 1. Channels go from one leg to another. We introduce a bipartite multigraph $B(V_1 \cup V_2, E_{12})$ to represent these channels in the spider, where V_1 and V_2, $V_1 = V_2$, are the set of legs, that is, each leg of the spider is represented as a vertex in V_1 and V_2. An edge $e_{ij} \in E_{12}$ if there is a channel which goes through legs i and j for $i \neq j$. Note that there may be more than one edges between two vertices in $B(V_1 \cup V_2, E_{12})$.

For example in Fig. 5.5(b), there are two edges between vertex 1 and vertex 2, as there are two channels going from leg 1 to leg 2. Now assigning wavelengths to the channels without causing wavelength conflict becomes the problem of coloring the edges of graph $B(V_1 \cup V_2, E_{12})$ such that the condition that two adjacent edges must be colored in two different colors. Since there are at most L channels going from leg i to leg j for $i \neq j$, multigraph $B(V_1 \cup V_2, E_{12})$ has the maximum degree L. Thus, $B(V_1 \cup V_2, E_{12})$ can be colored by using at most L colors. (Refer to [10]).

Case 2. Channels on a leg. Consider each of the legs separately. See Fig. 5.5(c). Note that each channel belonging to Case 1 is cut into two parts, one in a leg heading the body and the other in another leg leaving the body. We use the same method as discussed in the proof of Lemma 5.1 for assigning wavelengths to channels in a path. The channels heading the body of the spider are ordered from left to right by their destinations or the splitting point at the spider body, while the channels leaving the spider body are also ordered from left to right by their sources or the spider body. Ties are broken arbitrarily. The channels which have been assigned wavelengths in Case 1 retain the same wavelengths, and the rest of the channels in this leg will be assigned the wavelengths which are free. Thus, all channels (in both Case 1 and Case 2) can be assigned with at most L wavelengths.

For each of the channels that traverse different connected components in $G_S(V, E)$, it can be assigned different wavelengths. However, this will not cause wavelength conflict, because each vertex in S is assumed to have full wavelength conversion capability.

"Only if": By contradiction argument, we assume that one of the connected components of $G_S(V, E)$ is not a spider. There are only two possibilities of the component. We consider each of them separately as below.

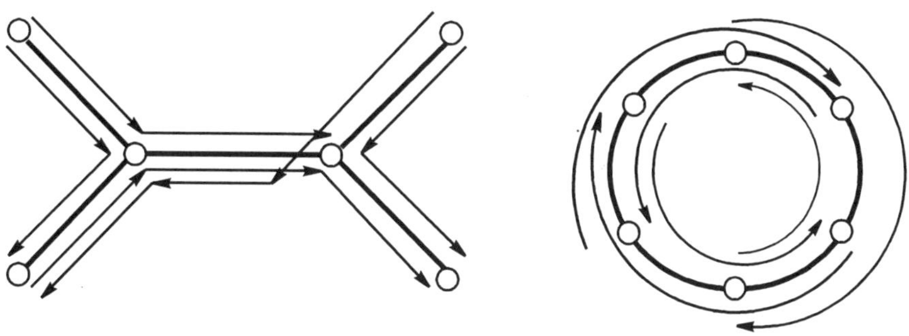

(a) (b)

Figure 5.6. (a) Five channels on a tree. (b) Six channels on a ring.

Case 1. There are two vertices having degrees greater than two. We construct an example in Fig. 5.6(a), where the maximal link load is two. It is not difficult to find that three wavelengths are needed for all the channels in this case. This contradicts the LWA for unidirectional channels.

Case 2. The component is a ring. Consider a case we constructed in Fig. 5.6(b), where the maximal link load is two. Again, three wavelengths are needed in this case. This leads to the same contradiction. ■

The following theorem [15] shows that it is $\mathcal{NP}$-hard to find the minimal sized set S that guarantees LWA.

THEOREM 5.2 *The problem of determining whether, for a given graph $G(V, E)$ and a positive integer $k > 0$, there is a subset of V with k vertices that guarantees LWA, is $\mathcal{NP}$-complete, even if $G(V, E)$ is a planar graph.*

PROOF For a given graph $G(V, E)$, it is easy to check in polynomial-time if a subset $S \subseteq V$ guarantees LWA since this can be done by checking that no $G_S(V, E)$ contains a cycle or more than one vertices with degrees greater than two. In order to show that our problem is $\mathcal{NP}$-hard, we will reduce the problem of planar 3-SAT [2] to it.

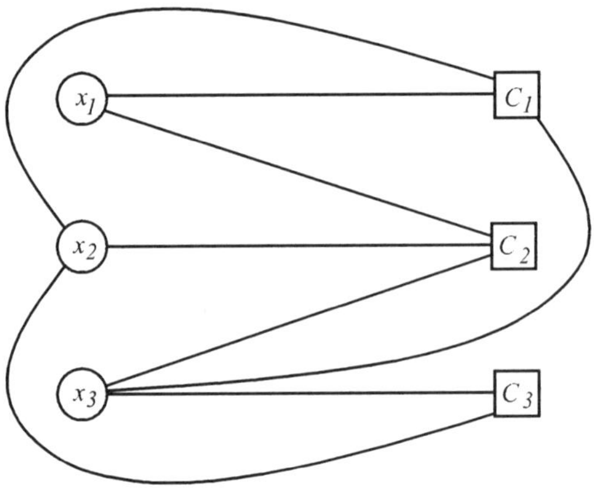

Figure 5.7. An instance of planar 3-SAT problem.

An instance I of this problem is a planar bipartite graph (or equivalently a set of clauses over a set of Boolean variables) . Its vertex-set is composed of two disjoint sets, one consists of Boolean variables and the other consists of clauses. There is an edges between a variable vertex x and a clause vertex C if and only if x or $\overline{x}$ is in the clause C. Here each clause includes the disjunction of at most three literals (i.e., a variable or the negation of a variable), and it is satisfied if and only if at least one of the literals is true. Fig. 5.7 shows an example of such an instance. The *planar 3-SAT problem* is to decide if there is a truth assignment satisfying all given clauses. It consists of three clauses over three Boolean variables,

$$C_1 = x_1 \vee \overline{x}_2 \vee \overline{x}_3, \; C_2 = \overline{x}_1 \vee x_2 \vee x_3, \text{ and } C_3 = x_2 \vee \overline{x}_3.$$

It is easy to see that it has a truth assignment, x_1 and x_2 are both true while x_3 can be either true or false.

Now we will show how to construct in polynomial-time an instance I' of our problem that has a desired set if and only if a instance I of the planar 3-SAT problem has a truth assignment. Let I be an instance of planar 3-SAT with m clauses $C_1, \cdots, C_m$ on n variables $x_1, \cdots, x_n$. Each clause C_i is a disjunction of three literals y_{i1}, y_{i2}, and y_{i3}. Each y_{ij} is some x_k or $\overline{x}_k$. But for the easiness of presentation, we will use distinct labels for y_{ij}, x_k and $\overline{x}_k$ in the construction of a graph. In addition, we will use the notation $x(y_{ij})$ to denote the vertex labelled by x_k if y_{ij} is x_k or the vertex labelled by $\overline{x}_k$ otherwise.

To define an instance I' of our problem, we need to define an integer $k > 0$ and a graph $G(V, E)$. Set $k \equiv 2m + n$, vertex-set V be

$$V \equiv \left(\cup_{i=1}^{n} \{x_i, \overline{x}_i, a_i, b_i, c_i, d_i\} \right) \cup \left(\cup_{i=1}^{m} \{y_{i1}, y_{i2}, y_{i3}\} \right),$$

and the edge-set E be

$$\begin{aligned} E \equiv \quad & \left(\cup_{i=1}^{n} \{(x_i, \overline{x}_i)\} \right) \\ \cup \quad & \left(\cup_{i=1}^{n} \{(x_i, a_i), (x_i, b_i), (\overline{x}_i, c_i), (\overline{x}_i, d_i)\} \right) \\ \cup \quad & \left(\cup_{i=1}^{m} \{(y_{i1}, y_{i2}), (y_{i2}, y_{i3}), (y_{i3}, y_{i1})\} \right) \\ \cup \quad & \left(\cup_{i=1}^{m} \{(y_{i1}, x(y_{i1})), (y_{i2}, x(y_{i2})), (y_{i3}, x y_{i3})\} \right) \end{aligned}$$

Fig. 5.8 illustrates the reduction from the instance I in Fig. 5.7 to the instance I'. Thus each clause C_i of I becomes a triangle of vertices y_{i1}, y_{i2} and y_{i3} with an edge from each such vertex to the vertex labelled by its corresponding variable or negated variable. In addition, each vertex labelled by a variable x_i has an edge to the vertex labelled by its negation and to the dummy vertices a_i and b_i. Similarly, the vertex labelled by $\overline{x}_i$ also has edges to dummy vertices c_i and d_i. Clearly, the size of $G(V, E)$ is polynomial in the size of I. Also, the graph $G(V, E)$ is obviously planar since the instance I was an instance of planar 3-SAT.

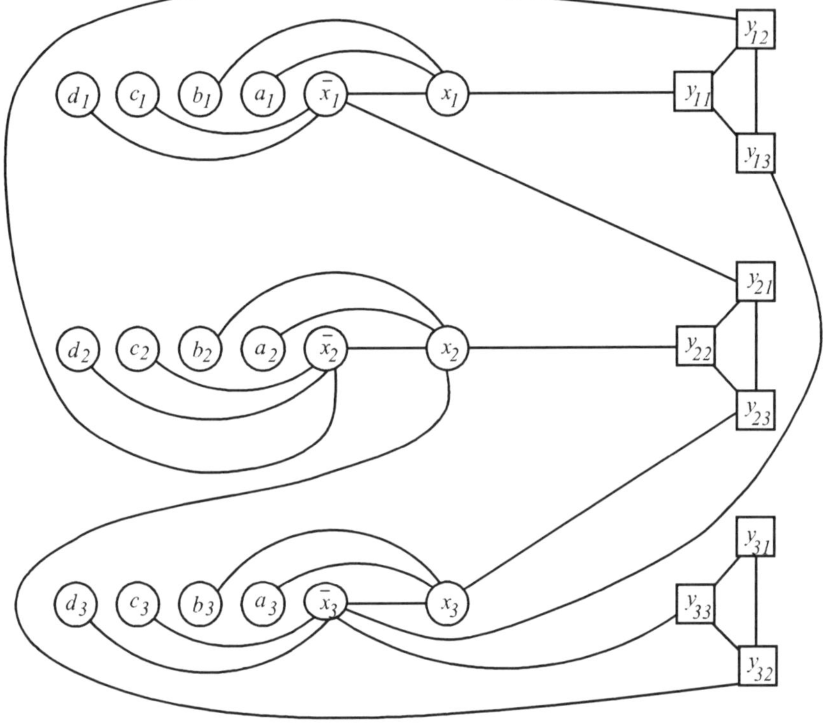

Figure 5.8. The reduction from the planar 3-SAT problem.

Now we prove that the instance I has a truth assignment of variables x_i if and only if the instance I' has a set of size k that guarantees LWA. Consider the following four conditions on a subset $S \subseteq V$.

(1) Exactly one of x_i or $\overline{x}_i$ is in S, for $1 \leq i \leq n$;

(2) Exactly two of y_{i1}, y_{i2} and y_{i3} are in S, for $1 \leq i \leq m$;

(3) If y_{ij} is not in S then $x(y_{ij})$ is in S;

(4) a_i, b_i, c_i and d_i are not in S, for $1 \leq i \leq n$.

We will show that a set S satisfies above four conditions if and only if S has $(2m+n)$ vertices guaranteeing load-wavelength assignability.

For "If" part. Suppose that S has $(2m + n)$ vertices and guarantees LWA. Note that each vertex labelled by some y_{ij}, x_k or $\overline{x}_k$ has degree at least 3 in $G(V, E)$, and so none of them can be in the same component of $G_S(V, E)$ since it must be a spider. Then for each clause C_i at least two of y_{i1}, y_{i2}, and y_{i3} must be in S. Also at least one of x_j or $\overline{x}_j$ must be in S. But since $|S| = 2m + n$, this means that exactly two of

y_{i1}, y_{i2}, and y_{i3} must be in S and no other vertices are in S. Thus conditions (1), (2) and (4) are satisfied. Similarly, due to the edges $(y_{ij}, x(y_{ij}))$, if y_{ij} is not in S then $x(y_{ij})$ must be in S else y_{ij} and $x(y_{ij})$ are in the same component of $G_S(V, E)$ and both have degrees at least 3 in the component. This contradicts that S guarantees LWA. Hence $G_S(V, E)$ is a spider, and condition (3) is also satisfied.

For "Only if" part. Suppose that set S satisfies conditions (1-4). Clearly, $|S| = 2m + n$ by conditions (1, 2, 4). Consider a component $G_S^j(V, E)$ of $G_S(V, E)$. Notice that vertices of G_S^j resulting from the splitting operation on a vertex in S have degree 1 in $G_S^j(V, E)$. Thus if $G_S^j(V, E)$ is not a spider graph then it must have at least one vertex labelled by some x_i, $\overline{x}_i$ or y_{st} that is not in S. Suppose the vertex labelled x_i is in $G_S^j(V, E)$ and not in S (a similar argument holds if the vertex labelled $\overline{x}_i$ is in $G_S^j(V, E)$). Then the vertex is adjacent to $a_i, b_i, \overline{x}_i$ and some number of vertices labelled by some y_{st}, where $x(y_{st})$ is x_i. By definitions a_i and b_i have degree 1, by condition (1) $\overline{x}_i$ must be in S and hence has degree 1 in $G_S^j(V, E)$, and by condition (3) the other vertices adjacent to x_i are also in S. Thus $G_S^j(V, E)$ must be a spider (in fact, it is a star since each leg has length 1). This contradicts the assumption. Suppose that the vertex labelled y_{st} is in $G_S^j(V, E)$ but not in S. Then conditions (2) and (3) implies that it is adjacent in $G_S^j(V, E)$ only to vertices with degree 1 in $G_S^j(V, E)$, and again this means that $G_S^j(V, E)$ is a spider (again it is actually a star). Hence S must guarantee LWA.

In the end we show that a desired assignment A for instance I can be found if and only if there is a set S_A of size $(2m + n)$ guaranteeing LWA. For $1 \leq i \leq n$, define t_i to be x_i or $\overline{x}_i$ depending on whether x_i is true or not respectively according to assignment A. In addition, define S_A to be the set of vertices of $G(V, E)$ containing those vertices labelled by each t_i and for each j, $1 \leq j \leq m$, two of those vertices labelled by y_{j1}, y_{j2}, and y_{j3} so that the one not placed in S_A evaluates to true according to A (there is always such a vertex since A is a desired assignment). Hence, $|S_A| = 2m + n$. Clearly S_A satisfies conditions (1, 2, 4). Suppose that the vertex labelled y_{ij} is not in S_A. Then y_{ij} must be true according to A since it is not in S_A. Thus the vertex labelled $x(y_{ij})$ is true, this implies that S_A satisfies condition (3). Therefore, S_A satisfies all conditions and hence it is a set of size $(2m + n)$ guaranteeing LWA.

Suppose that S is a set of size $(2m + n)$ that guarantees LWA. Then S satisfies conditions (1-4). Consider the truth assignment A_S that assigns true to each variable x_i such that the vertex labelled x_i is in S. Consider any clause C_i of instance I. Since S satisfies condition (2) we know that exactly one of the vertices labelled by the literals y_{i1}, y_{i2} and y_{i3} is not in S. Suppose that literal is y_{ij}. Then by condition (3), the vertex labelled by $x(y_{ij})$ is in S and hence is assigned true by S_A and so y_{ij} is true according to A_S. Hence in C_i there is at least one true literal and so A_S is a desired assignment. The proof is then finished. ■

The following lemma characterizes the property of the vertices in the set that guarantees LWA.

LEMMA 5.3 *If a graph $G(V, E)$ has a vertex with degree greater than two, then there exists a minimal sized subset of V that guarantees LWA and every vertex in the set has degree greater than two in $G(V, E)$.*

PROOF Let $v \in V$ be the vertex with degree greater than two, and S be a minimal sized subset of V that guarantees LWA. Now assume there is a vertex $s \in S$ with degree less than or equal to two. Let u be the closest vertex to s in a path between v and s, and u has degree greater than two (u can be v itself). We consider the following two cases of u.

Case 1. $u \in S$. Let $S' = S \setminus \{s\}$. It is obvious that each connected component of $G_{S'}(V, E)$ is still a spider, because s is a vertex of degree less than or equal to two. See Fig. 5.9(a).

Case 2. $u \notin S$. Let $S' = S \cup \{u\} \setminus \{s\}$. It is easy to see that each connected component of $G_{S'}(V, E)$ is still a spider, because substituting s with u only makes one component of $G_S(V, E)$ have a longer leg and another component is split into several paths. See Fig. 5.9(b).

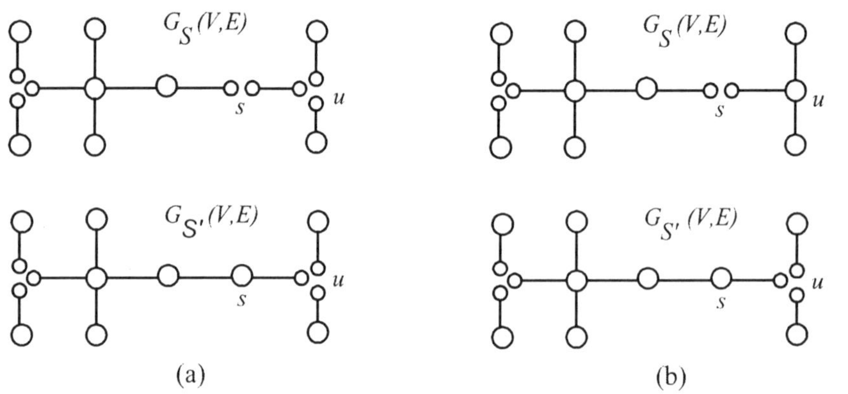

Figure 5.9. Two cases in the proof of Lemma 5.3: (a) Case 1 and (b) Case 2.

In either cases, S' can guarantee LWA because of Lemma 5.2. In Case 1, S' is a proper subset of S, this contradicts that S has the minimal size. In Case 2, the substituting operation can be repeated until all vertices in S having degrees greater than two. The desired subset can thus be obtained. The proof is then finished. ■

The significance of Lemma 5.3 is that when searching for the minimal sized S that guarantees LWA, we can ignore all vertices whose degrees are less than or equal to two in $G(V, E)$. Unfortunately, even we are able to reduce the Problem 5.1 in original graph $G(V, E)$ into a simple graph $G'(V', E')$ by removing from $G(V, E)$ all the vertices having degree less than or equal to two, it is still $\mathcal{NP}$-hard to find the minimal sized S in $G'(V', E')$. The following lemma transforms the Problem 5.1 for unidirectional channels to the vertex-cover problem. A *vertex-cover* (VC) of a graph $G(V, E)$ is a set of vertex $C \subseteq V$ such that each edge of G has at least one endpoint in C. The *vertex-cover problem* is to find a vertex-cover of the minimal size.

LEMMA 5.4 *If every vertex in graph $G(V, E)$ has degree greater than two, then a subset $S \subseteq V$ guarantees LWA if and only if S is a vertex-cover of $G(V, E)$.*

PROOF "If": Since S is a vertex-cover of $G_S(V, E)$, every connected component of $G_S(V, E)$ must be a spider (actually a star). From Lemma 5.2, we have that S guarantees LWA.

"Only if": Assume, by contradiction, that there are two vertices u and v in $G(V, E)$ such that edge $(u, v) \in E$ is not incident to any vertex in S. Then the connected component in $G_S(V, E)$ which contains edge (u, v) is not a spider, because it has two vertices (u and v) with degrees greater than two. This contradicts Lemma 5.2. The proof is then finished. ■

If we remove the condition of Lemma 5.4, then the claim of the lemma will not hold any more. See Fig. 5.10 for two simple counterexamples. In the tree network of Fig. 5.10(a), it is clear that one converter installed at the center (in black) can guarantee LWA; However, the minimum vertex-cover has eight vertices (in grey). In the ring network of Fig. 5.10(b), it is clear that one converter at (any) one vertex (in black) can guarantee LWA; However, the minimum vertex-cover contains half number of vertices in the network.

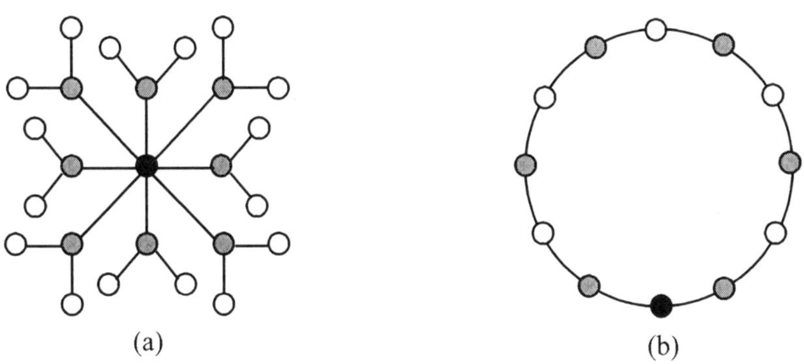

Figure 5.10. Two simple examples.

1.2 Algorithm Design and Performance Analysis

In this subsection we will develop some algorithms that locate a set guaranteeing LWA for unidirectional channels. (The case of bidirectional channels is trivial as shown in Section 5.1.1.)

1.2.1 General Networks

Given a graph $G(V, E)$, if no vertex in $G(V, E)$ has degree greater than two, then $G(V, E)$ is either a path or a ring. In the case of a path, there is no need to use any converter. In the case of a ring network, putting one converter at any vertex in the ring can achieve LWA. Consider the case that $G(V, E)$ has at least one vertex with degrees greater than two. According to Lemma 5.3, we do not need to consider degree-one or degree-two vertices. Thus, we first remove every degree-two vertex in V by substituting two edges incident to the vertex with one edge linking the two endpoints directly. Then, we remove every degree-one vertex by condensing it to the vertex adjacent to it. After these two operations we obtain an induced graph with fewer vertices and edges. Note that this induced graph may still have degree-two or degree-one vertices which are condensed vertices. Finally, we employ an algorithm for the vertex-cover problem on the induced graph, and obtain a vertex-cover C. The set C is able to guarantee LWA of the original graph. Fig. 5.11 illustrates of the proposed algorithm. At the first step, three degree-two vertices (in grey in Fig. 5.11(a)) are removed. At the second step, six degree-one vertices (in grey in Fig. 5.11(b)) in the reduced graph are condensed to their adjacent vertices, respectively. Finally, a vertex-cover of six vertices (in black in Fig. 5.11(c)) is found, which guarantees LWA. The proposed algorithm is formally presented as below.

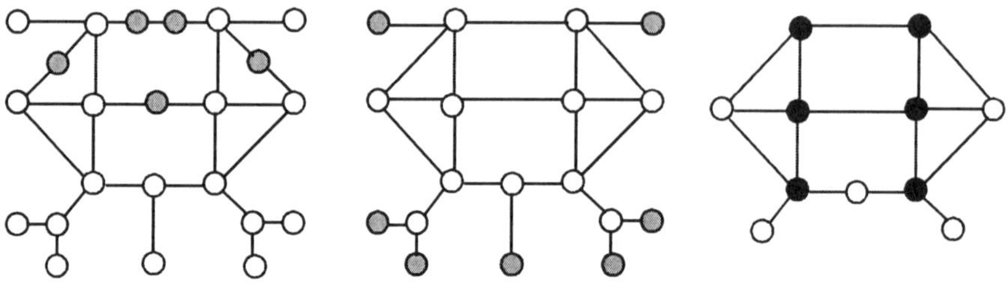

Figure 5.11. Algorithm 5.1: (a) remove degree-two vertices, (b) remove degree-one vertices, and (c) find a vertex-cover of the induced graph.

ALGORITHM 5.1 *Producing a Set Guaranteeing LWA*

Step 0 *Process simple cases*

if $G(V, E)$ is a ring **then**

return C including one vertex in V.
if $G(V, E)$ is a path **then**
return $C := \emptyset$.
if $G(V, E)$ is a spider **then**
return $C := \{v \in V | v$ has degree greater than two$\}$.

Step 1 *Removing degree-two vertices*
$V^2 := \{v \mid v$ is adjacent to exactly two vertices v_1 and $v_2\}$.
while $V^2 \neq \emptyset$ **do begin**
choose $v \in V^2$,
$V := V \setminus \{v\}$.
$E := E \cup \{(v_1, v_2)\} \setminus \{(v_1, v), (v, v_2)\}$,
$V^2 := V^2 \setminus \{v\}$.
end-while

Step 2 *Removing degree-one vertices*
$V^1 := \{u \mid u$ is adjacent to exactly one vertex $u_1\}$.
while $V^1 \neq \emptyset$ **do begin**
choose $u \in V^1$,
$V := V \setminus \{u\}$.
$E := E \setminus \{(u_1, u)\}$,
$V^1 := V^1 \setminus \{u\}$.
end-while

Step 3 *Generate a vertex-cover*
Generate a vertex-cover C of the reduced graph $G(V, E)$,
Return C.

The following theorem proves the validity of Algorithm 5.1. In the analysis, we do not need to consider the case that $G(V, E)$ is a spider or a ring, since the algorithm can find the optimal solution in these two simple cases.

THEOREM 5.3 *For any given graph $G(V, E)$, Algorithm 5.1 produces a set $C \subseteq V$ that guarantees LWA.*

PROOF Let $G(V_2, E_2)$ be the final graph obtained after Step 2. To prove the theorem, it suffices to show that every vertex-cover C of $G(V_2, E_2)$ guarantees LWA for original network $G(V, E)$. According to the way that $G(V_2, E_2)$ is constructed, it is not difficulty to verify that every connected component of $G_C(V, E)$ is a spider. Thus it follows from Lemma 5.2 that C guarantees LWA for original network $G(V, E)$. ■

The following theorem shows that Algorithm 5.1 can produce a 2-approximation solution to Problem 5.1 in an efficient way.

THEOREM 5.4 *Given a graph $G(V, E)$, Algorithm 5.1 produces a solution C to Problem 5.1 in time $O(|E| + |V|)$ satisfying $|C| \leq 2|C_{opt}|$, where C_{opt} is the optimal solution to the problem.*

PROOF It is obvious to see that Step 0-1-2 can be finished in time $O(|E|+|V|)$. In Step 3, the vertex-cover C of $G(V_2, E_2)$ can be found in time $O(|E|+|V|)$ in the following way.

Step 1. Construct a maximal matching M of $G(V_2, E_2)$ such that any pair of edges in M do not share an endpoint; Moreover, any edge in $E_2 \setminus M$ shares an endpoint with an edge in M.

Step 2. Construct C by including both endpoints of each edge in matching M.

Note that every edge in $E_2 \setminus M$ has at least one endpoint matched in M; Otherwise the edge could be added to M to provide a larger matching. This implies that every edge in E_2 has at least one endpoint that is matched and thus C is a vertex-cover. Moreover, it can be produced in time of $O(|E|+|V|)$. Hence Algorithm 5.1 can finish in time $O(|E|+|V|)$.

Now we prove $|C| \leq 2|C_{opt}|$. By Lemma 5.3 we can assume that $S \subseteq V_2$. In fact, we can further assume that C_{opt} is a vertex-cover of $G(V_2, E_2)$. If not, there exist two vertices $w_1 \in V_2$ and $w_2 \in V_2$ such that $(w_1, w_2) \in E_2$ with $w_1 \notin C_{opt}$ and $w_2 \notin C_{opt}$. According to the rules of Algorithm 5.1, w_1 and w_2 has degree greater than two in $G(V, E)$ (they may have degrees one or two in $G(V_2, E_2)$). Moreover, they are in one connected component of $G_S(V, E)$, because removing degree-one and degree-two vertices from $G(V, E)$ does not destroy its connectivity. This contradicts Lemma 5.2. To see $|C| \leq 2|C_{opt}|$, consider the edges in maximal matching M. To cover these edges we need at least $|M|$ vertices, since no two of them share a endpoint. This implies that the minimum vertex-cover has size at least $|M|$ and thus C contains exactly $2|M|$ vertices. Hence $|C_{opt}| \geq |M| = |C|/2$. The proof is then finished. ■

As Lemma 5.4 shows that the WCPP for LWA (i.e., Problem 5.1) is equivalent to the vertex-cover problem, which is believed unlikely to have approximation algorithms with a constant performance ratio less than two [12], Algorithm 5.1 for WCPP is believed to be the best possible. Furthermore, Algorithm 5.1 can find, in polynomial-time, the optimal solution to the WCPP in some special graphs, such as trees, meshes, toruses, and hyper-cubes, because the minimum vertex-cover can be computed efficiently in these cases.

1.2.2 Special Networks

In this subsection, we will discuss the WCPP for LWA (i.e., Problem 5.1) in WDM networks with special topologies including trees, rings, meshes, and hyper-cubes, which are widely used in parallel and distributed computing systems. The proposed theorems and methodology can also be applied to other interconnecting networks for parallel computing networks, such as Banyan networks and Shuffle networks.

We first study the case of bidirectional channels where all corollaries follow from Lemma 5.1.

COROLLARY 5.1 *To guarantee LWA for bidirectional channels on trees of n vertices, $\lfloor (n-2)/2 \rfloor$ converters are sufficient for all trees and necessary for some trees.*

PROOF "Sufficiency": Let n_i be the number of vertices with degree i, for $i = 1, 2$, and let n_3 be the number of vertices with degrees at least three. Clearly, $n = n_1 + n_2 + n_3$ and the total degree of n vertices is at least $(n_1 + 2n_2 + 3n_3)$. In addition, a tree of n vertices has $(n-1)$ edges and the total degree of n vertices equals $2(n-1)$. Therefore, $n - 1 \geq (n_1 + 2n_2 + 3n_3)/2$. This implies, $n_3 \leq (n - n_2 - 2)/2 \leq (n-2)/2$. According to Lemma 5.1, we know that $(n-2)/2$ converters are sufficient to guarantee LWA for bidirectional channels.

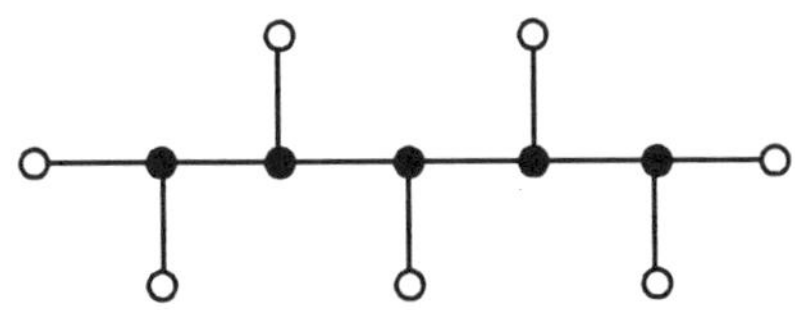

Figure 5.12. The optimal placement in the worst case of trees of n vertices.

"Necessity": For the tree, as shown in Fig. 5.12 that has a path of $(n+2)/2$ vertices and each of $(n-2)/2$ intermediate vertices in the path is adjacent to a degree-one vertex, it is obvious that $(n-2)/2$ converters are necessary to guarantee LWA for bidirectional channels. ■

COROLLARY 5.2 *To guarantee LWA for bidirectional channels on a ring, one converter is both sufficient and necessary.*

PROOF Let S be a singleton including one vertex. Clearly, $G_S(V, E)$ is a path. Hence according to Lemma 5.1 a set containing one vertex in V is both sufficient and necessary for guaranteeing LWA for bidirectional channels. ■

COROLLARY 5.3 *To guarantee LWA for bidirectional channels on the mesh of n vertices, $(n-4)$ converters are both sufficient and necessary.*

PROOF Except four degree-two vertices in the four corners all other vertices in a mesh have degree three or four. The conclusion follows immediately from Lemma 5.1. ■

COROLLARY 5.4 *To guarantee LWA for bidirectional channels on $(n \times n)$-torus or n-hypercube, for $n \geq 3$, n^2 or 2^n converters are both sufficient and necessary.*

PROOF Note that all vertices in $(n \times n)$-torus or n-hypercube have degrees at least three, for $n \geq 3$. The conclusion then follows from Lemma 5.1. ■

Now we study the case of unidirectional channels where the results follow from Lemma 5.2.

For tree networks, if no vertex has degree greater than two, then a tree becomes a path. According to Lemma 5.2 no converter is necessary for LWA. So we only consider the case that there exists at least one vertex with degree greater than two. Notice that Lemma 5.3 can not be applied to trees since all trees have vertices with degree one. According to Lemma 5.4, we do not need to consider degree-one or degree-two vertices when placing converters. Thus we can find the minimum set that guarantees LWA in the following way.

(1) Remove every degree-two vertex by substituting two edges incident to the vertex with one edge incident to two endpoints provided that they are not degree-two vertices.

(2) Remove every degree-one vertex by condensing it to the vertex adjacent to it.

(3) Find the minimum vertex-cover of the resultant tree.

Fig. 5.13 illustrates the above process. At the first step, three degree-two vertices (in grey in Fig. 5.13(a)) are removed. At the second step, eleven degree-one vertices (in grey in Fig. 5.13(b)) in the reduced network are condensed to five vertices that they are incident to, respectively (Fig. 5.13(c)). Finally a vertex-cover consisting of two vertices (in black in Fig. 5.13(d)) is found. Notice that the final tree, reduced from by removing degree-one and degree-two vertices in the original tree, may have some degree-one or degree-two vertices, which are condensed vertices. The proposed method is formally presented at below.

ALGORITHM 5.2 *Producing a Set Guaranteeing LWA in Trees*

Input A tree $G(V, E)$
Output A subset $C \subseteq V$ guaranteeing LWA
Step 1 *Remove degree-two vertices from* $G(V, E)$
  $V' := V$,
  $E' := E$.
  $V_2 := \{v \mid v$ is adjacent to exactly two vertices v_1 and $v_2\}$,
  $V' := V' \setminus V_2$.
  while $V_2 \neq \emptyset$ **do begin**
    choose $v \in V_2$,
    $E' := E' \cup \{(v_1, v_2)\} \setminus \{(v, v_1), (v, v_2)\}$,
    $V_2 := V_2 \setminus \{v\}$.
  end-while
Step 2 Remove degree-one vertices from $G(V', E')$
  $V_1 := \{u \mid u$ is adjacent to exactly one vertex $u_1\}$,
  $V' := V' \setminus V_1$.
  while $V_1 \neq \emptyset$ **do begin**
    choose $u \in V_1$,
    $E' := E' \setminus \{(u, u_1)\}$,

$V_1 := V_1 \setminus \{u\}$.
end-while

Step 3 *Generate a minimum vertex-cover of* $G(V', E')$
$C := \emptyset$
$\overline{V} := V'$,
$\overline{E} := E'$.
while $\overline{E} \neq \emptyset$ **do begin** (Loop ends until all edges in E' are covered)
$V_1 := \{v \in \overline{V} \mid v$ has degree one in $G(\overline{V}, \overline{E})\}$,
$C := C \cup \{u \in \overline{V} \setminus V_1 \mid u$ is adjacent to some $v \in V_1\}$,
$\overline{V} := \overline{V} \setminus \{v \in \overline{V} \mid v \in C$ or v is adjacent to some $u \in C\}$,
$\overline{E} := \{(u, v) \in \overline{E} \mid u$ and v are belong to $\overline{V}\}$.
end-while
return C.

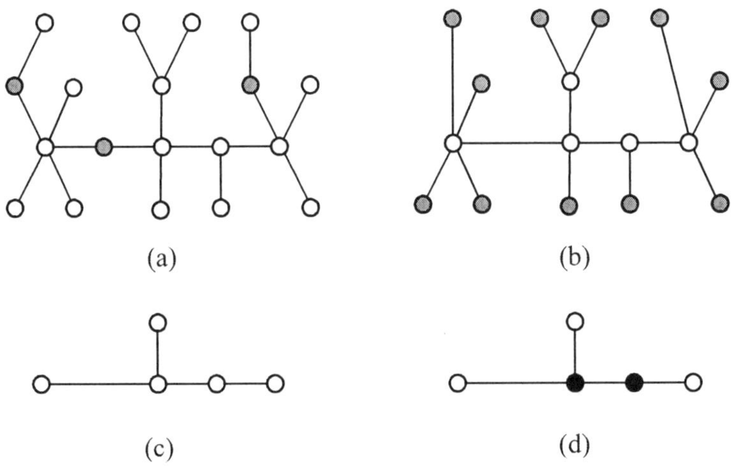

Figure 5.13. Algorithm 5.2: (a) remove degree-two vertices, (b) remove degree-one vertices, (c, d) find a vertex-cover of the induced tree.

The following theorem shows that the vertex-cover of the finally reduced tree is an optimal solution to the WCPP for LWA in the original tree. The proof is very similar to the proof of Theorem 5.3.

THEOREM 5.5 *For any given tree* $G(V, E)$, *Algorithm 5.2 finds an optimal solution to the WCPP for unidirectional channels in* $G(V, E)$ *in time* $O(|E| + |V|)$.

PROOF Let $G(V', E')$ be the final graph obtained after Step 2 and C be the returned vertex-cover of $G(V', E')$. According to Lemma 5.4, we know that there exists an optimal solution to the WCPP in $G(V, E)$ that is a subset of V'. In the following, we will prove that a subset S of V' guarantees LWA in $G(V,E)$ if and only if it is a vertex-cover of $G(V',E')$.

"If": By contradiction, we assume that $S \subset V'$ is a vertex-cover of $G(V',E')$ but it does not guarantee LWA in $G(V,E)$. According to Lemma 5.2, there exists a connected component of $G_S(V,E)$ which is not a spider. This means that there exist two vertices x and y which are not in S and have degrees greater than two in $G_S(V,E)$. Note there is a path in $G_S(V,E)$ between x and y and no vertex in the path is included in S. Moreover, vertices x and y are included in V', and removing degree-one and degree-two vertices from $G(V,E)$ does not destroy the connectivity between them. Thus x and y remain connected by the path in $G(V',E')$, which implies that all edges in the path are not covered by any vertex in S. This contradicts that S is a vertex-cover of $G(V',E')$.

"Only if": By contradiction again, we assume that there exists a subset $S \subset V'$ that guarantees LWA in $G(V,E)$ but it is not a vertex-cover. Then there are two vertices $x \in V'$ and $y \in V'$ such that $(x,y) \in E'$ but $x \notin S$ and $y \notin S$. Note that x and y have degrees greater than two in $G(V,E)$; otherwise they are removed away at Step 1 or Step 2. Hence the component of $G_S(V,E)$, which contains (x,y), is not a spider. This contradicts Lemma 5.2.

Now we prove that set C produced by Algorithm 5.2 is a minimum vertex-cover of $G(V',E')$. In the while-loop at Step 3, the vertices that are adjacent to degree-one vertices in current $\overline{V}$ are included in C. They can cover the edges incident to them. Thus these edges are removed from $\overline{E}$. In such way when $\overline{E}$ becomes empty, all edges in E' can be covered by some vertices in C. This means that C is a vertex-cover of $G(V',E')$. Then, we prove C is the minimal. Assume that C^* is a minimum vertex-cover of $G(V',E')$. If C^* includes some degree-one vertices in $V_1 \subset \overline{V}$, then we can replace them with the vertices that are adjacent to them. It is obvious that such modified C^* remains a minimum vertex-cover of $G(V',E')$. It is not difficult to see that the finally modified C^* is exactly the set of C, because this is the exact way how C is produced. Thus C is a minimum vertex-cover of $G(V',E')$. Therefore, C is an optimal solution to the WCPP in $G(V,E)$.

To see that Algorithm 5.2 finishes in time $O(|E|+|V|)$, note that the key operation in all three steps is to check the degree of each vertex in the current graph and include or exclude the vertex and its incident edges in some sets. The proof is complete. ■

COROLLARY 5.5 *To guarantee LWA for unidirectional channels on a ring, one converter is both sufficient and necessary.*

PROOF Let S be a singleton including one vertex. Clearly, $G_S(V,E)$ is a path. Hence according to Lemma 5.2 a set containing one vertex in V is both sufficient and necessary for guaranteeing LWA for unidirectional channels. ■

COROLLARY 5.6 *To guarantee LWA for unidirectional channels on trees of n vertices, $\lfloor (n-2)/4 \rfloor$ converters are sufficient for all trees and necessary for some trees.*

PROOF "Sufficiency": For a tree of n vertices, let n_3 be the number of vertices with degrees at least three. When Algorithm 5.2 runs with the tree, the final graph of $G(V', E')$ includes n_3 vertices and $(n_3 - 1)$ edges. Because each vertex can cover at least two edges, the minimum vertex-cover has size at most $\lceil (n_3 - 1)/2 \rceil$, which can be proved to be less than or equal to $\lfloor (n-2)/4 \rfloor$.

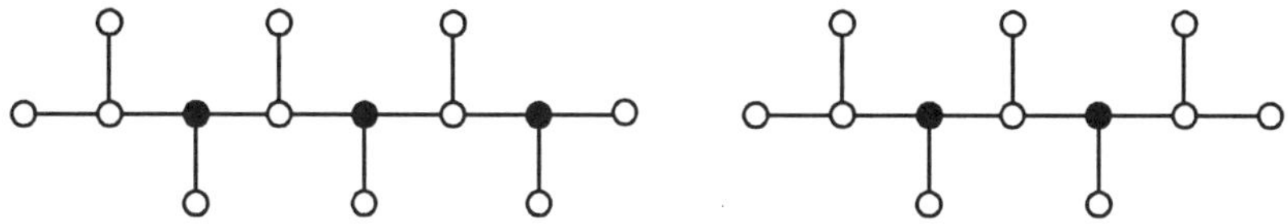

Figure 5.14. The optimal placement in the worst case of trees of n vertices.

"Necessity": It can be verified that $\lfloor (n-2)/4 \rfloor$ converters are necessary for trees, as shown in Fig. 5.14, that consists of a path and each vertex is adjacent to exactly three vertices. ∎

COROLLARY 5.7 *To guarantee LWA for unidirectional channels on* $(n \times n)$*-mesh,* $2\lceil n/2 \rceil \lfloor n/2 \rfloor$ *converters are both sufficient and necessary.*

PROOF For $(n \times n)$-mesh, according to Lemma 5.3 we can remove each of four degree-two vertices at the corners by substituting two edges incident to the vertex with one edge linking the two endpoints directly. By Lemma 5.4, a minimum vertex-cover of modified mesh is both sufficient and necessary for LWA. It is easy to see that a minimum vertex-cover is the set that includes every other vertex in each row and column (extra vertex must be added in even rows and columns for odd n), and thus the size of the cover is $n^2/2$ for even n and $(n^2 - 1)/2$ for odd n. See Fig. 5.15, where converters are optimally placed at the vertices in black. ∎

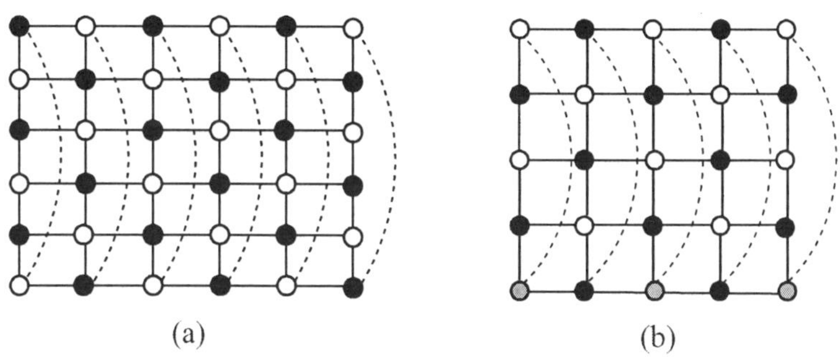

Figure 5.15. Optimal placement of converters on meshes and torus: (a) in the case of even n and (b) in the case of odd n.

COROLLARY 5.8 *To guarantee LWA for unidirectional channels on $(n \times n)$-torus, $n\lceil n/2 \rceil$ converters are both sufficient and necessary.*

PROOF According to Lemma 5.4, the minimum vertex-cover of $(n \times n)$-torus is both sufficient and necessary for LWA. It is easy to see that the minimum vertex-cover is the set that includes every other vertex in each row and column (extra vertex must be added for odd n), and thus the size of the cover is $n^2/2$ for even n and $n(n+1)/2$ for odd n. See Fig. 5.15. Converters are optimally placed at the vertices marked in black and grey. ■

COROLLARY 5.9 *To guarantee LWA for unidirectional channels on n-cube for $n \geq 3$, 2^{n-1} converters are both sufficient and necessary.*

PROOF According to Lemma 5.4, the minimum vertex-cover of n-cube is both sufficient and necessary for LWA. Construct two subsets of the vertex-set of n-cube $V(n) = \{i = (i_1, i_2, \cdots, i_n) \mid i_j = 0 \text{ or } 1, j = 1, 2, \cdots, n\}$ as follows: $C_1(2) = \{(0,0), (1,1)\}$ and $C_2(2) = \{(0,1), (1,0)\}$. For $n \geq 3$,

$$\begin{aligned} C_1(n) &= \{i, \text{ for } i \in C_1(n-1)\} \cup \{i + 2^{n-1}, \text{ for } i \in C_2(n-1)\}, \\ C_2(n) &= \{i, \text{ for } i \in C_2(n-1)\} \cup \{i + 2^{n-1}, \text{ for } i \in C_1(n-1)\}. \end{aligned}$$

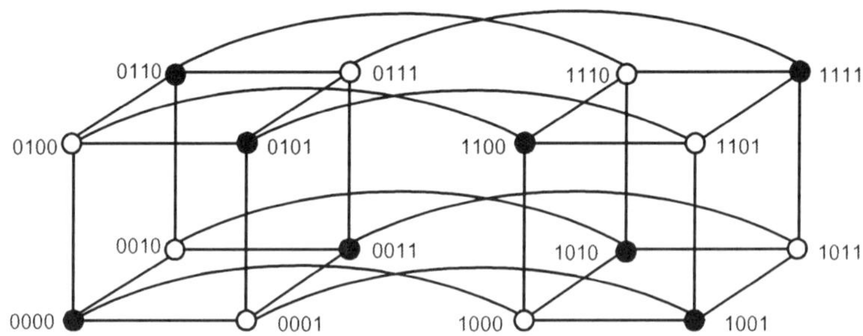

Figure 5.16. Optimal placement of converters on n-cube.

It is not difficult to verify (by mathematical induction) that for $n \geq 3$, $C_1(n)$ and $C_2(n)$ make an equal partition of 2^n vertices in $V(n)$, and they are two minimum vertex-covers of n-cube. See Fig. 5.16, where $C_1(n)$ consists of the vertices marked in black while $C_2(n)$ consists of the vertices marked in white. ■

The results obtained from the above discussion are summarized in Table 5.1.

2. Placement for Relaxed Load-Wavelength Assignability

In the preceding section we have discussed the WCPP for LWA. Our study shows that introducing wavelength converters and placing them properly can reduce the

Table 5.1. The number of converters required for load-wavelength assignability.

Networks	Unidirectional channels	Bidirectional channels
Stars	1	0
Ring of n vertices	1	1
Tree of n vertices	$\lfloor (n-2)/4 \rfloor$	$\lfloor (n-2)/2 \rfloor$
$(n \times n)$-Mesh	$2\lceil n/2 \rceil \lfloor n/2 \rfloor$	$n^2 - 4$
$(n \times n)$-torus	$n\lceil n/2 \rfloor$	n^2
n-Cube	2^{n-1}	2^n
General	$\exists$ 2-approximation	$\exists$ polynomial-time algorithm

number of wavelengths needed in a system as much as possible. Notice, however, that the optimal placement of wavelength converters for LWA may demand a large number of converters, especially when the density of the network is high, such as the networks of mesh and hyper-cube (refer to Table 5.1).

In this section, we will consider the problem of tradeoff between the number of wavelengths required and the number of wavelength converters placed in a system, that is, how to use a little bit more wavelengths than the network load in order to reduce the number of wavelength converters used.

2.1 Optimization Formulation and Theory

Given a graph $G(V, E)$ and a real number $\alpha \geq 1$, a subset $S \subseteq V$ is said to guarantee *Relaxed Load-Wavelength Assignability* (RLWA) with respect to α, denoted simply by α-RLWA, if, by configuring each vertex in S as a wavelength converter, the number of wavelengths needed to set up all possible sets of channels in the system is equal to α times the maximal link load. Clearly, α-RLWA is a natural generalization of LWA with $\alpha = 1$. In this section, we are interested in how many wavelength converters can be saved when α is relaxed to a small number, i.e, $1 < \alpha \leq 2$. This problem is formulated below as the WCPP for RLWA.

PROBLEM 5.2 *Converter Placement for Relaxed Load-Wavelength Assignability*

Instance A graph $G(V, E)$ and a real number $\alpha > 1$.
Solution A subset $S \subseteq V$ that guarantees α-RLWA.
Objective Minimizing the cardinality of set S, i.e., $|S|$.

From the results summarized in Table 1 of Chapter 4, we know that for tree networks, any set of channels with maximal link load L can be assigned by using at most $\frac{5}{3}L$ wavelengths for unidirectional channels, and $\frac{3}{2}L$ wavelengths for bidirectional channels. Our approach for Problem 5.2 is based on these results. The basic idea is the same as the one used in the preceding section. It decomposes any given network of general topology into a network of special topology, precisely, edge-disjoint trees.

The decomposition of a graph is done again by the splitting operation described in the preceding section, by which we are able to reduce Problem 5.2 to the feedback vertex set problem.

Given a graph $G(V, E)$, a subset $S \subseteq V$ is called a *Feedback Vertex Set* (FVS) if $G(V \setminus S, E)$, the induced subgraph of $G(V, E)$ on $V \setminus S$, has no cycle. *The feedback vertex set problem* is to find a feedback vertex set of minimal cardinality. It was proved in [8] that this problem is $\mathcal{NP}$-hard.

It is easy to see that placing wavelength converters for α-RLWA is equal to finding the minimum feedback vertex set of a network and configuring the vertices in the set as converter nodes. By splitting the vertices in the minimum set S in $G(V, E)$, the reduced graph $G_S(V, E)$ consists of a set of trees. Thus, the problem of assigning lightpaths, by using wavelengths at most α times network load, in the original graph is then reduced to the same problem in trees. The following lemma characterizes the structure of $G_S(V, E)$ in terms of an FVS of $G(V, E)$.

LEMMA 5.5 *Given a network $G(V, E)$ and a subset $S \subseteq V$, each connected component of $G_S(V, E)$ is a tree if and only if S is an FVS of $G(V, E)$.*

PROOF "Only if": Suppose that $G_S(V, E)$ consists of k connected components, each of which is a tree and denoted by $T_i(S), 1 \leq i \leq k$. Thus, we have $G_S(V, E) = \cup_{i=1}^{k} T_i(S)$. Let $E_i(S) = \{sv \mid s \in S$ is adjacent to v in $G(V, E)\}$ and $T_i'(S) = T_i(S) \setminus E_i(S), 1 \leq i \leq k$. Note that $T_i'(S)$ is a tree in $G(V \setminus S, E)$ for each i and $G(V \setminus S, E) = \cup_{i=1}^{k} T_i'(S)$. Therefore, S is an FVS of $G(V, E)$.

"If": If $S \subseteq V$ is an FVS of $G(V, E)$, then $G(V \setminus S, E)$ is a forest. Suppose there are k connected components in $G(V \setminus S, E)$, which are trees and denoted by $T_i(S)$, $1 \leq i \leq k$, i.e., $G(V \setminus S, E) = \cup_{i=1}^{k} T_i(S)$. Now let $E_i(S) = \{sv \mid v$ is a leaf of $T_i(S)$ and $s \in S$ is adjacent to v in $G(V, E)\}$ and $T_i'(S) = T_i(S) \cup E_i(S)$, $1 \leq i \leq k$. Because each vertex $s \in S$ is split into $\delta(s)$ different vertices, $T_i'(S)$ is a tree in $G_S(V, E)$, $1 \leq i \leq k$. Note that each edge $uv \in E$ is in either $T_i'(S)$ (when $u, v \notin S$) or $E_i(S)$ (when $u \in S$ or $v \in S$) for some i. Thus $G_S(V, E)$ consists of k trees. The proof is then finished. ■

The following theorem establishes the relationship between the set that guarantees α-RLWA for bidirectional channels and an FVS of $G(V, E)$.

THEOREM 5.6 *Given a network $G(V, E)$, a subset $S \subseteq V$ guarantees $\frac{3}{2}$-RLWA for bidirectional channels if and only if S is an FVS of $G(V, E)$.*

PROOF "If": If S is an FVS of $G(V, E)$, then according to Lemma 5.5 each connected component in $G_S(V, E)$ is a tree. Note that any set of bidirectional channels in a tree network with network load L can be assigned by using at most $\frac{3}{2}L$ wavelengths (refer to Section 3.1 of Chapter 3). Although the channel that traverses several connected components in $G_S(V, E)$ may be assigned different wavelengths, this will not cause any wavelength conflict, because each vertex in S is assumed to have full wavelength conversion capability.

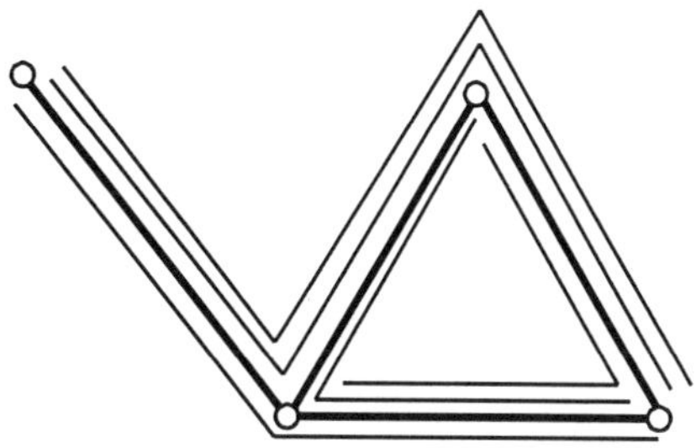

Figure 5.17. Five channels that produce maximal link load three require five wavelengths.

"Only if": We prove the theorem by contradiction. Assume that S is not an FVS of $G(V, E)$, but any set of bidirectional channels producing network load L in $G(V, E)$ can be assigned by using $\frac{3}{2}L$ wavelengths. Then by Lemma 5.5 there is a connected component in $G_S(V, E)$ which includes a cycle. Note that there exists a vertex in the component which is connected with the cycle but not on the cycle. Now consider a set of five bidirectional channels on the component as shown in Fig. 5.17. It can be verified that the network load is three but five wavelengths are needed, since each of those five channels shares a link with the other four channels. This contradicts $\frac{3}{2}$-RLWA. The proof is then finished. ■

The following theorem shows that the Problem 5.2 with $\alpha > 1$ for bidirectional channels is $\mathcal{NP}$-hard. Recall that Theorem 5.1 implies that the Problem 5.2 with $\alpha = 1$ (i.e., Problem 5.1) for bidirectional channels is polynomial-time solvable.

THEOREM 5.7 *The Problem 5.2 with $\alpha = 3/2$ for bidirectional channels is $\mathcal{NP}$-hard, and it has a 2-approximation algorithm of time $O(|E||V|)$.*

PROOF The $\mathcal{NP}$-hardness of this problem follows directly from Theorem 5.6 and the $\mathcal{NP}$-hardness of the feedback vertex set problem [8]. In addition, there is an $O(|E||V|)$-algorithm to produce a 2-approximation solution to the latter problem [1], which clearly is also a 2-approximation solution to the former problem. ■

The following theorem for unidirectional channels gives the parallel results of the above theorem.

THEOREM 5.8 *Given a graph $G(V, E)$, if a subset $S \subseteq V$ is an FVS of $G(V, E)$, then S can guarantee $\frac{5}{3}$-RLWA for unidirectional channels.*

PROOF Given a set of unidirectional channels routed in $G(V, E)$ that make maximal link load L, examine them in $G_S(V, E)$. Note that each connected component in $G_S(V, E)$ is a tree and the unidirectional channels in a tree network can be assigned by using $\frac{5}{3}L$ wavelengths (refer to Section 3.1 of Chapter 3). Although the channel that traverses several connected components in $G_S(V, E)$ may be assigned different wavelengths, this will not cause any wavelength conflict, because each vertex in S is assumed to have full wavelength conversion capability. ■

It follows immediately from the above theorem that given a graph $G(V, E)$ the $O(|E||V|)$-algorithm in [1] for the feedback vertex set problem can be used to produce a set that guarantees $\frac{5}{3}$-RLWA for unidirectional channels.

2.2 Algorithm Design and Performance Analysis

In this subsection we will develop an approximation algorithm that locate a set guaranteeing RLWA. The algorithm is applicable for both bidirectional and unidirectional channels.

2.2.1 General Networks

According to Theorem 5.6 and Theorem 5.8, given any network $G(V, E)$, finding the minimum set of converters in the network to guarantee α-RLWA is equivalent to finding the minimum FVS of $G(V, E)$. We now employ the 2-approximation algorithm, proposed in [1], to find the minimum FVS in $G(V, E)$. To present the algorithm we need some notations. A graph is said *clean* if it has no degree-one vertex. A cycle in a graph is said *semi-disjoint* if at most one vertex in the cycle has degree greater than two.

ALGORITHM 5.3 *Finding a Set Guaranteeing RLWA*

Input A graph $G(V, E)$
Output A feedback vertex set $F \subseteq V$
Step 1 *Construct an initial feedback vertex set F'*
  $F' := \emptyset$.
  while $V \neq \emptyset$ **do begin**
    while $V \neq \emptyset$ **do begin** // make (current) graph clean
      if V has degree-one vertex $v \in V$ **then**
        $V := V \setminus \{v\}$,
        $E := E \setminus \{(v, u) \in E\}$.
    end-while
    if $G(V, E)$ contains a semi-disjoint cycle C **then**
      $F' := F' \cup \{v \in C, v$ has degree greater than two$\}$,
      $E := E \setminus \{(v, u) \in E\}$.
    else choose $v \in V$ which has the largest degree,
    $F' := F' \cup \{v\}$,
    $E := E \setminus \{(v, u) \in E\}$.
  end-while
Step 2 *Remove redundant vertices from F'*
  $F := F'$.
  while $F' \neq \emptyset$ **do begin**
    choose vertex $v \in F'$ which was latest put into F',

$F' := F' \setminus \{v\}$.
if $F \setminus \{v\}$ is a *FVS* of original $G(V, E)$ **then**
$F := F \setminus \{v\}$. // v is redundant
end-while
return F.

The following theorem [1] proves the correctness and the approximation performance analysis of the above algorithm. The basic idea of its proof is to use a divide-and-conquer technique.

THEOREM 5.9 *For any given graph $G(V, E)$, Algorithm 5.3 outputs an FVS of $G(V, E)$ with cardinality at most two times that of the minimum FVS.*

2.2.2 Special Networks

In this subsection, we will discuss the WCPP for RLWA (i.e., Problem 5.2) in WDM networks with special topologies including trees, rings, meshes, and hypercubes. The following corollary comes directly from Theorem 5.6 and Theorem 5.8.

COROLLARY 5.10 *In tree networks there is no need to use any converter for guaranteeing $\frac{3}{2}$-RLWA for bidirectional channels and $\frac{5}{3}$-RLWA for unidirectional channels. In ring networks one converter is required to guarantee α-RLWA for any $\alpha \geq 1$.*

THEOREM 5.10 *To guarantee $\frac{3}{2}$-RLWA for bidirectional channels, or $\frac{5}{3}$-RLWA for unidirectional channels in an $(n \times n)$-mesh, $(n^2 - 2n + 2)/3$ converters are necessary and $(\frac{n^2+n+1}{3} + \frac{1}{4}\log 4(n-2))$ converters are sufficient.*

PROOF It was proved in [11] that the optimal solution to the feedback vertex set problem in $(n \times n)$-mesh has the cardinality at least $\frac{(n-1)^2+1}{3}$, and at the same time the inductive algorithm was proposed to produce an FVS whose cardinality is at most $(\frac{(n-1)^2+3n}{3} + \frac{1}{4}\log 4(n-2))$. Therefore, the theorem follows directly from Theorem 5.6 and Theorem 5.8. ■

Fig. 5.18 illustrates an optimal solution to the feedback vertex set problem in (17×17)-mesh, which consists of eighty-six vertices (in black). Note that removing all black vertices from the mesh makes the resulting graph a tree, or equivalently all cycles in the mesh include at least one black vertex. Moreover, the optimal solution consists of four optimal solutions to the same problem in (9×9)-mesh (four small square meshes divided by dashed lines).

In fact, for small n we can find the optimal solutions to the feedback vertex set problem in $(n \times n)$-mesh whose size is less than the upper bound given in Theorem 5.10. See the results included in Table 5.2 at the end of the next section.

THEOREM 5.11 *To guarantee $\frac{3}{2}$-RLWA for bidirectional channels, or $\frac{5}{3}$-RLWA for unidirectional channels in an n-cube, $\lceil \frac{(n-2)2^{n-1}+1}{n-1} \rceil$ converters are necessary and 2^{n-1} converters are sufficient.*

Figure 5.18. An optimal solution to the feedback vertex set problem in (17×17)-mesh that includes four optimal solutions to the problem in (9×9)-mesh.

PROOF "Sufficiency": Consider the binary representation of the vertex-labels of n-cube, and define the *weight* of a vertex as the number of 1's in its binary representation. Now partition all vertices into two sets V_1 and V_2 so that V_1 (V_2) contains the vertices with odd (even) weight. See Fig. 5.19, where $n = 4$. Clearly, n-cube is a bipartite graph $B(V_1, V_2; E)$. Thus, V_1 (consisting of vertices in black) and V_2 (consisting of vertices in white) are two FVS's of n-cube.

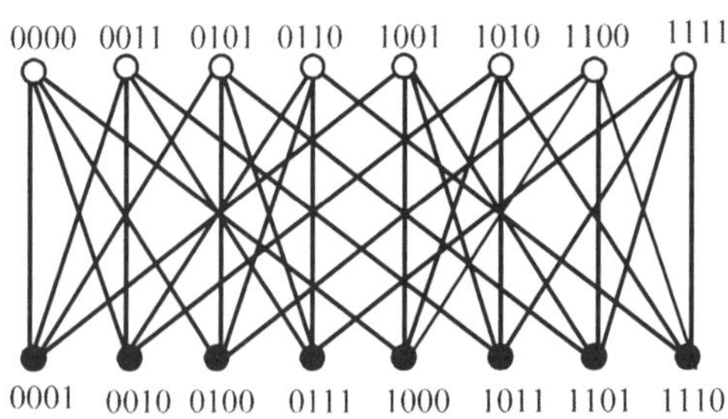

Figure 5.19. Two FVS's in an n-cube.

"Necessity": Consider a solution S to the feedback vertex set problem in an n-cube. Due to Lemma 5.5, $G_S(V, E)$ is a forest. Note that $G_S(V, E)$ has $n2^{n-1}$ edges and $(2^n + (n-1)|S|)$ vertices, since each vertex in $S \subset V$ (in n-cube) becomes n vertices in $G_S(V, E)$. Thus

$$n2^{n-1} + 1 \leq 2^n + (n-1)|S|,$$

which produces $|S| \geq \lceil \frac{(n-2)2^{n-1}+1}{n-1} \rceil$. The proof is then finished. ■

The following theorem gives the optimal solutions to the feedback vertex set problem in small n-cubes.

THEOREM 5.12 *To guarantee $\frac{3}{2}$-RLWA for bidirectional channels, or $\frac{5}{3}$-RLWA for unidirectional channels in an n-cube, three, six, and fourteen converters are both necessary and sufficient for $n = 3$, $n = 4$, and $n = 5$, respectively.*

PROOF According to Theorem 5.6 and Theorem 5.8 it suffices to prove that the optimal solutions to the feedback vertex set problem in 3-cube, 4-cube, and 5-cube have cardinalities of three, six, and fourteen, respectively. We consider these three cubes one by one.

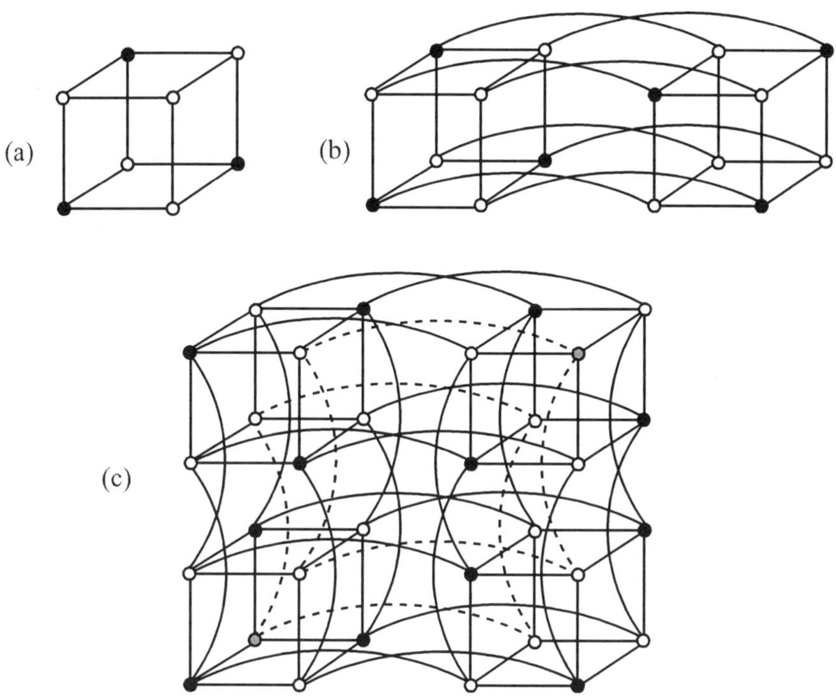

Figure 5.20. Optimal solutions of FVS problem in: (a) 3-cube, (b) 4-cube, and (c) 5-cube.

3-cube: Note that there are six faces, each of which is a 4-cycle, and each vertex is on three of them. If an FVS has two vertices, then they must be in a diagonal position. But there exists a 6-cycle which consists of other six vertices. Hence, the FVS of minimum cardinality has at least three vertices. Fig. 5.20(a) displays an FVS that includes three vertices (in black).

4-cube: Note that 4-cube consists of two vertex-disjoint 3-cubes. Thus, any FVS has at least six vertices. Fig. 5.20(b) shows an FVS which contains exactly six vertices (in black), three in each of two 3-cubes.

5-cube: Note that 5-cube consists of two vertex-disjoint 4-cubes, and it can be verified that the feedback vertex set problem in 4-cube has an unique solution (under isomorphic transformation through vertex-labelling), which is given in Fig. 5.20(b). However, such two solutions, in the total of twelve vertices (in black), can not make an FVS of 5-cube, because there are two vertex-disjoint 4-cycles (marked by dashed lines in Fig. 5.20(c)). To destroy these two cycles, one vertex (in grey) in each of two cycles needs to be included in an FVS. Therefore, the optimal solution to the feedback vertex set problem in 5-cube has fourteen vertices. The proof is then finished. ■

The results obtained from the above discussion are summarized in Table 5.2.

Table 5.2. The number of converters required for the relaxed load-wavelength assignability.

Networks	Unidirectional channels with $\alpha = 5/3$	Bidirectional channels with $\alpha = 3/2$
Tree	No wavelength converter is needed.	
Ring	One wavelength converter is needed.	
Mesh	$(n^2 - 2n + 2)/3 \leq \ \leq (n^2 + n + 1)/2 + \frac{1}{4} \log 4(n-1)$	
Cube	At most 2^{n-1} wavelength converters are needed.	
General	There is a 2-approximation algorithm.	

3. Comparison Study of LWA and RLWA

In this section, we study the difference between the number of wavelength converters required for α-RLWA and that for LWA.

3.1 General Networks

The study on general networks is done through simulations. For each generated network, the FVS-algorithm (i.e., Algorithm 5.2) and the VC-algorithm (as described in the proof of Theorem 5.4) are applied to find the FVS and VC of the network, respectively. The simulation results are illustrated in Fig. 5.21-23, from which we draw the following conclusions.

1 The sizes of VC and FVS of networks increases with the number of nodes in networks (see Fig. 5.21). However, the increasing rate of VC is faster than that of FVS.

2 Networks of bigger sizes have relatively larger VC and FVS (see Fig. 5.22). VC's contain 40%-50% of network nodes for small networks and 90% of network nodes for large networks. However, as a contrast, the percentages of FVS are 10%-20% and 50%, respectively.

3 The size of FVS is 40%-60% smaller than that of VC (see Fig. 5.23) depending on network sizes. This means that relaxing from LWA to RLWA can save about half number of converters, and the saving for small networks is more significant than for large networks.

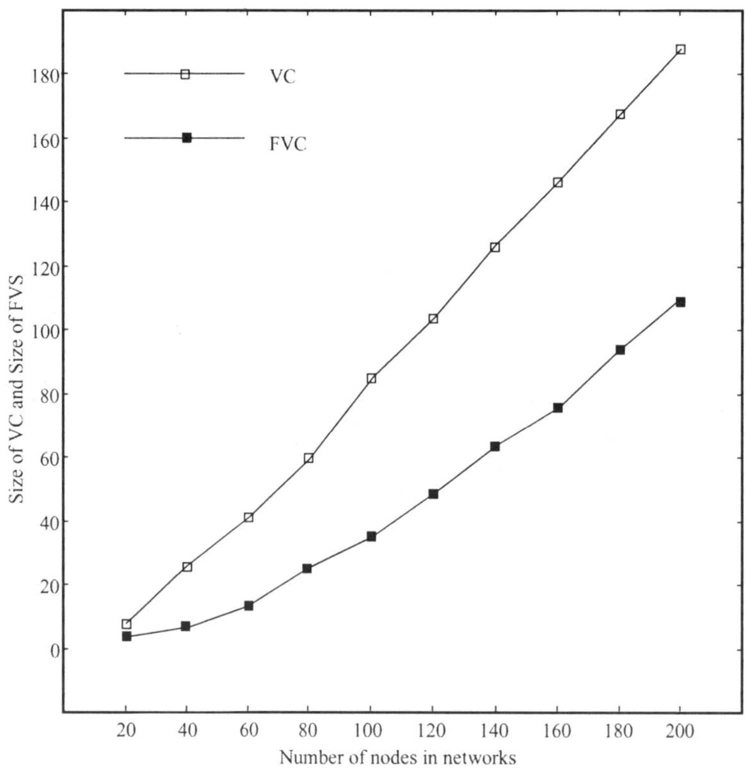

Figure 5.21. The sizes of VC and FVS against the size of networks.

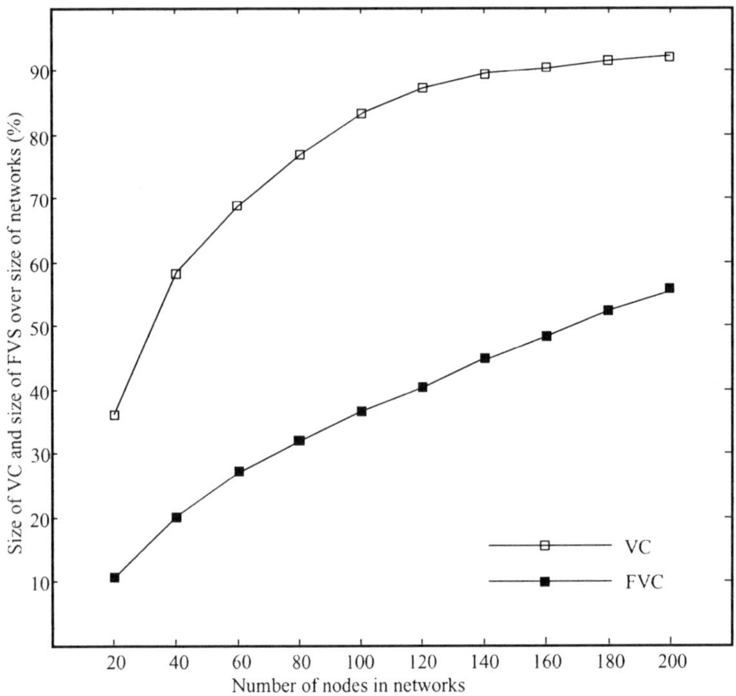

Figure 5.22. The percentages of VC and FVS in the network.

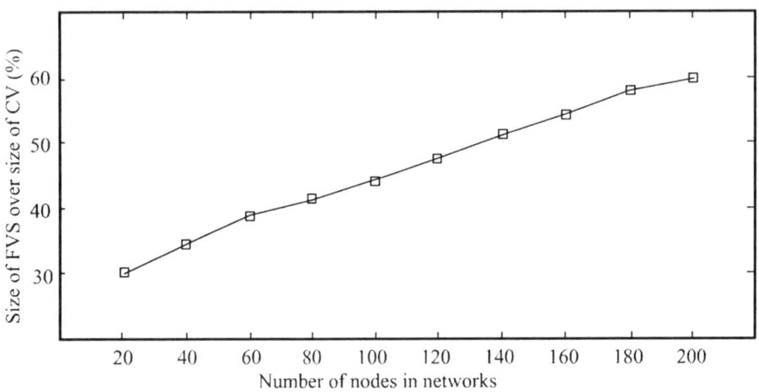

Figure 5.23. The ratio of VC over FVS.

3.2 Special Networks

In order to compare the numbers of wavelength converters required for LWA with that for RLWA, we now summarize the obtained results in Table 5.3 putting them side by side.

Table 5.3. The comparison between LWA and RLWA.

Network	Unidirectional channels		Bidirectional channels	
Topologies	LWA	$\frac{5}{3}$-RLWA	LWA	$\frac{3}{2}$-RLWA
Star	0	0	1	0
Tree of n vertices	$\lfloor (n-2)/4 \rfloor$	0	$\lfloor (n-2)/2 \rfloor$	0
Ring of n vertices	1	1	1	1
$(n \times n)$-mesh	$2\lceil n/2 \rceil \lfloor n/2 \rfloor$	$\leq n(n+2)/3$	$n^2 - 4$	$\leq n(n+2)/3$
4×4-mesh	8	4	12	4
5×5-mesh	12	6	21	6
6×6-mesh	18	10	32	10
7×7-mesh	24	13	45	13
8×8-mesh	32	18	60	18
9×9-mesh	40	22	77	22
10×10-mesh	50	27	96	27
11×11-mesh	60	34	117	34
12×12-mesh	72	41	140	41
n-cube	2^{n-1}	$\leq 2^{n-1}$	2^n	$\leq 2^{n-1}$
3-cube	4	3	8	3
4-cube	8	6	16	6
5-cube	16	14	32	14

By observing the results in Table 5.2, we can see a significant saving of wavelength converters when LWA is relaxed to $\frac{3}{2}$-RLWA for bidirectional channels and $\frac{5}{3}$-RLWA for unidirectional channels, respectively.

- On $(n \times n)$-mesh of small size ($n \leq 12$) about 70% and 50% of wavelength converters can be saved, respectively.
- On $(n \times n)$-mesh of large size two third and one third of wavelength converters can be saved, respectively.
- On n-cube the saving of the wavelength converters for bidirectional channels is about 60% when n is small and 50% when is large. But for unidirectional channels the saving is negligible.

4. Discussion

In Section 1 of this chapter we have studied the WCPP for LWA. The obtained results have significant impact to the design of multihop WDM networks. Firstly, by achieving LWA, the number of wavelengths needed in a system is made minimal, because the low bound of the number of wavelengths required is equal to the maximal link load. Secondly, with LWA, network applications can feel free to maximally utilize the network bandwidth without worrying about the availability of wavelengths, so long as the maximal link load is kept less or equal to the number of wavelengths employed. Thirdly, by using the minimal number of converters to achieve the LWA, it reduces the hardware cost of a network.

In Section 2 of this chapter, we have discussed the WCPP for α-RLWA of small α. Our study shows that when there are more than load number of wavelengths available in the network, a considerable number of wavelength converters can be saved. This result is very important for the following reasons. Firstly, they can help understand the relationship between the number of wavelengths required and the placement of converters. Secondly, they can be used to guide the placement of converters at the design of a WDM network. Thirdly, they can help determine the maximal traffic load that the network can support, given the network topology and converter placement.

In this chapter we assumed that a wavelength converter has *full wavelength conversion* capability, that means, a wavelength on any input port can be converted to any other wavelength on output ports. In this case, communication channels may be connected regardless of their wavelengths. So full wavelength conversion provides the most efficient use of wavelengths, but it is also the most expensive to implement. There are other three possible types of wavelength conversion at the routing nodes as follows.

1 *No conversion* means that no wavelength shifting is allowed. In this case, only channels with the same wavelengths can be connected.

2 *Limited conversion* provides a compromise solution. In this case, a wavelength on an input port can be shifted to only some of wavelengths on output ports (due to the limit on the range of wavelength shifting). Presumably, a small amount of wavelength conversion capability will be easier or cheaper to implement than full wavelength conversion capability and will be able to provide better wavelength efficiency than no wavelength conversion capability at all.

3 *Fixed conversion* is very restricted form of limited conversion and a general form of no conversion. In this case, at each node a channel can be connected to exactly one predetermined channel on all other links.

Ramaswami et al [13] proposed ring networks with fixed wavelength conversion capability within the nodes that can guarantee $(L + 1)$-assignability and ring networks with a small additional amount of wavelength conversion capability within the nodes that achieves LWA. They also proposed a star network with fixed wavelength conversion capability at its hub that guarantees load-wavelength assignability.

Karasan and Ayanoglu [7] studied the performance benefits of using wavelength converters in WDM networks. They introduced two metrics to quantify the performance gain (enhancement) with respect to the reduction in blocking probability and the increase in maximum utilization, compared to networks without wavelength converters. The *blocking probability gain* is defined as the ratio of blocking probabilities of the network without converters and with converters, and the *utilization gain* is defined as the ratio of maximum offered loads of the network without converters and with converters. In general, there are many factors that may affect the performance gain, such as the network topology (in particular, network size and connectivity). Their analysis shows that fully connected networks make one extreme of using converters, where there is no gain with the shortest path routing. Another extreme case is the ring networks, where the gain is also relatively small. The intermediately connected networks have the largest gain. Their studies match the obtained results in this chapter. For fully connected networks (complete graphs) a wavelength converter is needed to be placed at almost every node. For the ring networks, one wavelength converter can guarantee LWA. For mesh (torus) or hypercube networks, about one half number of nodes in the networks need to be placed wavelength converters to guarantee LWA. Thus the network that requires more wavelength converters for LWA can get more performance benefits from using wavelength conversions.

Another objective of the problem of placing wavelength converters in WDM networks is to minimize the blocking probability of communication channels or maximize system throughput. Under such formulation of the wavelength converter placement problem, the optimal placement is dependent on many factors other than the network topologies, such as the traffic model or distribution, number of wavelengths available on fibers, number of fibers on links, routing and wavelength assignment algorithms. These problems will be addressed in Chapter 7.

References

[1] V. Bafna, P. Berman, and T. Fujito, A 2-approximation algorithm for the undirected feedback vertex set problem, *SIAM Journal on Discrete Mathematics*, 12 (3) (1999), 289-297.

[2] S. A. Cook, The complexity of theorem-proving procedures, *Proceedings of the 3rd Annual ACM Symposium on Theory of Computing* (STOC), (1971), 151-158.

[3] J. M. H. Elmirghani and H. T. Moutfah, All-optical wavelength conversion technologies and applications in WDM networks, *IEEE Communication Magazine*, 38 (3) (2000), 86-92.

[4] F. Gavril, Algorithms for minimum coloring, maximum clique, minimum covering by cliques, and maximum independent set of a chordal graph, *SIAM Journal on Computing*, 1 (1972), 180-187.

[5] X.-H. Jia, D.-Z. Du, X.-D. Hu, H.-J. Huang, and D.-Y. Li, Optimal placement of wavelength converters in WDM networks for parallel and distributed computing systems, *Proceedings of the 4th International Conference on Algorithms and Architectures for Parallel Processing*, (2000), 548-559.

[6] X.-H. Jia, D.-Z. Du, X.-D. Hu, H.-J. Huang, and D.-Y. Li, Placement of Wavelength Converters for Minimal Wavelength Usage in WDM Networks, *Proceedings of IEEE Conference on Computer Communications* (INFOCOM), 2002.

[7] E. Karasan and E. Ayanoglu, Effects of wavelength routing and selection algorithms on wavelength conversion gain in WDM optical networks, *IEEE/ACM Transactions on Networking*, 6 (2) (1998), 186-196.

[8] R. M. Karp, Reducibility among combinatorial problems, in R. E. Miller and J. W. Thatcher (eds.), *Complexity of Computer Computations*, Plenum Press, New York, 85-103.

[9] J. Kleinberg and A. Kumar, Wavelength conversion in optical networks, *Journal of Algorithms*, 38 (1) (2001), 25-50.

[10] D. König, Über graphen und ihre anwendung auf determinantentheorie und mengenlehre, *Mathematische Annalen*, 77 (1916), 453-465.

[11] F. L. Luccio, Almost exact minimum feedback vertex set in meshes and butterflies, *Information Processing Letters*, 66 (2) (1998), 59-64.

[12] B. Monien and E. Speckenmeyer, Ramsey numbers and an approximation algorithm for the vertex cover problem, *Acta Informatica*, 22 (1985), 115-123.

[13] R. Ramaswami and G. Sasaki, Multiwavelength optical networks with limited wavelength conversion, *IEEE/ACM Transactions on Networking*, 6 (6) (1998), 744-754.

[14] C. Savage, Depth first search and the vertex cover problem, *Information Processing Letters*, 14 (1982), 233-235.

[15] G. Wilfong and P. Winkler, Ring routing and wavelength translation, *Proceedings of the 9th Annual ACM-SIAM Symposium on Discrete Algorithms* (SODA), (1998), 333-341.

Chapter 6

MINIMIZATION OF BLOCKING PROBABILITY

In an all-optical WDM network, the route of a pair of communication nodes usually consists of multiple hops. If a transmission can occupy the same wavelength on every hop, it can then remain in optical form within the network. Otherwise, it encounters wavelength conflict and it has to be blocked. In single-hop systems of WDM networks, the blocking probability can be reduced through proper wavelength assignment. In multihop systems of WDM networks, the blocking probability can be reduced by using wavelength converters at network nodes to avoid wavelength conflict (so that the wavelength continuity constraint is relaxed). That is, when a transmission encounters a wavelength conflict on a hop, we can use a wavelength converter to convert its wavelength to another one, so that it can remain in optical form on this hop.

In this chapter we study two approaches for reducing the overall blocking probability in WDM networks. One is to assign wavelengths properly to communication channels in a single-hop system (where no wavelength converter is equipped at any node in the network). The other is to place a given number of wavelength converters properly in a multi-hop system of WDM networks.

The rest of this chapter is organized as follows. Section 1 discusses wavelength assignment to the communication channels such that the overall blockings in the system is minimized. The problem is solved by transforming it into the maximum weight k-cut problem. In addition, the relationship between the number of wavelengths and the overall blocking probability in the system is investigated. Section 2 discusses the problem of how to place a given number of wavelength converters so that the overall blocking probability is minimized. The problem is formulated as a problem of nonlinear programming with integral variables, and it is further solved by using a branch-and-bound method. Section 3 concludes this chapter.

1. Wavelength Assignment Approach

In this section, we focus on single-hop systems. In a single-hop system, each pair of communication nodes should have an all-optical channel configured and a wavelength is assigned to the channel. Due to the limited number of available wavelengths in the system, a wavelength is usually shared by many channels. If two channels share a common physical link and are assigned with the same wavelength, a blocking will occur when both of them request for a connection in the same period of time (i.e., a connection request arrives before the other one closes). In this case, the connection request arrived later is blocked and has to be rejected. One of the important goals for designing a single-hop system is to prevent the blockings to occur.

1.1 Optimization Formulation and Theory

Let $C = \{r(s,d) \mid s \in V \text{ and } d \in V\}$ denote the set of channels in the system. $r(s,d)$ is also used to denote the route of the channel, which is a set of physical links which constitute the path between source s and destination d. In this section we assume bidirectional channels. Let $p_r(s,d)$ be the probability that the channel between s and d requests for connection, and $q_r(s,d) \equiv 1 - p_r(s,d)$ the probability that it does not. We assume that the routes and the probabilities of requesting for connections of channels are given *in prior*. Fig. 6.1(a) illustrates a WDM network with the ring topology. There are ten channels connecting all vertex pairs in the network. Their routes are the shortest paths in terms of hops.

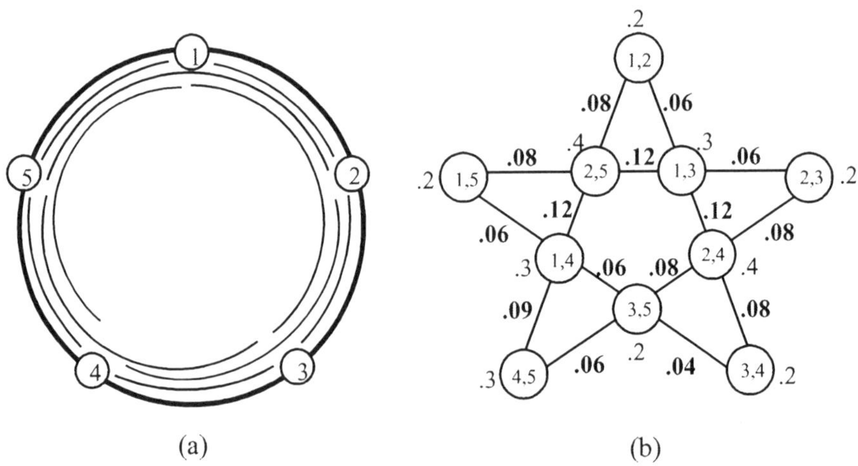

Figure 6.1. (a) Ring network $G(V, E)$ with routed channels over it. (b) Obtained auxiliary graph $G_a(V_a, E_a)$.

In all-optical WDM networks, each channel is assigned with a wavelength. A channel may share a wavelength with others due to the limited number of available wavelengths. Let ω be the number of available wavelengths in the system. Then we have C_i denote the set of channels that are assigned with wavelength i, $1 \leq i \leq \omega$ and $C = C_1 \cup \cdots \cup C_\omega$. A *blocking* will occur if two channels $r(s,d) \in C_i$ and $r(s',d') \in C_i$, for some i, share a common physical link and request for connections in the same time. In such a case one channel has to be blocked. Here, we are interested in the *overall blocking*, the expected number of channels blocked in the system. In order to give this concept a more formal definition, we introduce an auxiliary graph $G_a(V_a, E_a)$ as follows.

ALGORITHM 6.1 *Constructing Auxiliary Graph*

Step 1 The vertex set V_a is the set of channels $C = \{r(s,d) \mid s \in V \text{ and } d \in V\}$.
Step 2 There is an edge in E_a between two vertices $r(s,d) \in V_a$ and $r(s',d') \in V_a$ if and only if they share a link in network $G(V,E)$, that is, $r(s,d) \cap r(s',d') \neq \emptyset$.

In graph $G_a(V_a, E_a)$, if there is no edge between two vertices, it means that their corresponding routes do not share any physical link. They can thus be assigned with the same wavelength without causing any blocking. If there is an edge between two vertices, it means that their corresponding routes share a physical link. Assigning them with the same wavelength will cause a blocking if they request for connections in the same time. Fig. 6.1(b) is the auxiliary graph of the network in Fig. 6.1(a). The weight of a vertex represents the probability of the channel requesting for connections. For example, there is an edge between channels $r(1,2)$ and $r(1,3)$, because they share link $(1,2)$ in the ring.

For any subset $S \subseteq V_a$, let $G_a(S)$ denote the induced subgraph of $G_a(V_a, E_a)$ whose vertex-set is S and $E_a(S)$ its edge-set. For the example in Fig. 6.1(b), suppose that two wavelengths are available, $\omega = 2$, and consider 2-partition of C as below.

$$\begin{aligned} C_1 &= \{(1,5),(1,2),(2,5),(2,3),(2,4)\}, \\ C_2 &= \{(1,4),(4,5),(3,5),(3,4),(1,3)\}. \end{aligned}$$

Fig. 6.2(a) and Fig. 6.2(b) illustrate $G_a(C_1)$ and $G_a(C_2)$, respectively. In addition, let $I(S)$ denote the maximum independent set of $G_a(S)$. An *independent set* of a graph is a subset of vertex-set such that there is no edge between any pair of vertices in the set. The probability that all channels in $S \subseteq C_i$ request for connections at the same time is

$$P_r(S) = \Bigg(\prod_{r(s,d) \in S} p_r(s,d) \Bigg) \Bigg(\prod_{r(s,d) \notin S} q_r(s,d) \Bigg). \tag{6.1}$$

When this event occurs, among $|S|$ channels only $|I(S)|$ of them can be satisfied and the rest of $(|S| - |I(S)|)$ channels have to be blocked. Thus the overall blockings

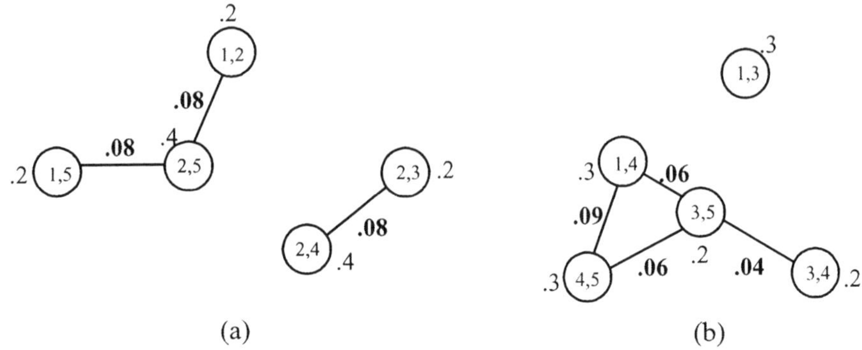

Figure 6.2. The induced subgraphs, (a) $G_a(C_1)$ and (b) $G_a(C_2)$.

of channels in C_i is

$$\text{Block}(C_i) = \sum_{S \subseteq C_i} (|S| - |I(S)|) P_r(S). \tag{6.2}$$

Since the blocking of channels in C_i is independent from the channels in C_j, for $i \neq j$, this means that as long as two channels are in different C_i's blocking will not occur even if they request for connections at the same time. Thus the overall blockings of the entire system is:

$$\text{Block}(C_1, \cdots, C_i) = \sum_{i=1}^{\omega} \sum_{S \subseteq C_i} (|S| - |I(S)|) P_r(S). \tag{6.3}$$

Observe that to evaluate the overall blockings in the system we need to determine the maximum independent sets $I(S)$. Clearly, $S = I(S)$ for $|S| = 1$; For $|S| = 2$, $S = I(S)$ if two vertices (that are channels) in S do not share a common physical link, other $I(S)$ is any one of two vertices in S. Unfortunately, however, finding the maximum independent set is $\mathcal{NP}$-hard [5]. In fact, there is no polynomial-time approximation algorithm with any constant performance ratio for this problem [8]. Accordingly, we have to consider the approximation of Block(C_i). Note that for small $p_r(s,d)$, Block(C_i) in equation (6.2) is dominated by the terms associated with $|S| = 2$. Thus we have

$$\begin{aligned} \text{Block}(C_i) &\approx \sum_{S \subseteq C_i, |S|=2} P_r(S) \\ &= \sum_{(r(u,v), r(x,y)) \in E_a(C_i)} p_r(u,v) p_r(x,y) \prod_{r(s,d) \neq r(u,v), r(x,y)} q_r(s,d) \end{aligned}$$

Now by ignoring production of $q_r(s,d)$ for $r(s,d) \neq r(u,v), r(x,y)$, which is close to one, we get the approximation of Block(C_i) as follows.

$$B(C_i) \equiv \sum_{(r(u,v),r(x,y))\in E_a(C_i)} p_r(u,v)p_r(x,y), \tag{6.4}$$

$$B(C_1,\cdots,C_\omega) \equiv \sum_{i=1}^{\omega} B(C_i) \tag{6.5}$$

We define the *Minimum Overall Blocking Problem* (MOBP) as follows.

PROBLEM 6.1 *Minimum Overall Blocking Problem*

Instance A network $G(V,E)$, a set of channels $r(s,d)$ with their routes and probabilities $p_r(s,d)$ to be requested, and a number of wavelengths ω.
Solution A wavelength assignment to $r(s,d)$.
Objective Minimizing the overall blockings $B(C_1,\cdots,C_\omega)$ as in equation (6.4-5).

To solve the above problem we reformulate it as follows. First we give a weight to every edge in E_a that equals the multiplication of the weights of the two endpoints in V_a, which represents the probability of the two channels requesting for connection in the same time (the case of blocking). Notice that the weight of an edge in C_i represents the blocking probability of the two channels corresponding to the two endpoints of the edge. The overall blockings in equation (6.5) can be rewritten as:

$$B(C_1,\cdots,C_\omega) = \sum_{i=1}^{\omega} WT(E_a(C_i)) \equiv \sum_{i=1}^{\omega} \sum_{e\in E_a(C_i)} WT(e). \tag{6.6}$$

In the example of Fig. 6.1(b), edge $(r(1,2), r(1,3))$ has weight $WT((1,2),(2,3))$ $= p_r(1,2)p_r(1,3) = 0.2 \times 0.3 = 0.06$. In the example of Fig. 6.2(a, b),

$$\begin{aligned}
B(C_1) &= WT((1,2),(2,5)) + WT((2,5),(1,5)) + WT((2,4),(2,3)) \\
&= 0.24, \\
B(C_2) &= WT((1,4),(4,5)) + WT((1,4),(3,5)) + \\
&\quad WT((4,5),(3,5)) + WT((3,5),(3,4)) = 0.25, \\
B(C_1,C_2) &= B(C_1) + B(C_2) = 0.49.
\end{aligned}$$

Thus Problem 6.1 can be transformed to the problem of partitioning the vertex-set V_a into ω disjoint sets, $C_1,\cdots,C_\omega$, so that the overall blocking (6.6) is minimized.

PROBLEM 6.1' *Minimum Overall Blocking Problem*

Instance An auxiliary graph $G_a(V_a,E_a)$ and a positive integer ω.
Solution An ω-partition of V_a, $C_1,\cdots,C_\omega$.
Objective Minimizing the overall blockings $B(C_1,\cdots,C_\omega)$ as in equation (6.6).

THEOREM 6.1 *Minimum Overall Blocking Problem is $\mathcal{NP}$-hard.*

PROOF We consider the decision version of the wavelength assignment problem (Problem 3.2). It is to decide if a set of channels can be assigned with a given number of wavelengths without causing any wavelength conflict. It can be reduced to the special case of the decision version of Problem 6.1, where the bound on the overall blocking is set to zero (i.e., no wavelength conflict is allowed). Since the wavelength assignment problem is $\mathcal{NP}$-hard (refer to Theorem 3.1), the minimum overall blocking problem is $\mathcal{NP}$-hard as well. ■

In order to develop an approximation algorithm to solve Problem 6.1', we relate it with the maximum weight k-cut problem. For an edge-weighted graph $G(V, E)$ and a k-partition of V, $V_1, \cdots, V_k$, a *k-cut* is a subset of E consisting of edges whose two endpoints are in two different sets V_i and V_j, $i \neq j$. The *weight* of a k-cut is the sum of weights of all edges in the cut. The *maximum weight k-cut problem* is to find the k-partition that produces a k-cut with maximum weight.

The following theorem states the relationship between the minimum overall blocking problem in $G(V, E)$ and the maximum weight k-cut problem in $G_a(V_a, E_a)$.

THEOREM 6.2 *Given a set of channels C, a k-partition of C is the optimal solution to the minimum overall blocking problem in $G(V, E)$ if and only if it is the optimal solution to the maximum weight k-cut problem in $G_a(V_a, E_a)$ with $k = \omega$.*

PROOF Since the minimum overall blocking problem is transformed into the problem of partitioning V_a into ω subsets such that equation (6.6) is minimized, we need to prove that finding a k-partition of V_a which minimizes equation (6.6) is equivalent to finding the optimal solution to the maximum weight k-cut problem in $G_a(V_a, E_a)$ with $k = \omega$.

Consider an ω-partitioning of V_a, $C_1, \cdots, C_\omega$. From equation (6.6), we have

$$B(C_1, \cdots, C_\omega) = \sum_{i=1}^{\omega} WT(E_a(C_i)).$$

Let $WT(C_1, \cdots, C_k)$ denote the weight of a k-cut and $WT(E_a)$ the total weight of all edges in E_a. From the definition of a k-cut, we have:

$$WT(C_1, \cdots, C_\omega) = WT(E_a) - \sum_{i=1}^{\omega} WT(E_a(C_i)).$$

Therefore, minimizing the overall blocking $B(C_1, \cdots, C_\omega\}$ in $G(V, E)$ is equivalent to maximizing the k-cut in $G_a(V_a, E_a)$ with $k = \omega$. ■

Fig. 6.3 illustrates the optimal solutions to MOBP and the maximum k-cut problem of the example in Fig. 6.1(b). Fig. 6.3(a) shows the optimal solution to the MOBP in the case of $\omega = k = 2$. It consists of two sets, one contains three channels $\{r(2,4), r(2,5), r(4,5)\}$ (marked by bold circles) and the other contains the rest of

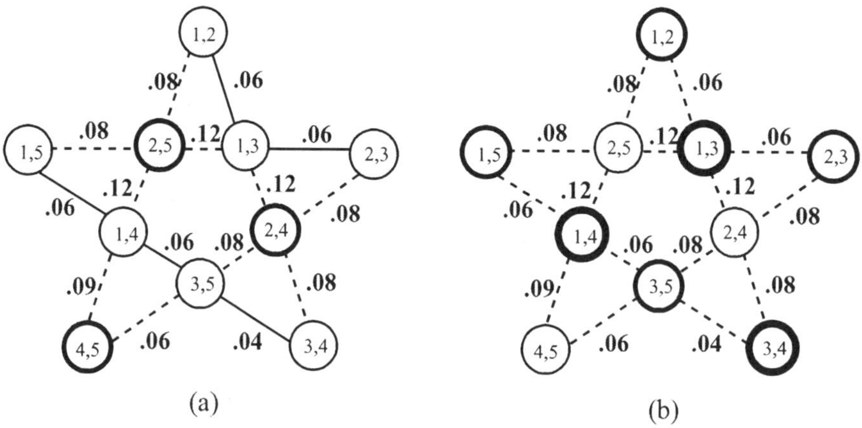

Figure 6.3. The optimal solutions to the minimum overall blocking problem and the maximum weight k-cut problem with $k = \omega$: (a) in the case of $\omega = 2$ and (b) in the case of $\omega = 3$.

the channels. The overall blocking is 0.28. The optimal solution to the maximum weight k-cut problem consists of ten edges (marked by dash lines) and has weight 0.91. Fig. 6.3(b) shows the optimal solution to the MOBP in the case of $\omega = k = 3$ in a similar way, where the overall blocking is reduced to zero. The weight of maximum k-cut, in this case, is the weight of all edges. It is worth noting that finding the minimal number of wavelengths which makes the blocking zero is equivalent to the problem of finding a minimal number of sets such that there is no edge between any two vertices in the same set. The former is the wavelength assignment problem while the latter is the vertex coloring problem (refer to the discussion in Section 3.3).

1.2 Algorithm Design and Performance Analysis

In the preceding subsection, we transform the minimum overall blocking problem to the maximum weight k-cut problem. The latter is $\mathcal{NP}$-hard [9] and it has an approximation algorithm proposed in [4] that could produce a k-cut whose weight is at least $1/(1-1/k+2\ln k/k^2)$ times of the weight of maximum k-cut. However, this method is based on solving a semi-definite programming relaxation, thus complicated to be implemented. In the following, we will apply a simple heuristic which also has a guaranteed performance ratio and is easy to be carried out in practice.

The basic idea is to generalize the local search method for solving the maximum cut problem [13], which is a special case of k-cut problem with $k = 2$. The algorithm is summarized as the following.

ALGORITHM 6.2 *Finding a k-partition*

Step 1 Generate an initial k-partition of V_a: $V_1, \cdots, V_k$.
Step 2 Move a vertex from V_i to V_j if the weight of k-cut is strictly increased.
Step 3 Repeat Step 2 until no further improvement can be made.

The following theorem shows that Algorithm 6.2 has a small constant approximation performance ratio.

THEOREM 6.3 *Given any auxiliary $G_a(V_a, E_a)$ and positive integer $k \geq 2$, let $WT_{LS}(G_a, k)$ be the weight of k-cut of the k-partition produced by Algorithm 6.2 and $WT_{Opt}(G_a, k)$ the weight of the maximum k-cut. Then $WT_{LS}(G_a, k) \geq (1 - 1/k)WT_{Opt}(G_a, k)$.*

PROOF For $V' \subseteq V_a$, let $V'(v)$ be the set of edges $(v, u) \in E_a$ such that $u \in V'$, and $WT(V'(v))$ the total weight of edges in $V'(v)$. Suppose that $V_1, \cdots, V_k$ is the k-partition output by Algorithm 6.2, and $WT(V_1, \cdots, V_k)$ is the weight of the k-cut that it produces. We introduce a function $i(v)$ to indicate which partition a node v belongs to when Algorithm 6.2 terminates. For example, $i(v) = j$ indicates that v belongs to V_j. Then, we have

$$WT(V_1, \cdots, V_k) = \frac{1}{2} \sum_{v \in V_a} \Big(\sum_{i \neq i(v)} WT(V_i(v)) \Big). \tag{6.7}$$

The ratio 1/2 comes from the fact that the weight of each edge in the k-cut is counted twice. Now note that

$$WT(V_{i(v)}(v)) \leq WT(V_j(v)), \quad \text{for any } j \neq i(v),$$

which implies

$$\sum_{i \neq i(v)} WT(V_i(v)) \geq \sum_{i \neq j} WT(V_i(v)), \ \ \text{for any } j \neq i(v), \tag{6.8}$$

because, according to Steps 2-3 of Algorithm 6.2, vertex v should be moved from $i(v)$ to j whenever inequality (6.8) is not true for some $j \neq i(v)$. Summing up inequality (6.8) over j, we obtain

$$k \sum_{i \neq i(v)} WT(V_i(v)) \geq (k-1) \sum_{i=1}^{k} WT(V_i(v)). \tag{6.9}$$

Combining inequality (6.9) with equality (6.7), we obtain

$$\begin{aligned} WT(V_1, \cdots, V_k) \ &\geq \ \frac{1}{2} \sum_{v \in V_a} \Big(\frac{k-1}{k} \sum_{i=1}^{k} WT(V_i(v)) \Big) \\ &= \ \frac{k-1}{k} \sum_{e \in E_a} WT(e) = \Big(1 - \frac{1}{k}\Big) WT(E_a) \end{aligned} \tag{6.10}$$

It is obvious that the weight of the maximum k-cut is not greater than $WT(E_a)$. Hence the theorem follows from inequality (6.10). The proof is then finished. ■

Now we consider the time complexity of Algorithm 6.2. To simplify the notation, we define

$$\Theta \equiv \min_{v \in V} \min_{V' \subseteq V} \left| WT(V'(v)) - WT(V_a \setminus V'(v)) \right|,$$

which is the minimum gain that Algorithm 6.2 can obtain. The following theorem shows that although Algorithm 6.2 is simple, it is not a polynomial-time algorithm [14].

THEOREM 6.4 *Given any auxiliary $G_a(V_a, E_a)$ and positive integer $k \geq 2$, it takes time $O\Big(k|V_a|^2 WT(E_a)/\Theta\Big)$ for Algorithm 6.2 to find a local optimal solution.*

PROOF For each move of vertex v from V_i to V_j, the weight of k-cut is increased by

$$WT(V_i(v)) - WT(V_j(v)) \geq \Theta,$$

and the weight of the maximum k-cut is at most $WT(E_a)$. Thus, there are at most Θ/θ moving operations. In addition, it takes at most $O(k|V_a|^2)$ time to find vertex $v \in V_a$ that can increase the weight of k-cut by moving it from V_i to V_j. Thus, the time-complexity of Algorithm 6.2 is bounded above by $O\Big(k|V_a|^2 WT(E_a)/\Theta\Big)$. ■

The following theorem gives the time-complexity of solving Problem 6.1 by the proposed method, we call *max-cut algorithm*, that integrates Algorithm 6.1 with Algorithm 6.2.

THEOREM 6.5 *The max-cut algorithm produces an approximate solution to Problem 6.1 in time $O\Big(\omega|V|^4 WT(E_a)/\Theta\Big)$.*

PROOF Since there are at most $O(|V|^2)$ channels in the given graph $G(V, E)$, the auxiliary graph has vertex-set of size at most $O(|V|^2)$. It takes time at most $O(|V|^4)$ to check if two channels share a link in $G(V, E)$. Thus, $G_a(V_a, E_a)$ can be constructed in time $O(|V|^4)$ by Algorithm 6.1. To sum up this time with the time required for Algorithm 6.2 in Theorem 6.4, we obtain the total time $O(\omega|V|^4 WT(E_a)/\Theta)$ to produce an approximate solution by the max-cut algorithm. ■

1.3 Simulation Study

The purpose of the simulation is to evaluate the performance of the proposed max-cat algorithm under various network situations. We compare it with a simple wavelength assignment method that does not consider the possible wavelength conflict and overall blocking directly. Given a k-partition of $C = C_1 \cup C_2 \cup \cdots \cup C_k$, let $P_r(C_i) = \sum_{r(s,d) \in C_i} p_r(s,d)$ for each i and $P_{max}(C) = \max_{1 \leq i \leq k} P_r(C_i)$. The *balanced assignment* is to allocate each channel $r(s,d)$ to a set C_i (all channels in C_i share a wavelength) such that the differences among $P_r(C_i)$ is as small as possible.

In other words, the goal is to find a balanced assignment minimizing $P_{max}(C)$. It is easy to see that the balanced assignment assigns wavelengths in such a way that the expected number of channels that each wavelength serves is as close as possible. Since the balanced assignment avoids using one wavelength to serve a large expected number of channels, it clearly avoids a lots of blockings in the network. Thus this simple method can be used as a heuristic for the minimum overall blocking problem.

However, the following theorem shows that finding the optimal balanced assignment is also $\mathcal{NP}$-hard.

THEOREM 6.6 *Given a graph $G(V, E)$, a set C of channels with their routes $r(s, d)$ and probabilities $p_r(s, d)$ to be requested for connections, and a number k of wavelengths available. The problem of finding a k-partition of $C = C_1 \cup C_2 \cdots \cup C_k$ to minimize $P_{max}(C)$ is $\mathcal{NP}$-hard.*

PROOF The balanced assignment problem can be considered as the minimum scheduling problem with identical machines, which was proved to be $\mathcal{NP}$-hard [5]. In fact, set C of channels is a set J of jobs, each available wavelength w_i is a machine M_i, probability $p_r(s, d)$ is the length $l(s, d)$ for executing job $j(s, d)$. The objective of minimum scheduling is to assign jobs to machines such that the makespan $\max_{1 \leq k \leq k} \sum_{j(s,d) \in M_i} l(s, d)$ is minimized. ■

In the simulation we will use the following method, that is based on the *list scheduling* with largest processing time rule [7] for the minimum scheduling problem.

ALGORITHM 6.3 *Balanced Wavelength Assignment*

Step 1 Sort the channels in decreasing order of their probabilities;
Step 2 Take a channel in the order and put it into C_i if $P_r(C_i)$ is minimal.
Step 3 Repeat Step 2 until every channel is assigned a wavelength.

The following theorem shows that Algorithm 5.3 produces a near-optimal balanced assignment.

THEOREM 6.7 *Given a graph $G(V, E)$, a set C of channels with their routes $r(s, d)$ and probabilities $p_r(s, d)$ to be requested for connections, and a number k of wavelengths available. Algorithm 5.3 finds an assignment $C = C_1 \cup C_2 \cdots \cup C_k$ satisfying $P_{max}(C) \leq (\frac{4}{3} - \frac{1}{3k}) P_{max}(C^*)$, where $C^* = C_1^* \cup C_2^* \cdots \cup C_k^*$ is the optimal balanced assignment.*

PROOF Let p_{min} be the probability of the last channels that was considered. Consider the following two cases separately.

Case 1. $p_{min} > \frac{1}{3} P_{max}(C^*)$. It is easy to verify that each wavelength may have been assigned to at most two channels and Algorithm 5.3 outputs the optimal balanced assignment.

Case 2. $p_{min} \leq \frac{1}{3}P_{max}(C^*)$. Suppose that $P_r(C_h) = \max_{i\leq i\leq k} P_r(C_i)$ and let $r(u,v)$ be the last channel put into C_h. Note that $P_r(C_i) \geq P_r(C_h) - p_r(u,v)$ for all i due to the rule of Step 2 in Algorithm 5.3. This implies that

$$P_r(C) \equiv \sum_{1\leq i\leq k} P_r(C_i) \geq k(P_r(C_h) - p_r(u,v)) + p_r(u,v). \tag{6.11}$$

Note that $p_r(u,v) \geq p_{min}$. Thus inequality (6.11) leads to

$$P_r(C_h) \leq \frac{P_r(C)}{k} + \frac{k-1}{k} p_{min}. \tag{6.12}$$

Since $k \max_{1\leq i\leq k} P_r(C_i^*) \geq P_r(C^*) = P_r(C)$ and $p_{min} \leq \frac{1}{3}\max_{1\leq i\leq k} P_r(C_i^*)$, from inequality (6.12) we have

$$\begin{aligned} P_r(C_h) = \max_{1\leq i\leq k} P_r(C_i) &\leq \max_{1\leq i\leq k} P_r(C_i^*) + \frac{k-1}{3k} \max_{1\leq i\leq k} P_r(C_i^*) \\ &= (\frac{4}{3} - \frac{1}{3k}) \max_{1\leq i\leq k} P_r(C_i^*) \end{aligned}$$

The proof is then finished. ■

In all the simulations, we first produce a balanced assignment by using Algorithm 6.3 and then apply Algorithm 6.2 to optimize the overall blockings.

In the simulation the routes of channels between pairs of nodes are the shortest paths in terms of distance. The probabilities of channels requesting for connections are randomly distributed among closed interval $[0, 1]$.

Fig. 6.4 shows the overall blockings versus the number of wavelengths in NSFnet. Fig. 6.5-6 and Fig. 6.7-8 are the cases for the randomly generated networks. The network graph used in Fig. 6.5-6 have 414 edges (as an emulation of sparse networks), while the graph used in Fig. 6.7-8 have 714 edges (as an emulation of dense networks). In Fig. 6.5-8, there are two pairs of curves, representing the average connection probabilities 0.65 with 0.5 and 0.35 with 0,5, respectively. From these figures, the following observations can be made.

1 The overall blockings decrease exponentially as the number of wavelengths increases.

2 Comparing the two pairs of curves in Fig. 6.5-6 (or Fig. 6.7-8), it can be seen the overall blockings decrease significantly as the decrease of network traffic load (i.e., the probabilities of channels for connections).

3 Comparing Fig. 6.5-6 with Fig. 6.7-8, we can see that the system has less overall blockings when the density of the network is higher (i.e., the network has more edges). This is because that two routes would have less chance to share a common link when the network has more links. As an extreme case where there is a link

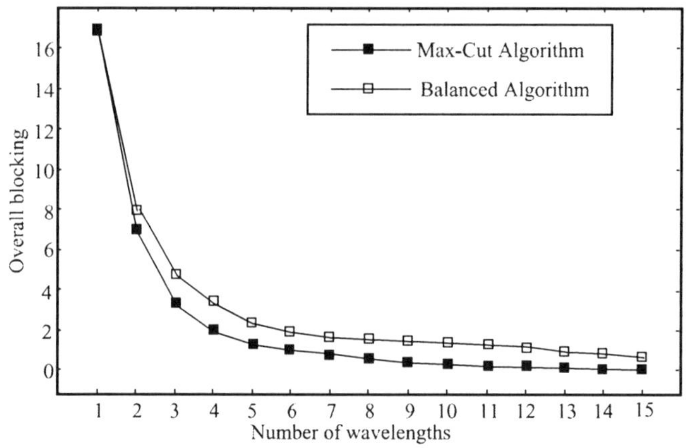

Figure 6.4. The overall blocking against the number of wavelengths in NSFnet.

between every pair of nodes, the blocking in the network is zero, because each channel has a dedicated link as the route.

4 The gap between the balanced algorithm and the max-cut algorithm is more significant when the number of wavelengths is larger than four. When the number of wavelengths is too small, there is no much space for optimization. When the number of wavelengths is large enough, the overall blockings in the system by using the proposed algorithm are getting very close to zero. From the simulation data produced, the overall blocking in the system drops below 0.1 when the number of wavelengths reaches 40.

In order to demonstrate the improvement of the max-cut algorithm over the balanced method, we introduce an *improvement ratio*. It is the overall blockings produced by balanced assignment over the overall blockings produced by the max-cut algorithm. The ratio is at least one (where no improvement is made), and the greater it is, the more effective the max-cut algorithm is. Fig. 6.9 shows the improvement ratios in NSFnet, while Fig. 6.10 shows the ratios in a dense network with 714 edges and Fig. 6.11 in a sparse network with 414 edges. It can be observed from the figures.

1 The improvement ratio increases sharply as the increase of wavelengths. This is because the overall blockings incurred by using the max-cut algorithm get close to zero quickly when the number of wavelengths is large enough, which makes the ratio very big.

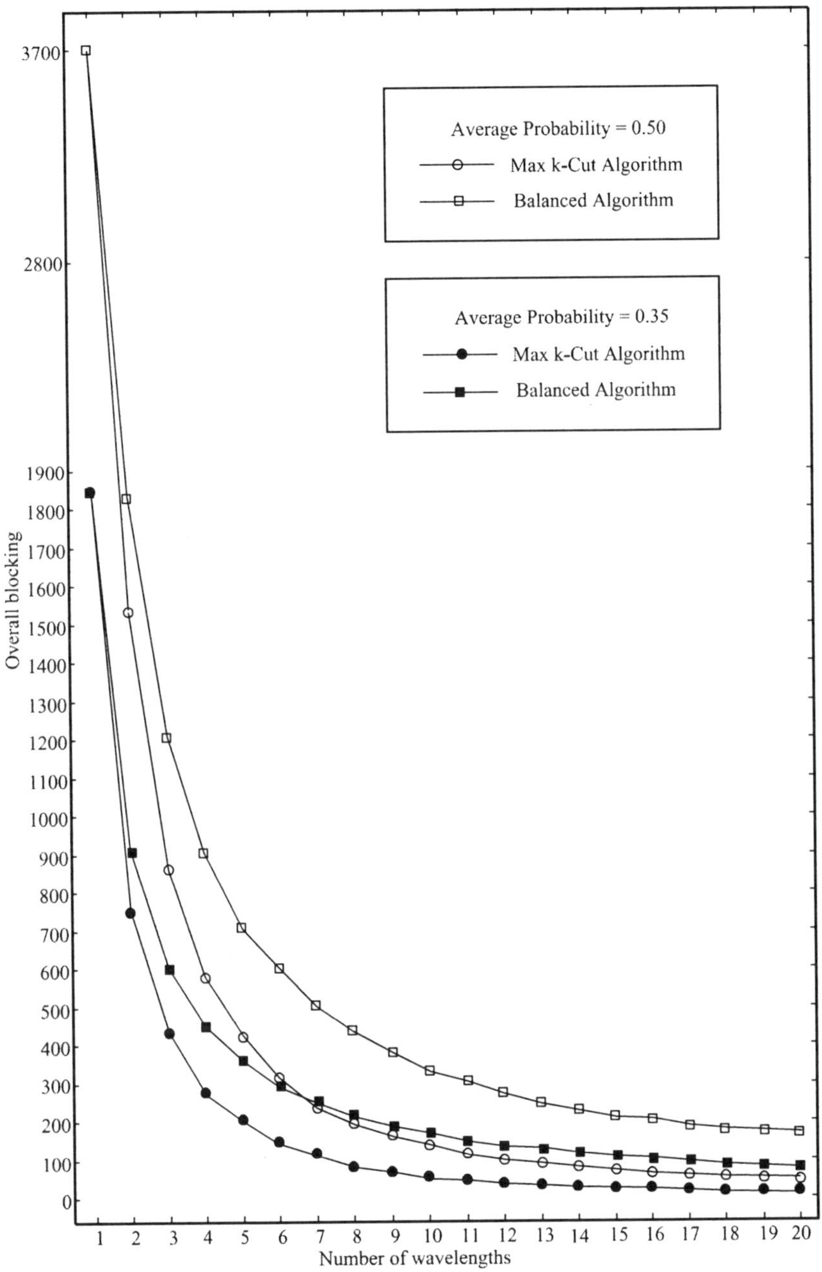

Figure 6.5. The overall blocking against the number of wavelengths in a network of 414 edges.

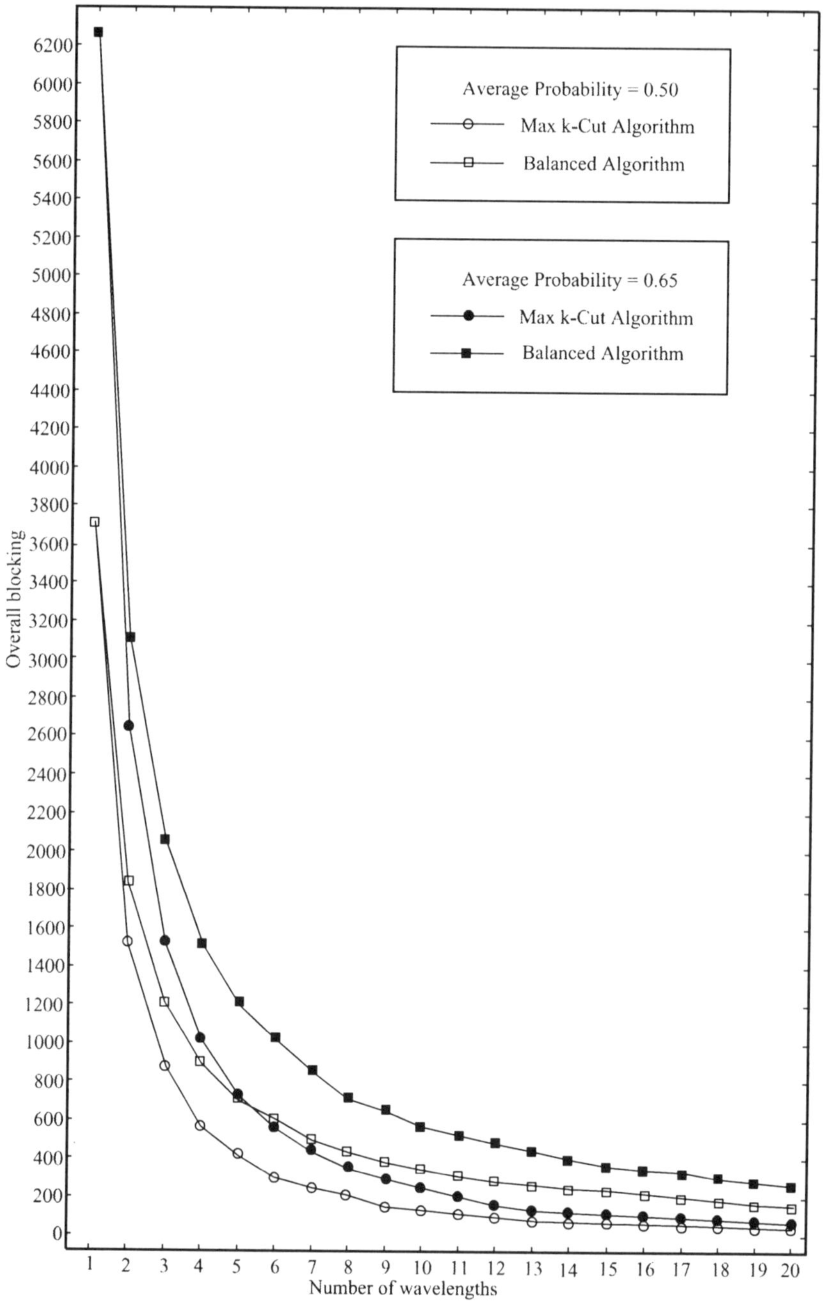

Figure 6.6. The overall blocking against the number of wavelengths in a network of 414 edges.

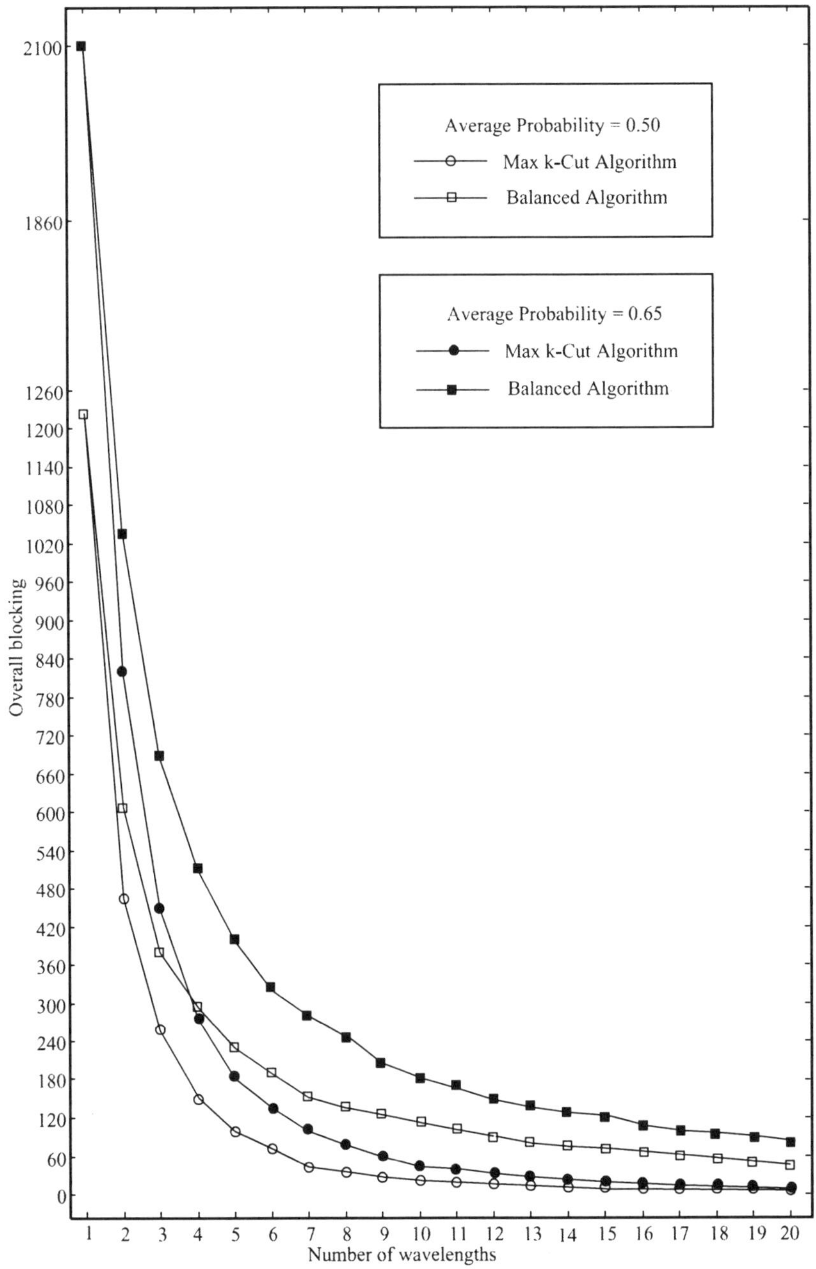

Figure 6.7. The overall blocking against the number of wavelengths in a network of 714 edges.

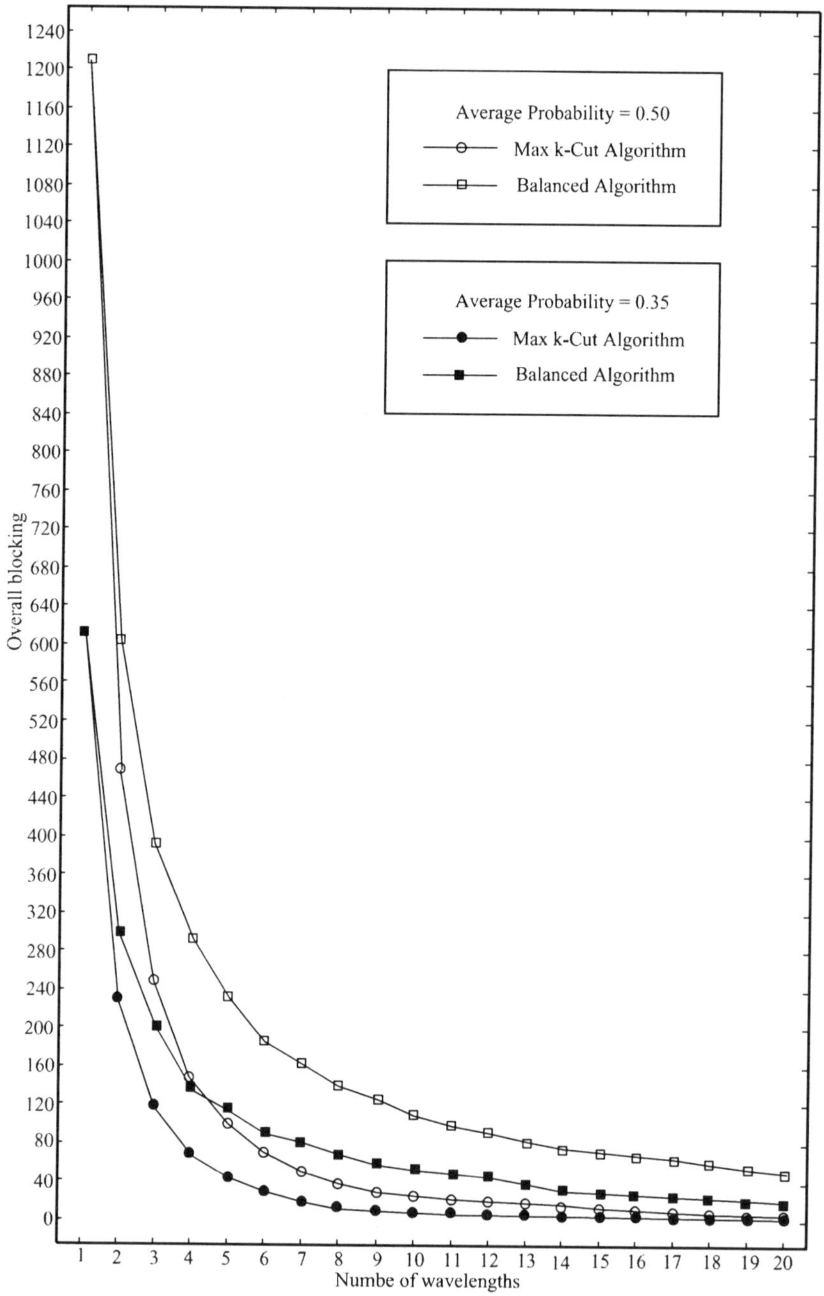

Figure 6.8. The overall blocking against the number of wavelengths in a network of 714 edges.

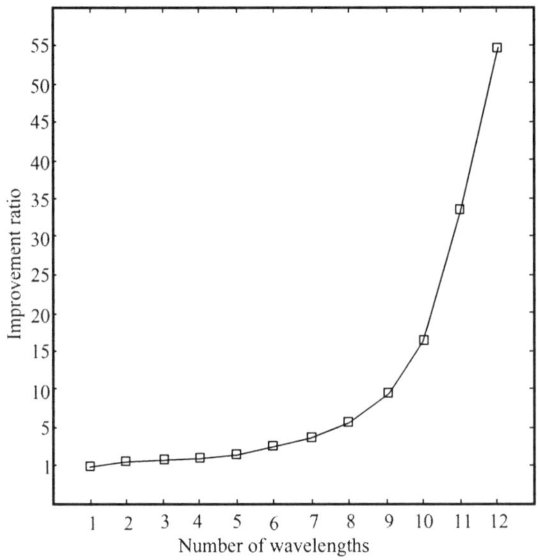

Figure 6.9. The improvement ratio in NSFnet.

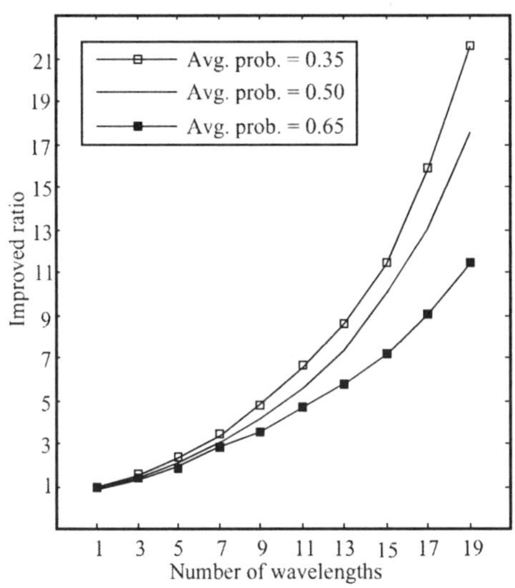

Figure 6.10. The improvement ratio in a network with 714 edges.

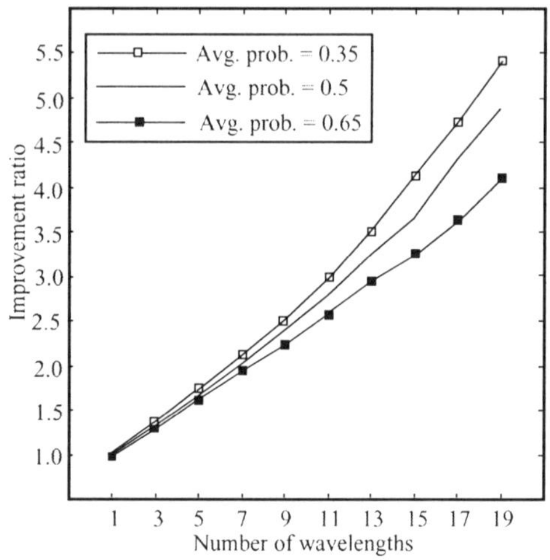

Figure 6.11. The improvement ratio in a network with 414 edges.

2 The improvement of the max-cut algorithm is more significant with smaller connection probabilities or with higher network density (comparing Fig. 6.10 with Fig. 6.11).

3 The improvement ratios for NSFnet and two generated networks display the very similar pattern. This implies that the simulation model well describes the real situation and the simulation data are stable.

2. Wavelength Converter Placement Approach

In this section we will study how to minimize the overall blocking through placing a given number of wavelength converters in WDM networks. We will first express the overall blocking probability as a polynomial function of binary variables, which indicate that a converter is placed at a node or not; We then propose a brand-and-bound method to find the minimum of the polynomial function and get the optimal placement.

2.1 Optimization Formulation and Theory

To model the overall blocking probability of a WDM network where some wavelength converters are equipped, we introduce the following notations.

(i,j): the directed link from node v_i to node v_j.
P_{ij}: the channel route from node v_i to node v_j, that is a path in the network between two nodes.
ω_{ij}: the number of wavelengths on link (v_i, v_j). Here, we assume $\omega_{ij} = \omega$. for all $v_i, v_j \in V$.
λ_{st}: the end-to-end traffic from node v_i to node v_j.
λ_{st}^{ij}: the amount of traffic λ_{st} between node s and node t that goes through : link (v_i, v_j).
λ_{st}^{i}: the amount of traffic λ_{st} flowing through node v_i.
λ_i: the total amount of traffic flowing through v_i. Here $\lambda_i = \sum_{\forall s,t} \lambda_{st}^i$.
ρ_{ij}: the load per wavelength over link (v_i, v_j) representing the probability that a wavelength on (v_i, v_j) is occupied. Here $\rho_{ij} = \frac{1}{\omega_{ij}} \sum_{s,t \in V} \lambda_{st}^{ij}$.
q_{ij}: the probability that a wavelength on link (v_i, v_j) is available. Here we assume $q_{ij} = 1 - \rho_{ij}$ for all wavelengths.

Now consider a channel between node s and node t. Suppose that its route is P_{st} and some converters are placed in path P_{st}. A *segment* of P_{st} is the set of links between two consecutive converter nodes or between an end-node of P_{st} and the converter in the path closet to it. An i-segment is a segment consisting of i links.

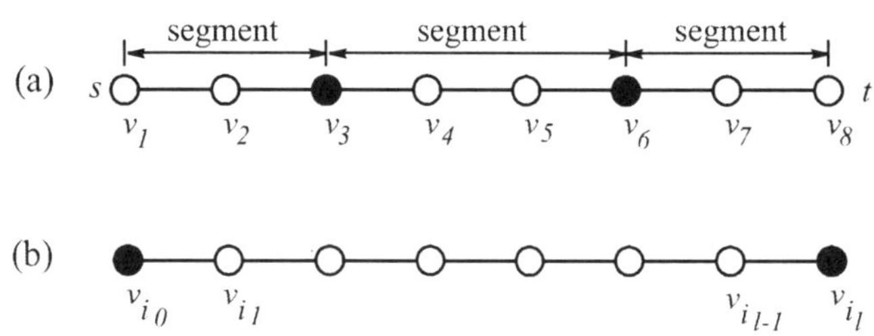

Figure 6.12. (a) Segments of a channel route. (b) A path consisting of l links.

Clearly, placing k converters in P_{st} produces $k+1$ segments. In the example of Fig. 6.12(a), two converters (black nodes) are placed at node v_3 and node v_6, respectively, that divide the path into three segments.

Suppose that the i-th segment $P_{st}(i)$ of P_{st} consists of l links, $(v_{i_0}, \cdots, v_{i_l})$. See Fig. 6.12(b). Note that a channel is blocked in segment $P_{st}(i)$ if and only if there is no wavelength that is available on every link in the segment. Thus the probability of a successful channel connection can go through this segment is

$$S_p(P_{st}(i)) = 1 - (1 - q_{i_1 i_2} q_{i_2 i_3} \cdots q_{i_{l-1} i_l})^{\omega} \tag{6.13}$$

In addition, a channel over P_{st} can be put through successfully if and only it can be done successfully on every segment of P_{st}. Therefore, the probability of a successful channel connection on P_{st} is

$$S_p(P_{st}) = \prod_{i=1}^{k+1} S_p(P_{st}(i)) \tag{6.14}$$

Clearly, the probability of a successful channel connection is dependent on the number of converters placed in its route and their locations. The more converters are placed, the shorter the length of each segment will be, that implies the larger successful probability.

In order to formulate $S_p(P_{st})$ as a function of converter placement, we introduce the *state vector* of n binary variables $(x_1, x_2, \cdots, x_n)$ to indicate a convert placement in the network. That is, x_i is one if a converter is placed at node i and zero otherwise.

We first consider a simple case that the link load over all links in the network is the same, i.e., $\rho_{ij} = \rho$ and $q_{ij} = 1 - \rho$ for all i, j. Suppose that P_{st} consists of d links, and for simplicity of notations, we assume $P_{st} = v_1 v_2 \cdots v_d v_{d+1}$ with $v_1 = s$ and $v_{d+1} = t$. For the simplicity of presentation, we denote $1 - x_i$ by $\overline{x}_i$ for i. The following lemma gives the expression of $S_p(P_{st})$ as a function of $(x_1, x_2, \cdots, x_n)$ under a fixed converter placement $(x_1, x_2, \cdots, x_n)$.

LEMMA 6.1 *When* $d = 1$, $S_p(P_{st}) = 1 - (1-q)^{\omega}$; *When* $d = 2$, $S_p(P_{st}) = (1-(1-q)^{\omega})^{2x_2}(1-(1-q^2)^{\omega})^{\overline{x}_2}$.

PROOF The case of $d = 1$ is obvious. We consider the case of $d = 2$, i.e., $P_{st} = v_1 v_2 v_3$. The value of $S_p(P_{st})$ depends on whether there is a converter at middle node v_2. If there is no converter placed at v_2 (i.e., $x_2 = 0$), then $S_p(P_{st}) = 1 - (1-q^2)^{\omega}$; Otherwise a converter is placed at v_2 (i.e., $x_2 = 1$), path P_{st} is divided into two segments of length 1. From equations (6.13-14), we have

$$S_p(P_{st}) = S_p(P_{st}(1)) S_p(P_{st}(2)) = (1-(1-q)^{\omega})^2.$$

It can be verified that

$$S_p(P_{st}) = \begin{cases} 1-(1-q^2)^{\omega}, & \text{if } x_2 = 0; \\ (1-(1-q)^{\omega})^2, & \text{if } x_2 = 1. \end{cases}$$

This is the desired equation. ■

For the case of $d > 2$, the explicit expression of $S_p(P_{st})$ is complicated. Let

$$S_1 = x_2 + \sum_{j=2}^{d-1} x_j x_{j+1} \text{ and } S_d = \prod_{j=2}^{d} \overline{x}_2,$$

and for $2 < i < d$,

$$S_i = x_{i+1} \prod_{j=2}^{i} \overline{x}_j + x_{d+1-i} \prod_{j=d+2-i}^{d} \overline{x}_j + \sum_{j=2}^{d-2} x_{j-i+2} \overline{x}_{j-i+3} \cdots \overline{x}_{j+1} x_{j+2}.$$

LEMMA 6.2 *For any positive integer* $d > 2$, $S_p(P_{st}) = \prod_{i=1}^{d}(1-(1-q^i)^\omega)^{S_i}$.

PROOF Nodes in P_{st} placed with converters divide P_{st} into a number of segments, which have various lengths varying from 1 to d. In fact, P_{st} may have d 1-segments, $(d-1)$ 2-segments, $\cdots$, and one d-segment. In the following we consider them separately.

First, we consider possible 1-segments. Clearly, the successful probability of channel connection on each of them is $1-(1-q)^\omega$. When link (v_1, v_2) is an 1-segment, that means a converter is placed in node v_2, i.e., $x_2 = 1$. Then $1-(1-q)^\omega$ is included in $S_p(P_{st})$ as $(1-(1-q)^\omega)^{x_2}$. When link (v_d, v_{d+1}) is an 1-segment, that means a converter is placed in node v_d, i.e., $x_d = 1$. Then $1-(1-q)^\omega$ is included in $S_p(P_{st})$ as $(1-(1-q)^\omega)^{x_d}$. In general, when link (v_i, v_{i+1}) is an 1-segment for $1 < i < d$, that means two converters are placed at nodes v_i and v_{i+1}, respectively. i.e., $x_i = x_{i+1} = 1$. Then $1-(1-q)^\omega$ is included in $S_p(P_{st})$ as $(1-(1-q)^\omega)^{x_i x_{i+1}}$. Thus, the term associated with 1-segments in $S_p(P_{st})$ is

$$(1-(1-q)^\omega)^{x_2}(1-(1-q)^\omega)^{x_2x_3}\cdots(1-(1-q)^\omega)^{x_{d-1}x_d}(1-(1-q)^\omega)^{x_d},$$

which equals $(1-(1-q)^\omega)^{S_1}$.

Secondly, we consider possible 2-segments. Clearly, the successful probability of channel connection on each of them is $1-(1-q^2)^\omega$. When links (v_1, v_2) and (v_2, v_3) make a 2-segment, that means a converter is placed at node v_3 but not at node v_2, i.e., $x_3 = 1$ and $x_2 = 0$. Then $1-(1-q^2)^\omega$ is included in $S_p(P_{st})$ as $(1-(1-q^2)^\omega)^{\overline{x}_2 x_3}$. When links (v_{d-1}, v_d) and (v_d, v_{d+1}) make a 2-segment, that means a converter is placed in node v_{d-1} but not at node v_d, i.e., $x_{d-1} = 1$ and $x_d = 0$. Then $1-(1-q^2)^\omega$ is included in $S_p(P_{st})$ as $(1-(1-q)^\omega)^{x_{d-1}\overline{x}_d}$. In general, when links (v_i, v_{i+1}) and (v_{i+1}, v_{i+2}) make a 2-segment for $1 < i < d-1$, that means two converters are placed at nodes v_i and v_{i+2} but not at node v_{i+1}, respectively. i.e., $x_i = x_{i+2} = 1$ and $x_{i+1} = 0$. Then $1-(1-q^2)^\omega$ is included in $S_p(P_{st})$ as $(1-(1-q)^\omega)^{x_i\overline{x}_{i+1}x_{i+2}}$. Thus, the term associated with 2-segments in $S_p(P_{st})$ is

$$(1-(1-q^2)^\omega)^{\overline{x}_2 x_3}(1-(1-q^2)^\omega)^{x_{d-1}\overline{x}_d}\prod_{i+2}^{d-2}(1-(1-q^2)^\omega)^{x_i\overline{x}_{i+1}x_{i+2}},$$

which equals $(1-(1-q^2)^\omega)^{S_2}$.

Thirdly, we consider possible i-segments for $2 < i < d$. By using the similar argument, we know that the term associated with i-segments in $S_p(P_{st})$ is $(1-(1-q^i)^\omega)^{S_i}$.

Finally, we consider the possible d-segment, which is the case that no converter is placed at any intermediate node of P_{st}, i.e., $x_2 = x_3 = \cdots = x_d = 0$. Thus the term associated with d-segments in $S_p(P_{st})$ is

$$(1-(1-q^d)^\omega)^{\overline{x}_2\cdots\overline{x}_d} = (1-(1-q^d)^\omega)^{S_d}.$$

Note that $S_p(P_{st})$ is just the production of all these terms, which is the probability of a successful channel connection on all possible segments. The proof is then finished. ■

For the general case where traffic loads on links are non-uniform. By using the same method as in the proof of Lemmas 6.1-2 and substituting q with corresponding q_{ij}, we can prove the following lemma.

LEMMA 6.3 *When* $d = 1$, $S_p(P_{st}) = 1 - (1 - q_{12})^{\omega}$; *When* $d = 2$, $S_p(P_{st}) = (1 - (1 - q_{12})^{\omega})^{2x_2}(1 - (1 - q_{23})^{\omega})^{x_2}(1 - (1 - q_{12}q_{23})^{\omega})^{\overline{x}_2}$.

The arithmetical average and geometric average are the two common ways to measure the overall probabilities of successful channel connections between node pairs in the network. In the following, for the sake of easy treatment, we adopt the geometric average, that is

$$\left(\prod_{\forall s,t} S_p(P_{st})^{\lambda_{st}}\right)^{(\sum_{\forall s,t}\lambda_{st})^{-1}}.$$

For the simplicity of presentation, we take logarithm of above value and introduce a function of n binary variables

$$F(x_1, \cdots, x_n) \equiv (\sum_{\forall s,t} \lambda_{st})^{-1} \sum_{\forall s,t} \lambda_{st} \log S_p(P_{st}). \tag{6.15}$$

Thus the problem of how to minimize the overall blocking using a given number of wavelength converters in the network can be formulated as follows.

PROBLEM 6.2 *Minimum Overall Blocking Problem*

Instance A network $G(V, E)$, a set of channels with their routes and traffics λ_{st} between node pairs s and d, and ω wavelengths and k converters.
Solution A placement of converters at nodes of the network.
Objective Minimizing the overall blockings $F(x_1, \cdots, x_k)$ as in equation (6.15).

Problem 6.2 can be simply defined as the following optimization problem.

$$\text{Max} \quad F(x_1, \cdots, x_n) \tag{6.16}$$
$$\text{s.t.} \quad \textstyle\sum_{i=1}^{n} x_i = k, \tag{6.17}$$
$$x_i \in \{0, 1\}, i = 1, \cdots, n. \tag{6.18}$$

The objective (6.15-16) of the above problem is to find an assignment of n binary variables, under the constraint (6.17) that at most k converters can be placed in the network, such that the geometric average of the probabilities of all successful channel connections in the network is maximized. From the expression (6.13-14) of $S_p(P_{st})$ and Lemmas 6.1-3, we know that the objective function (6.15) is a polynomial function of binary variables.

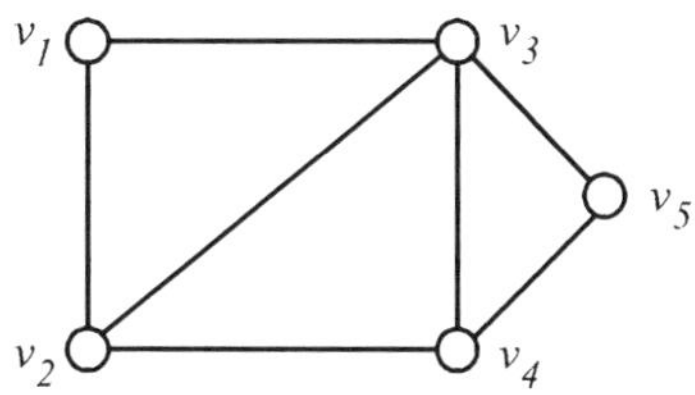

Figure 6.13. A simple network of five nodes.

Now to illustrate our mathematical model, we consider a simple example taken from [17]. It is a cross-connected WDM network of five nodes interconnected by point-to-point bidirectional fiber links with a mesh topology as shown in Fig. 6.13. We assume uniform load case that $\lambda_{ij} = 0.05$ for any node pair (v_i, v_j), and $\omega = 10$. The channel routes between node pairs, which are constructed as shortest paths, are given in Table 6.1. The traffic loads ρ_{ij} on links are given in Table 6.2. The probabilities $S_p(P_{st})$ of successful channel connections between pair of nodes are given in Table 6.3.

Table 6.1. Channel routes between each pair of nodes.

	v_1	v_2	v_3	v_4	v_5
v_1		$v_1 \to v_2$	$v_1 \to v_3$	$v_1 \to v_3 \to v_4$	$v_1 \to v_3 \to v_5$
v_2	$v_2 \to v_1$		$v_2 \to v_3$	$v_2 \to v_4$	$v_2 \to v_4 \to v_5$
v_3	$v_3 \to v_1$	$v_3 \to v_2$		$v_3 \to v_4$	$v_3 \to v_5$
v_4	$v_4 \to v_2 \to v_1$	$v_4 \to v_2$	$v_4 \to v_3$		$v_4 \to v_5$
v_5	$v_5 \to v_3 \to v_1$	$v_5 \to v_4 \to v_2$	$v_5 \to v_3$	$v_5 \to v_4$	

In Table 6.3, $s_1 = 1 - 0.97^\omega$, $s_2 = 1 - 0.98^\omega$, $s_3 = 1 - 0.99^\omega$, $s_4 = 1 - 0.9994^\omega$, $s_5 = 1 - 0.9996^\omega$. It can be verified that the objective function (6.15) is as follows.

$$\frac{1}{20}\Big((2 + x_2 + 2x_3 + x_4)\log s_1 + (8 + x_2 + 4x_3 + 3x_4)\log s_2$$
$$+4\log s_3 + (4 - x_2 - 2x_3 - x_4)\log s_4 + (2 - x_3 - x_4)\log s_5\Big)$$
$$= \frac{1}{20}\Big(\log(s_1^2 s_2^8 s_3^4 s_4^4 s_5^2) + x_2 \log\frac{s_1 s_2}{s_4} + x_3 \log\frac{s_1^2 s_2^4}{s_4^2 s_5} + x_4 \log\frac{s_1 s_2^3}{s_4 s_5}\Big)$$

Note that variables x_1 and x_5 do not appear in the above equation. This is because that they are not used as intermediate nodes in any routes of the channels. In addition,

Table 6.2. Traffic loads over each link.

	v_1	v_2	v_3	v_4	v_5
v_1		0.01	0.03		
v_2	0.02		0.01	0.02	
v_3	0.02	0.01		0.02	0.02
v_4		0.03	0.01		0.02
v_5			0.02	0.02	

Table 6.3. Probabilities of successful channel connection between each pair of nodes.

	v_1	v_2	v_3	v_4	v_5
v_1		s_3	s_1	$(s_1 s_2)^{2x_3} s_4^{\overline{x}_4}$	$(s_1 s_2)^{2x_3} s_4^{\overline{x}_4}$
v_2	s_2		s_3	s_2	$s_2^{2x_4} s_5^{\overline{x}_4}$
v_3	s_2	s_3		s_2	s_2
v_4	$(s_1 s_2)^{2x_2} s_4^{\overline{x}_2}$	s_1	s_3		s_2
v_5	$s_2^{2x_3} s_5^{\overline{x}_3}$	$(s_1 s_2)^{2x_4} s_4^{\overline{x}_4}$	s_2	s_2	

the objective function in this simple case is linear, since channels routes between all pair of nodes are within two hops, i.e., $d \leq 2$. As this is a small network and the expression is relatively simple, the optimal converter placement can be found by considering all possible placements. For instance, it can be verified that for $k = 2$ the optimal converter placement is to put two converters at node v_3 and node v_4, respectively.

2.2 Algorithm Design and Analysis

In the preceding section, we formulate the minimum overall blocking problem (Problem 6.2) as the minimization of a polynomial function of binary variables (6.16-8). In this subsection we will propose a branch-and-bound method to find the optimal solution to the problem. The basic idea is to find the optimal converter placement through searching a tree. Each vertex in the tree represents the *state* of a converter placement, which corresponds 0-1 assignments to some of n binary variables $\{x_1, \cdots, x_n\}$. We call a state $\{x_1, \cdots, x_n\}$ *feasible* if $\sum_{i=1}^{n} x_i = k$. This means that a feasible state represents a feasible converter placement satisfying constraint (6.17). For two states $\{x_1, \cdots, x_n\}$ and $\{x'_1, \cdots, x'_n\}$, $\{x'_1, \cdots, x'_n\}$ is called a *sub-state* of $\{x_1, \cdots, x_n\}$ if $x'_i \leq x_i$ for $i = 1, \cdots, n$. Moreover, a sub-state $\{x'_1, \cdots, x'_n\}$ of state $\{x_1, \cdots, x_n\}$ is called a *direct sub-state* if $\sum_{i=1}^{n} x'_i = \sum_{i=1}^{n} x_i - 1$. For a simple example of $n = 5$, state $\{1,1,0,0,0\}$ is a sub-state of $\{1,1,1,0,1\}$ and $\{1,1,0,0,1\}$ is a direct sub-state of $\{1,1,1,0,1\}$, while $\{1,1,0,0,0\}$ is a feasible state when $k = 2$.

2.2.1 Branching Method

The search tree is constructed as follows (see Fig. 6.14): The primary level, 0-level, has one vertex that is the root of the tree. The root state has all 1-components, i.e., $\{1, \cdots, 1\}$, indicating that every node in the network is placed with a converter. The next level of the tree, 1-level, consists of the direct sub-states of the root, whose components are all "1"s except one component. Inductively, for $i \geq 1$ each state (or vertex in the tree) at i-level is a direct sub-state at $(i-1)$-level. It has $(n-i)$ 1-components and i 0-components, indicating a placement of $(n-i)$ converters in the network. The tree has total of $(n-k)$ levels since there are at most k converters. The end level, $(n-k)$-level, consists of feasible states.

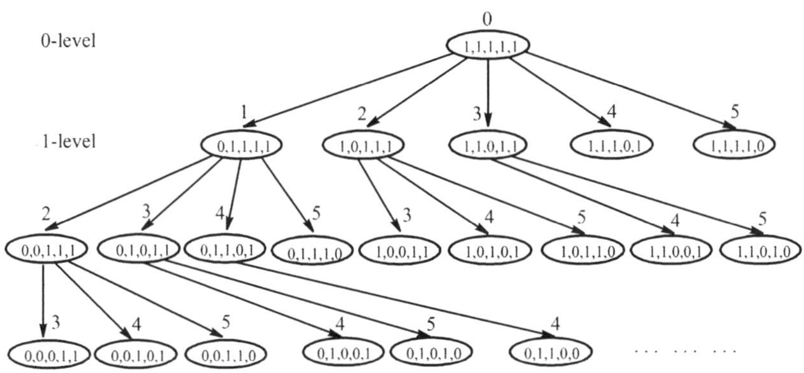

Figure 6.14. An example of search tree of $n = 5$ and $k = 2$.

To avoid duplicate states and computations, we label each state (i.e., a vertex in the tree) as follows (see Fig. 6.14). The root vertex is labelled as 0, and then its 1-*th* child (on the leftmost side), obtained by shifting its 1-*th* component to "0", is labelled as 1. Generally, its *i-th* child, whose *i-th* component is shifted to "0" while the rest of components remain unchanged, is labelled as i. Inductively, for a vertex in the tree with label l, its children are labelled by (from left to right) $l+1, l+2, \cdots, n$. Its *(l+i)-th* child inherits all its component assignments except changing its *(l+i)-th* component to "0". For the example in Fig. 6.14, the vertex labelled 3 at 1-level whose state is $\{1,1,0,1,1\}$ has two children $\{1,1,0,0,1\}$ and $\{1,1,0,1,0\}$ labelled 4 and 5, which are generated by changing its 4-*th* and 5-*th* components to "0", respectively.

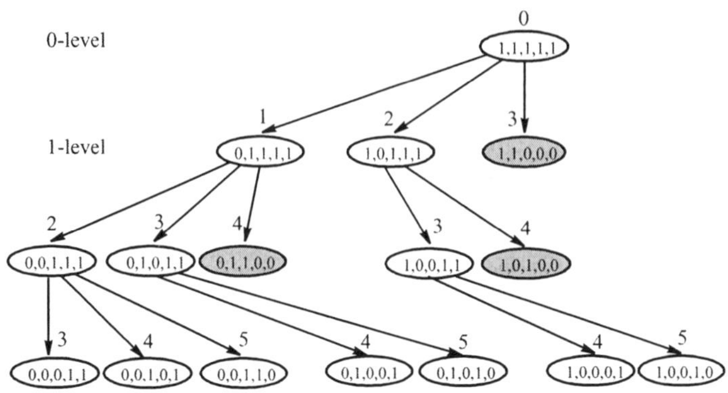

Figure 6.15. The resulting search tree obtained by applying the second pruning rule.

2.2.2 Bounding Method

We can apply the depth first search to the tree constructed as above. Since the method needs to search all vertices at all levels in the tree in the worst case, we now propose some pruning methods that make searching more efficient.

First note a simple but important fact. When one converter is added to the network, while the existing converter locations remain unchanged, the probabilities of successful channel connections will not decrease. This can be easily understood from the formulation of $S_p(P_{st})$ and can be formally expressed as: If $\{x'_1, \cdots, x'_n\}$ is a sub-state of $\{x_1, \cdots, x_n\}$, then $F(x'_1, \cdots, x'_n) \leq F(x_1, \cdots, x_n)$.

Now we denote by $X^* = (x^*_1, \cdots, x^*_n)$ the current best feasible state found during the searching process. Since any sub-state of X^* has a smaller success probability than $F(X^*)$ and thus is unable to make a better solution than X^*. In other words, the optimal state will not be in the sub-tree rooted from X, it thus can be pruned off. This leads to the first pruning rule.

PRUNING RULE 1. *If $F(X) \leq F(X^*)$, then the branch rooted from X can be pruned.*

According to the branching method for generating child states in the tree, when the children of a state at l-level are generated, the first l components of its state vector will remain unchanged. If there are already k 1-components among its first l components, this means, k converters have been already placed in the network, we need only to consider the state whose last $(n-l)$ components are all "0". Moreover, due to labelling rule, any of its siblings with label $l' > l$ should also have more than k

"1"s among its first l components. Therefore, all its siblings whose labels are greater l should be pruned too. This leads to the second pruning rule.

PRUNING RULE 2. *If a state X of label l satisfies $\sum_{i=1}^{l} x_i \geq k$, then prune all children of X and all the sibling states of X whose labels are greater than l. Meanwhile, change state X by making the last $(n - l)$ components "0".*

For the example in Fig. 6.14, state $\{1,1,0,1,1\}$ at 1-level is labelled 3 and has two "1"s in its first three components. For the case of $k = 2$, its children and its sibling states $\{1,1,0,0,1\}$ labelled 4 and $\{1,1,0,1,0\}$ labelled 5 are pruned off. At the same, the rest of its components (starting from the fourth) should be made to "0", which become a new state $\{1,1,0,0,0\}$ at 1-level. After the second pruning rule, we obtain a new tree whose leave nodes are all feasible states. See Fig. 6.15, where the second pruning rule is applied to the states marked in grey.

Combining the branch-and-bound methods, we develop the following algorithm. It starts searching from the root vertex $X_0 = (1, 1, \cdots, 1)$ with an initial feasible state X^* (when the algorithm terminates, X^* gives an optimal converter placement). The outer if-statement corresponds the first pruning rule, that means, any state X with label l and its children will not be considered if $F(X) \leq F(X^*)$. This process is $search(X, l)$. The middle if-statement follows the second pruning rule, that means, for a child with label $(l + i)$, if its first $(l + i)$ components contains k "1"s, then its last $(n - l - i)$ components will be set to "0". This process is $set0(X_{l+i}, l + i)$. Note that the rest of children will not be considered.

ALGORITHM 6.4 *Placing Wavelength Converters*

Generate an initial state X^*.
search(X^0,0) with $X^0 = \{1, 1, \cdots, 1\}$.
do *search*(X,l) as follows. // search sub-tree rooted at X with label l
if X is a feasible state **then**
  if $F(X) > F(X^*)$ **then** $X^* := X$; // a better feasible state found
else-if $F(X) > F(X^*)$ **then**
    for $(i = 1; |V| - l; i++)$ **do** // generate X's $|V| - l$ children
      generate X's child X_{l+i};
      if $\sum_{j=1}^{l+i} x_j = k$ **then**
        do $set0(X_{l+i}, l + i)$;
        quit **for-loop**
      do *search*($X_{l+i}, l + i$); // search X's child
    end-for
return X^*.

One of factors that affect the performance efficiency of the algorithm is how to choose an initial state X^*, since when the initial state is very close to the optimal state, most of the branches of the tree will be pruned. Some previous studies, e.g. [12], has discovered that placing converters at the nodes with high average congestion will result in a low blocking probability. We can use the heuristic proposed in [12] to find a good initial state. We compute λ_i for all network nodes and sort them in descending order with respect to λ_i. A reasonable placement is to place given k converters at the first k nodes (with higher values of λ_i). Another option for getting a good initial state X^* is to solve Problem 6.2 by using a general algorithm such as [3].

2.3 Simulation Study

In this subsection, we apply the proposed method (i.e., Algorithm 6.4) to two examples, a path and NSFnet. In a path network of 11 nodes, we assume the same traffic load between any pair of nodes, $\lambda_{ij} = 5$ and $\omega = 5$. The obtained results are given in Table 6.4 and Table 6.5, respectively. In these tables, the first column indicates the number of given converters in the networks. The second and third columns give the initial feasible states that Algorithm 6.4 uses and the optimal states that it outputs. A state is simply represented by a binary number whose i-th bit corresponds i-th component x_i of state vector $\{x_1, \cdots, x_n\}$. The fourth columns are the optimal values of the objective function $F(X)$ defined in equation (6.15). Note that they are negative because they are the logarithm values of the probabilities of successful channel connections. The fifth columns shows the corresponding probabilities by relieving the logarithm. The sixth columns record the number of states that the algorithm searches before it finds the optimal states. The last columns show the efficiency of the algorithm which are defined as

$$E \equiv \frac{\binom{|V|}{k} - \alpha}{\binom{|V|}{k}} \times 100\%,$$

where $\binom{|V|}{k}$ is the number of states that the exhaustive enumerate needs to search, and α is the number of searched states given in the sixth columns. In fact, E is the percentage of searches saved by our algorithm over the exhaustive search method.

From the experiment and simulation study, we make the following discoveries.

1 The heuristic proposed in [12] that places converters at highly congested nodes does not produce the optimal placement in most of the cases. In fact, in the study we first use this heuristic to generate the initial states (placements) from where we search the optimal placements. From the obtained results in [12] we can see that in most of time the initial solutions are different from the optimal solutions that found by Algorithm 6.4.

2 The probabilities of successful channel connections increase as the number of converters increase. However, as the value of k reaches a number large enough, which is about 20-30% of the network nodes, the probabilities would reach a stable and satisfactory level that is about as high as 98%. We notice that this percentage heavily depends on the traffic load in real network environment. Moreover,

Table 6.4. The results of the path network of 11 nodes.

k	Initial state X_0	Optimal state X_{opt}	Optimum $F(X_{opt})$	Probability $10^{F(X_{opt})}$	States α	Efficiency E (%)
1	00000100000	00000100000	-0.050359	0.890515	8	27.3
2	00001100000	00001010000	-0.024287	0.945612	19	65.5
3	00001110000	00010101000	-0.015058	0.965921	24	85.5
4	00011110000	00011101000	-0.010069	0.977081	28	91.5
5	00011111000	00011111000	-0.005081	0.988369	11	97.6
6	00111111000	00111111000	-0.003750	0.991403	9	98.1
7	00111111100	00111111100	-0.002419	0.994446	8	97.6
8	00111111110	00111111110	-0.002345	0.994614	9	94.5
9	01111111110	01111111110	-0.002272	0.994783	3	96.4
10	11111111110	11111111110	-0.002272	0.994783	2	90.9
11	11111111111	11111111111	-0.002272	0.994783	1	0.00

Table 6.5. The results of the NSFnet network.

k	Initial state X_0	Optimal state X_{opt}	Optimum $F(X_{opt})$	Probability $10^{F(X_{opt})}$	States α	Efficiency E (%)
1	00010000000000	00010000000000	-0.028255	0.937012	10	28.6
2	00010001000000	00010000010000	-0.018072	0.959241	19	79.1
3	00010011000000	00010011000000	-0.012552	0.971512	28	92.3
4	00010011010000	00010011010000	-0.009453	0.978468	23	97.7
5	00010011010100	00011011010000	-0.007947	0.981868	29	98.6
6	10010011010100	00011011010100	-0.006694	0.984704	21	99.3
7	10010011011100	10011011010100	-0.005490	0.987438	17	99.5
8	10011011011100	10011011011100	-0.005032	0.988481	9	99.7
9	10011011011101	10011011011101	-0.004864	0.988863	10	99.5
10	10111011011101	10011111011101	-0.004862	0.988868	15	98.5
11	10111011011111	10111111011101	-0.004861	0.988871	14	96.2
12	10111111011111	10111111011111	-0.004860	0.988872	13	85.7
13	10111111111111	10111111111111	-0.004860	0.988872	2	92.9
14	11111111111111	11111111111111	-0.004860	0.988872	1	0.00

further increase of k does not cause the proportional increase of the probabilities. This means that there is no need to place too many converters in networks.

3 The efficiency of Algorithm 6.4 is extremely high. From the results in the last columns of Table 6.4 and Table 6.5, we can see that the savings of the number

of searches over the exhaustive search method are over 90% for $k > 2$. We, however, notice that when $k = 1$ or 2, the exhaustive search is very efficient.

4 From the third columns of Tables 6.4 and 6.5, we can see that the optimal placement of k converters can be derived from the optimal placement of $(k - 1)$ converters in most of cases (but not in every case). This means that the optimal placement of k converters in a network can be produced by keeping the optimal placement of $(k-1)$ converters unchanged and then finding the optimal location for the k-th converter.

The fourth observation in the above leads to a very efficient heuristic as follows. Search all the nodes in a network to find the optimal placement of one converter, and then search all nodes except the one where the first converter is located and find the optimal placement of the second converter; Repeat this process until all k converters are placed in the network. Clearly, it takes time of $O(k|V|)$ for this simple heuristic to find an approximate solution.

3. Discussion

In Section 1 of this chapter, we have studied how to minimize the overall blocking through assigning wavelengths properly. The problem is transformed to the maximum weight k-cut problem. An efficient approximation algorithm with guaranteed performance ratio has been proposed. The obtained results show that the overall blockings in single-hop systems can be considerably reduced via proper wavelength assignment assuming that the routes of all connections are known.

Mokhtar et al [11] adopted a more general formulation in which they considered all possible paths (as candidates of routes) between a source-destination pair and incorporate network state information into the routing decision. Their approach performs routing and wavelength assignment algorithms to optimize the blocking performance, where the arrival process of connection requests is assumed to be a Poisson process and connection holding times are exponentially distributed.

Birman proposed a model in [2] to compute the end-to-end blocking probability in all-optical networks without wavelength conversion. In this model, the blocking probabilities under two routing schemes were analyzed: fixed routing and least loaded routing. This model was later improved by Sridharan et al [15]. The models were proposed in [1, 10] for the analysis of blocking probabilities in a system with or without wavelength converters. In particular, Barry and Humblet [1] studied the blocking probabilities on a path and discussed the effects of some network parameters to the blocking probabilities, such as the path length, switching size, etc.

In Section 2 of this chapter we have studied how to minimize the overall blocking probability of connections through placing a given number of wavelength converters. A general model is established in which the overall blocking probability is expressed as a polynomial function of binary variables indicating the possible locations of wavelength converters. Although it is quite complicated, it provides an exact formulation of the problem and makes a global optimization possible.

Subramaniam et al. [16] considered the same problem in a path. They adopted a simple traffic model in which connection requests arrive and depart from the network as a random point process. They showed that uniformly spaced converters yield the optimal performance for an end-to-end request on a path when the link loads are uncorrelated and uniform. They provided the optimal converter placement based on dynamic programming for the case of that the link loads are nonuniform or other requests are considered.

Xiao et al. [18] addressed the same problem but in a different way. Their approach begins with the assumption that every node in the network is equipped with a wavelength converter. Under the assumption it first records the utilization statistics of wavelength converters by simulations. It then optimizes the placement of a given number of wavelength converters in the network. The following two objectives are considered.

(1) Maximizing the sum or product of total utilizations of all nodes so that the overall utilization of wavelength converters can be improved. As a result, the overall blocking probability can be smaller and the mean quality of service is better.

(2) Maximizing the minimum value of total utilizations of all nodes so that the placement of wavelength converters to the nodes could be more fair. As a result, the maximum of the blocking probabilities experienced at all nodes could be minimized.

References

[1] R. A. Barry and P. A. Humblet, Models of blocking probability in all-optical networks with and without wavelength changers, *IEEE Journal on Selected Areas in Communications*, 14 (5) (1996), 858-867.

[2] A. Birman, Computing approximate blocking probabilities for a class of all-optical networks, *IEEE Journal in Selected Areas in Communications*, 14 (5) (1996), 852-857.

[3] W.-W. Fang, T.-J. Wu and J.-P. Chen, An algorithm of global optimization for rational functions with rational constraints, *Journal of Global Optimization*, 18 (2000), 211-218.

[4] A. Frieze and M. Jerrum, Improved approximation algorithms for MAX k-cut and MAX bisection, *Algorithmica*, 18 (1) (1997), 67-81.

[5] M. R. Garey and D. S. Johnson, *Computers and Intractability: A Guide to the Theory of NP-Completeness*, W. H. Freeman, San Francisco, CA, 1979.

[6] F. Gavril, Algorithms for minimum coloring, maximum clique, minimum covering by cliques, and maximum independent set of a chordal graph, *SIAM Journal on Computing*, 1 (1972), pp. 180-187.

[7] R. L. Graham, Bounds for certain multiprocessing anomalies, *Bell System Technical Journal*, 45 (1966), 1563-1581.

[8] M. M. Halldorsson, Approximating the minimum maximal independence number, *Information Processing Letters*, 46 (1993), 169-172.

[9] R. M. Karp, Reducibility among combinatorial problems, in R. E. Miller and J. W. Thatcher (eds.), *Complexity of Computer Computations*, Plenum Press, New York, 85-103.

[10] M. Kovacevic and A. S. Acampora, Benefits of wavelength translation in all-optical clear-channel networks, *IEEE Journal of Selected Areas in Communications*, 14 (5) (1996), 868-880.

[11] A. Mokhtar and M. Azizoglu, Adaptive wavelength routing in all-optical networks, *IEEE/ACM Transactions on Networking*, 6 (2) (1998), 197-206.

[12] B. Mukherjee, *Optical communication networks*, New York: McGraw-Hill, 1997.

[13] S. Poljak, Integer linear programs and local search for max-cut, *SIAM Journal on Computing*, 24 (4) (1995), 822-839.

[14] A. A. Schaffer and M. Yannakakis, Simple local search problems that are hard to solve, *SIAM Journal on Computing*, 20 (1) (1991), 56-87.

[15] A. Sridharran and K. N. Sivarajan, Blocking in all-optical networks, in *Proceedings of IEEE Conference on Computer Communications* (INFOCOM), (2000).

[16] S. Subramaniam, M. Azizoglu, and A. K. Somani, On optimal converter placement in wavelength-routed networks, *IEEE/ACM Transactions on Networking*, 7 (5) (1999), 754-766.

[17] S. Thiagarajan and A. K. Somani, An efficient algorithm for optimal wavelength converter placement on wavelength-routed networks with arbitrary topologies, in *Proceedings of IEEE Conference on Computer Communications* (INFOCOM), (1999), 916-923.

[18] G. Xiao and Y. W. Leung, Algorithms for allocating wavelength converters in all-optical networks, *IEEE/ACM Transactions on Networking*, 7 (4) (1999), 545-557.

Chapter 7

MINIMIZATION OF WAVELENGTH CONVERSIONS

In multihop WDM networks where wavelength converters are equipped at some routing nodes, a connection between two communication nodes consists of one or more lightpaths. A wavelength conversion is required at the joint of two lightpaths if they use different wavelengths. Wavelength conversion can be done either electronically or optically. In electronic approach, optic-electronic-optic conversion is required and it causes long delay compared with the transmission of light signals. In all optical approach, the optical signal is allowed in the optical domain throughout the conversion process, however shifting wavelength channels from one to another makes routing/switching complicated. Therefore, in either case, it is desirable to minimize the number of wavelength conversions used to reduce the conversion delay and workload of routing nodes.

In this chapter, we will focus on how to route the connections and assign wavelengths to them so that the overall number of wavelength conversions occurred in the system is as minimal as possible. Here we assume that the wavelength converters are placed at some routing nodes using the methods proposed in Chapter 5 or Chapter 6.

The rest of this chapter is organized as follows. Section 1 discusses wavelengths assignment to a set of given lightpaths such that the overall number of wavelength conversions in the system is minimized, where the routes of lightpaths and the number of available wavelengths are given. Section 2 studies the setup of a broadcast connection with minimal number of wavelength conversions. The basic approaches used in these works are to transform these problems into some typical covering problems, the minimum clique-partition problem in Section 1 and the vertex-cover and set-covering problems in Section 2. Section 3 concludes this chapter.

1. Wavelength Assignment for Unicast Connections

In Chapter 3 we have discussed the routing and wavelength assignment problem that aims at minimizing the number of wavelengths used in a system. However, in the design of a real multihop system, the number of wavelengths that are allowed for use is often a given parameter, because this number depends on the hardware setting of optical switches in the network. In this section, we will consider how to, given the number of available wavelengths in the system, assigns wavelengths to lightpaths such that the overall number of wavelength conversions in the whole system is minimized.

1.1 Optimization Formulation and Theory

Given a physical topology of network $G(V, E)$. Denote a unicast communication by a pair (s, d), where $s, d \in V$. Let $r(s, d)$ denote the route from node s to node d, which consists of one or several lightpaths, and let $|r(s, d)|$ denote the number of lightpaths in $r(s, d)$. Let $\lambda_{s,d}$ denote the amount of traffic, in terms of the number of connections, from s to d. Both $\lambda_{s,d}$ and $r(s, d)$ are given in the system.

Fig. 7.1(a) is a simple physical topology of a ring network of six nodes, on which six lightpaths are configured. Fig.7.1(b) is the corresponding logical topology. There is an edge between two nodes in the logical topology if and only if there is a lightpath between the two nodes. (Refer to Chapter 4 for the discussion on how to design a logical topology.)

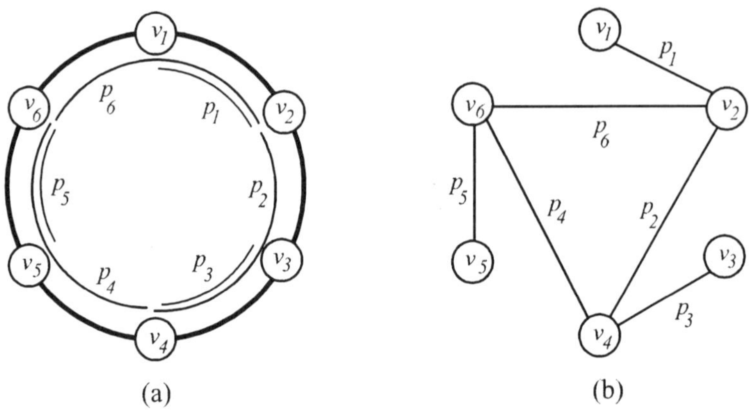

Figure 7.1. An example: (a) lightpaths on a ring network and (b) the logical topology.

Suppose that the traffic between node pairs are given as in Table 7.1, where the traffic is symmetry (the traffic from node v_i to node v_j is equal to the traffic from node v_j

to node v_i). Suppose that the route between two nodes is the shortest path, in terms of the least number of lightpaths, in the logical topology of Fig. 7.1(b), which are given as in Table 7.2. For instance, the route for connection from node v_1 to node v_2 uses one lightpath, i.e., $r(v_1, v_2) = p_1$, while the route for connection from node v_1 to node v_6 consists of two lightpaths via node v_2, i.e. $r(v_1, v_6) = p_1p_6$.

Table 7.1. Traffic between node pairs.

Node	v_1	v_2	v_3	v_4	v_5	v_6
v_1		3	2	2	3	2
v_2	3		2	2	2	2
v_3	2	2		3	2	1
v_4	2	2	3		2	2
v_5	3	2	2	2		3
v_6	2	2	1	2	3	

Table 7.2. Routes between node pairs.

Node	v_1	v_2	v_3	v_4	v_5	v_6
v_1		p_1	$p_1p_2p_3$	p_1p_2	$p_1p_6p_5$	p_1p_6
v_2	p_1		p_2p_3	p_2	p_6p_5	p_6
v_3	$p_1p_2p_3$	p_2p_3		p_3	$p_3p_4p_5$	p_3p_4
v_4	p_1p_2	p_2	p_3		p_4p_5	p_4
v_5	$p_1p_6p_5$	p_6p_5	$p_3p_4p_5$	p_4p_5		p_5
v_6	p_1p_6	p_6	p_3p_4	p_4	p_5	

When the route of a connection between two nodes consists of more than one lightpath, two adjacent lightpaths in the route use different wavelengths, a wavelength conversion is needed at the joint of the two lightpaths. Let $x(s, d)$ denote the number of wavelength conversions from s to d along route $r(s, d)$. The overall number of wavelength conversions in the system can be represented as:

$$\overline{W} \equiv \sum_{s,d \in V} \lambda_{s,d} x(s, d). \tag{7.1}$$

Then the *Minimum Wavelength Conversion Problem* (MWCP) we will study in this section can be formulated as follows.

PROBLEM 7.1 *Minimum Wavelength Conversion Problem*

Instance A network $G(V, E)$ and traffic between node pairs, a set W of ω wave-, lengths and a set P of lightpaths and their routes.

Solution An assignment of wavelengths to the lightpaths in P without causing any wavelength conflict.

Objective Minimizing the overall number of wavelength conversions $\overline{W}$ as defined in equation (7.1).

The following theorem shows that it is difficult to find an optimal solution to MWCP.

THEOREM 7.1 *Problem 7.1 is $\mathcal{NP}$-hard.*

PROOF The problem of determining whether lightpaths in P can be assigned by using wavelengths in W without causing any wavelength conflict is equivalent to the vertex coloring problem, which is $\mathcal{NP}$-complete (refer to the discussion on Problem 3.1 in Section 1 of Chapter 3). Thus the decision version of problem 7.1 is $\mathcal{NP}$-completer as well. ■

Our approach for solving MWCP is to transform it to the maximum clique-partition problem. Given a graph $G(V, E)$, a *clique* C in $G(V, E)$ is a complete subgraph of $G(V, E)$ (there is an edge between every pair of vertices in the subgraph). A *clique-partition* $\mathcal{C}$ is a set of cliques Cs such that every vertex in V is included in one of the cliques. For edge-weighted graph $G(V, E)$, the weight of a clique is the total weights of all edges in the clique. The weight of a clique-partition is the total weights of all cliques in the partition. The *Maximum Clique-Partition Problem* (MCPP) is to find a clique-partition whose weight is maximal.

Since the objective of MWCP is to minimize the overall wavelength conversions between lightpaths, we need to compute the transit traffic between two lightpaths. Let $\lambda(p_i, p_j)$ denote the traffic from lightpath p_i to lightpath p_j. $\lambda(p_i, p_j)$ can be easily computed from $\lambda_{s,d}$ and $r(s, d)$, where $s, d \in V$ and $p_i, p_j \in P$. Table 7.3 shows the transit traffic between any pair of lightpaths in Fig.7.1(b), which is computed based on Table 7.1 and Table 7.2.

To transform the minimum wavelength conversion problem into the maximum clique-partition problem, we introduce an auxiliary edge-weighted graph $G_a(U_a, E_a)$ as follows.

ALGORITHM 7.1 *Constructing an Auxiliary Graph*

Step 1 The vertex set U_a is P. A node $u_i \in U_a$ represents lightpath p_i in P.

Step 2 For $u_i, u_j \in U_a$, an edge $(u_i, u_j) \in E_a$ if and only if lightpaths p_i and p_j do not share any physical link in network $G(V, E)$, i.e., $p_i \cap p_j = \emptyset$.

Step 3 The weight of edge $(u_i, u_j) \in E_a$, denoted by $wt(u_i, u_j)$, is $\lambda(p_i, p_j)$.

Fig. 7.2 is the auxiliary graph constructed from the network in Fig. 7.1(a). The weight of an edge is marked at the edge if the weight is non-zero, and the weight is

Table 7.3. Transit traffic between pairs of lightpaths.

Lightpath	p_1	p_2	p_3	p_4	p_5	p_6
p_1		4	0	0	0	0
p_2	4		0	0	0	0
p_3	0	0		3	0	0
p_4	0	0	3		0	0
p_5	0	0	0	0		5
p_6	0	0	0	0	5	

zero if no number is associated with an edge (referring to the transit traffic in Table 7.3). For example, there is an edge between lightpaths p_1 and p_2, because they do not share any link in the network of Fig. 6(a). Moreover, since the traffic flowing from p_1 to p_2 consists of the traffic of node pairs (v_1, v_3) and (v_1, v_4), the weight of edge (p_1, p_2) is $\lambda(p_1, p_2) = \lambda_{1,3} + \lambda_{1,4} = 4$.

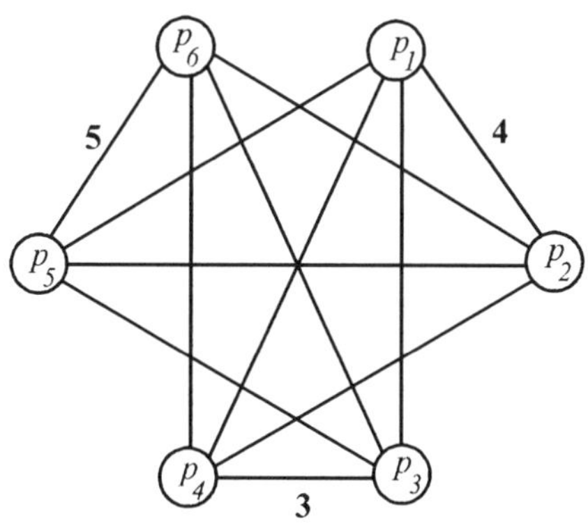

Figure 7.2. The auxiliary graph $G_a(U_a, E_a)$.

Graph $G_a(U_a, E_a)$ represents the relationships among the lightpaths in P. If there is an edge between two vertices in the graph, it means that their corresponding lightpaths do not share any physical links. It is safe to assign them the same wavelengths according to the wavelength conflict rule. Furthermore, a clique in the graph means there is no common physical link shared by any pair of lightpaths represented by the vertices in the clique. Thus, it is safe to assign the same wavelengths to the vertices in a clique. Let $wt(C)$ and $wt(K)$ denote the weight of a clique C and the weight of a clique-partition K of $G_a(U_a, E_a)$, respectively. From the features of $G_a(U_a, E_a)$, we have the following lemmas.

LEMMA 7.1 *For a clique C in $G_a(U_a, E_a)$, by assigning the vertices in C the same wavelength, the number of wavelength conversions that can be saved is $wt(C)$.*

PROOF For an edge $(u_i, u_j) \in E_a$, $wt(u_i, u_j)$ represents the transit traffic between lightpaths p_i and p_j. If p_i and p_j are assigned the same wavelength, there is no need of wavelength conversions for the traffic flowing between p_i and p_j. Thus, the number of conversions saved in this case is $wt(u_i, u_j)$. Similarly, if all vertices in C are assigned with the same wavelength, the total number of conversions that can be saved is $\sum_{(u_i,u_j)\in C} wt(u_i, u_j) = wt(C)$. ■

LEMMA 7.2 *K is a clique-partition of $G_a(U_a, E_a)$ if and only if lightpaths in P can be assigned by using k wavelengths, where k is the number of cliques in K.*

PROOF "Only if": Suppose $K = \{C_1, \cdots, C_k\}$ is a clique-partition of $G_a(U_a, E_a)$. We assign all the vertices in C_i the same wavelength w_i, $1 \leq i \leq k$. Since there is an edge between every pair of vertices in C_i, which means that they do not share any link in network $G(V, E)$, such an assignment will not cause wavelength conflict. Thus lightpaths can be assigned by using k wavelengths.

"If": Suppose that lightpaths in P can be assigned by using k wavelengths, w_1, $\cdots, w_k$. Let C_i be the set of vertices in $G_a(U_a, E_a)$ that are assigned wavelength w_i, $1 \leq i \leq k$. Since all vertices in C_i are assigned with the same wavelength, according to wavelength conflict rule, there is no shared link between any pair of lightpaths represented by the vertices in C_i. This implies that there is an edge between any pair of vertices in C_i, and C_i is thus a clique. Therefore, $\{C_1, \cdots, C_k\}$ makes up a clique-partition of $G_a(U_a, E_a)$ with size k. ■

THEOREM 7.2 *$G_a(U_a, E_a)$ has a clique-partition of size k if and only if lightpaths in P can be assigned by using k wavelengths and the number of wavelength conversions is*

$$\overline{W} = \sum_{s,d\in V} \lambda_{s,d}\Big(|r(s,d)| - 1\Big) - wt(K). \tag{7.2}$$

PROOF "Only if": Suppose that $K = \{C_1, C_2, \cdots, C_k\}$ is a clique-partition of $G_a(U_a, E_a)$. According to Lemma 7.2, lightpaths in P can be assigned by using k wavelengths.

Now, consider the number of wavelength conversions in the system. We assign wavelength w_i to the vertices in clique C_i, $1 \leq i \leq k$. By Lemma 7.1, $wt(C_i)$ is the number of the conversions that can be saved in clique C_i, $1 \leq i \leq k$. To sum up the conversions in all cliques in K, the total number of conversions that can be saved in the system is

$$\sum_{1\leq i\leq k} wt(C_i) = wt(K). \tag{7.3}$$

Since the number of lightpath joints in a route $r(s,d)$ is $|r(s,d)|-1$, the total number of lightpath joints in the routes of all connections in the system is

$$\sum_{s,d \in V} \lambda_{s,d}\Big(|r(s,d)| - 1\Big). \tag{7.4}$$

By deducting equation (7.3) from equation (7.4), we obtain the number of wavelength conversions required as in equation (7.2).

"If": Suppose the lightpaths in P can be assigned by using k wavelengths. According to Lemma 7.2, there should be a clique-partition of $G_a(U_a, E_a)$ and its size is k. The relationship between $\overline{W}$ and $wt(K)$ can be obtained in the same way as in the proof of "Only if" part. The proof is then finished. ■

In equation (7.2), $\sum_{s,d \in V} \lambda_{s,d}\Big(|r(s,d)| - 1\Big)$ is a constant. Thus minimizing $\overline{W}$ is equivalent to maximizing $wt(K)$. Theorem 1 states that finding a wavelength assignment for a set of lightpaths P in network $G(V,E)$ which minimizes the number of conversions is equivalent to finding a clique-partition of $G_a(U_a, E_a)$ whose weight is the maximal.

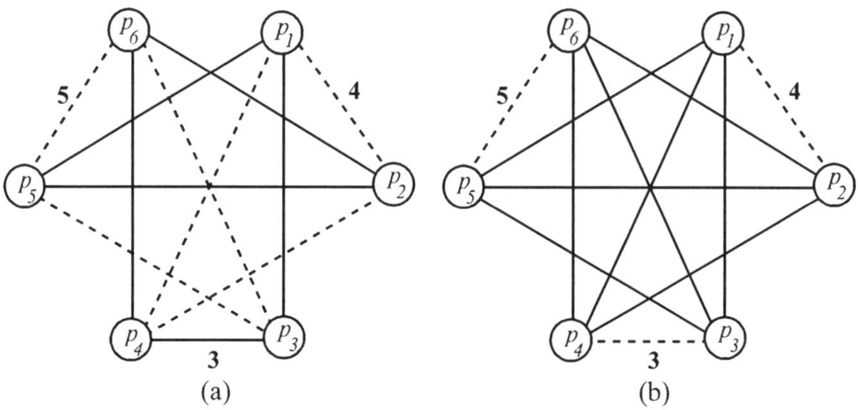

Figure 7.3. Optimal wavelength assignment of Problem 7.1: (a) $\omega = 2$ and (b) $\omega = 3$.

Fig. 7.3 gives two different wavelength assignments for the graph in Fig. 7.2. Fig. 3 (a) shows a clique-partition consisting of two cliques (two triangles in dashed lines). One contains lightpaths p_1, p_2 and p_4, with weight 4, and the other contains p_3, p_5, and p_6, with weight 5. p_1, p_2 and p_4 can be assigned with one wavelength, and p_3, p_5, and p_6 with another. Therefore, in total two wavelengths are used and 9 conversions can be saved. Fig. 7.3(b) shows another wavelength assignment using three wavelengths. It has a clique-partition consisting of three cliques (three edges in dashed lines). In this case 12 conversions are saved. From this example, we can see that by using more wavelengths, more wavelength conversions may be saved.

1.2 Algorithm Design and Analysis

In the preceding section we have proved that the minimal wavelength conversion problem is equal to the maximum clique-partition problem. In this section, we will first propose a greedy heuristic for the maximum clique-partition problem, and then discuss how to deal with the wavelength constraint.

We first consider how to find a clique of maximum weight. Since even for unweighted case of this problem, it was proved in [1] that there will be no algorithm that can produce a $|V|^{1/2-\epsilon}$-approximation solution for any $\epsilon > 0$. Hence we will use a simple greedy heuristic as follows.

ALGORITHM 7.2 *Finding a Clique*

Step 1 Find edge $(x, y) \in E_a$ that has the maximum weight. Denote the set U_a of vertices adjacent to both x and y by $U_a(x, y) = \{v \in U_a \mid (v, x) \in E_a$ and $(v, y) \in E_a\}$. Tie breaks in favor of the edge (x, y) that $U_a(x, y)$ includes more vertices if more than one edges have the same maximum weight.

Step 2 Find the maximum clique including (x, y) from the vertex set $U_a(x, y)$. Tie breaks in favor of the clique that includes more vertices if more than one cliques have the same maximum weight.

In the above we can simply use the exhaustive search at Step 2 to find the maximum clique. Its efficiency is reasonable, because the search space is upper bounded by the degrees of selected vertices x and y.

We now can find a clique-partition of $G_a(U_a, E_a)$ by repeatedly using algorithm Algorithm 7.2 as follows.

ALGORITHM 7.3 *Finding a Clique-Partition*

Step 1 Produce a clique C in $G_a(U_a, E_a)$ by using Algorithm 7.2.

Step 2 Remove C from $G_a(U_a, E_a)$ along with edges incident to vertices in C.

Step 3 Repeat Step 1-2 until all vertices in U_a are included in one of the produced cliques.

Consider the example in Fig. 7.4 continued from Fig. 7.3. We first choose edge (p_5, p_6) since it has the greatest weight 5 and obtain $V(p_5, p_6) = \{p_i \mid i = 1, 2, \cdots, 6\}$. Then we find a maximum clique $\{p_2, p_5, p_6\}$ (in dashed lines) that includes the most number of vertices as shown in Fig. 7.4(a). After that the clique is removed from the graph and a subgraph is obtained as shown in Fig. 7.4(b). In the resulting graph, we choose edge (p_3, p_4) since it has the greatest weight 3 and obtain $V(p_3, p_4) = \{p_1, p_3, p_4\}$. In the end we find a maximum clique $\{p_1, p_3, p_4\}$. As a result Algorithm 7.3 outputs a clique-partition of two cliques having weight 8. Observe that it is not an optimal solution (comparing it with the solution in Fig. 7.3(3)).

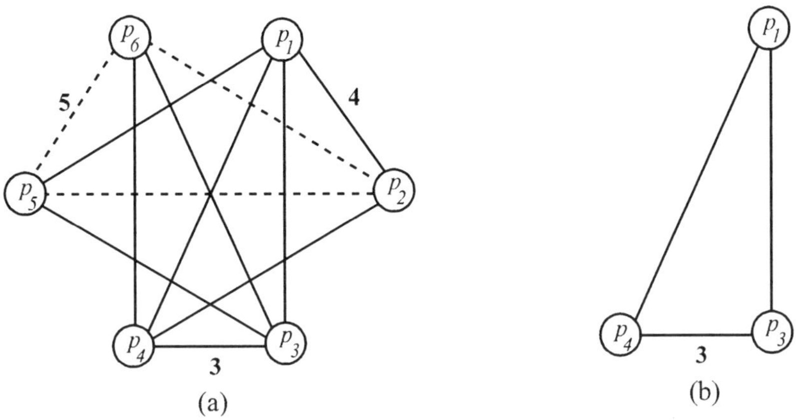

Figure 7.4. Finding a clique-partition: (a) find a clique in $G_a(U_a, E_a)$, and (b) find a clique in the remaining graph of $G_a(U_a, E_a)$.

Now we integrate the above discussed routines into a complete algorithm for solving the minimum wavelength conversion problem (Problem 7.1).

ALGORITHM 7.4 *Assigning Wavelengths to Lightpaths*

Step 1 Construct the auxiliary graph $G_a(U_a, E_a)$ by using Algorithm 7.1.
Step 2 Find a clique-partition of $G_a(U_a, E_a)$ by using Algorithm 7.3.
Step 3 Assign a wavelength to the lightpaths in each clique of the clique-partition.

THEOREM 7.3 *Algorithm 7.4 makes a wavelength assignment in time $O(|P|^2|E| + |P|^4 + 2^D|P|)$, where D is the maximal degree of vertices in $G_a(U_a, E_a)$.*

PROOF In Algorithm 7.4, Step 1 can be done in time $O(|P|^2|E|)$. This is because $G_a(U_a, E_a)$ has $|P|$ vertices (representing lightpaths). For each pair of vertices in U_a, it needs to go through the routes of two lightpaths to check if they share a common physical link. It thus takes at most time $O(|E|)$ to find if there is an edge between them. Since there are in total $|P|(|P| - 1)/2$ pairs of vertices, it takes time $O(|P|^2|E|)$ to find all edges in E_a.

Step 2 can be finished in time $O((|U_a| + 2^D)|P|)$. It takes time $O(|E_a|)$ to choose the edge having the greatest weight in $G_a(U_a, E_a)$, and it takes time $O(2^{2D-1})$ to find the maximum clique from the selected edge and the vertices adjacent to its end vertices, where D is the maximal vertex degree in $G_a(U_a, E_a)$. Since a clique-partition of $G_a(U_a, E_a)$ contains at most $O(|P|)$ cliques, it takes time $O(|P|(|P| + 2^{2D}))$ to find a clique-partition in $G_a(U_a, E_a)$.

Step 3 can be done in time $O(|P|)$. To sum up the time required in each step, the time-complexity of Algorithm 7.4 is $O(|P|^2|E| + |E_a|(|P| + |E_a|) + |P|(|P| +$

$2^{2D}) + |P|) \leq O(|P|^2|E| + |P|^4 + 4^D|P|)$, because $|U_a| \leq (|P| + |E_a|)$ and $|E_a| \leq |P|^2$. The proof is then finished. ■

At the end, we need to consider how to take care of the wavelength constraint. Note that the wavelength assignment produced by using Algorithm 7.4, which aims at minimizing the number of wavelength conversions, may require wavelengths more than ω. This is because the proposed algorithm always chooses the clique of the maximum weight (thus more wavelength conversions can be saved), but it ignores the size of the clique. This strategy may produce a clique with the greatest weight but having few vertices. Consequently, it may need more cliques to constitute a partition of the vertex-set of the given graph. Therefore, a clique-partition with the maximal weight may contain a large number of cliques, which results in a large number of wavelengths required.

In order to make an assignment by using no more than ω wavelengths, the size of a clique (i.e. the number of vertices in a clique) should be considered, as well as its weight. To achieve a tradeoff between the size and weight of a clique, when constructing auxiliary graph $G_a(U_a, E_a)$ we assign each vertex in U_a a weight of value $\alpha \geq 0$. In this case, the weight of clique C of $G_a(U_a, E_a)$, denoted by $wt_\alpha(C)$, is the total weights of both edges and vertices in C. That is, $wt_\alpha(C) \equiv \alpha|C| + wt(C)$. Algorithm 7.2 for finding a clique can be modified as follows.

ALGORITHM 7.5 *Finding a Clique*

Step 1 Choose the edge $(x, y) \in E_a$ or the vertex $z \in U_a$ that has the maximum (x, y) weight. If an edge was chosen, then follow Algorithm 7.2. If a vertex z was chosen, denote the set of vertices in U_a that is adjacent to z by $U_a(z) = \{v \in U_a \mid (v, z) \in E_a\}$. Tie breaks in favor of the vertex z that $U_a(z)$ includes more vertices if more than one vertices have the same maximum weight.

Step 2 Find the maximum clique from the vertex set $U_a(z)$. Tie breaks in favor of the clique that includes more vertices if more than one cliques have the same maximum weight.

When such modified heuristic Algorithm 7.5 is integrated with heuristic Algorithm 7.3, it will result in the following two effects:

(1) When the value of α becomes large enough (for example, greater than the maximum weight of edges), the weight of a clique C in $G_a(U_a, E_a)$ will be dominated by the size of clique C. Thus, finding a clique of maximal weight in $G_a(U_a, E_a)$ will be equivalent to finding a clique of maximum size in $G_a(U_a, E_a)$. In this case, each clique in the clique-partition will have a large size, which results in a small number of cliques in the clique-partition. This would lead to an assignment with a small number of wavelengths.

(2) When the value of α becomes small enough (for example, less than the minimal weight of nonzero edges), the weight of a clique C in $G_a(U_a, E_a)$ will be dominated

by the edge-weight of clique C. In this case, each clique in the clique-partition will have a great weight of edges. The number of wavelength conversions will become small. However, the size of a clique (i.e., the number of vertices in the clique) could be small in this case, which would lead to a large number of cliques in the clique-partition. That is, it requires more wavelengths.

From the above discussion, we can see that a clique-partition tends to contain a small number of cliques for a large α. In other words, by increasing the value of α, it will produce an assignment that requires less number of wavelengths. Therefore, we can use a binary search method to find the right value of α, which can produce a near-optimal wavelength assignment under the wavelength constraint ω. Let α_l and α_u denote the lower and upper bounds of weight α, respectively. The following is the binary searching algorithm.

ALGORITHM 7.6 *Assigning Wavelengths under Constraint*

Step 1 $\alpha_l := 0, \alpha_u := max_{(u_i,u_j)\in E_a}\{wt(u_i,u_j)\}, \alpha := \alpha_l$.
Step 2 Find a clique-partition K of $G_a(U_a, E_a)$.
Step 3 If $|K| > W$ then $\alpha_l := \alpha, \alpha := \alpha_l + (\alpha_u - \alpha_l)/2$;
If $|K| < W$ then $\alpha_u := \alpha, \alpha := \alpha_l + (\alpha_u - \alpha_l)/2$.
Step 4 Repeat Step 2-3 until $|K| = \omega$ or $\alpha_l = \alpha_u$.

Introducing the weight α helps us to find a wavelength assignment that causes small number of conversions and uses no more than ω wavelengths.

1.3 Simulation study

The aim of the proposed algorithm is to assign ω wavelengths to a set of lightpaths, such that the number of wavelength conversions in the system is minimized. In the discussion of preceding subsection, we have seen that the number of wavelengths in use and the number of conversions required are the two closely coupled factors. To demonstrate the performance of Algorithm 7.6, we simulate it under various wavelength requirements. In the simulation, we let α to vary from its lower bound, $\alpha = 0$, to its upper bound, $\alpha = max_{(u_i,u_j)\in E_a}\{wt(u_i,u_j)\}$. When $\alpha = 0$, Algorithm 7.6 aims at maximizing the weight of the clique-partition, which achieves the maximal saving of conversions. When $\alpha = max_{(i,j)\in E_a}\{wt(i,j)\}$, Algorithm 7.6 aims at maximizing the size of each clique (i.e., assigning the same wavelength to as many lightpaths as possible), which is reduced to the wavelength assignment by using the minimal number of wavelengths as in Chapter 3.

In the simulations, we apply Algorithm 7.6 to find a wavelength assignment, and compute the number of conversions saved in the system by using equation (7.3). Simulations are carried out on both the NSFnet and randomly generated networks.

For the simulation on the NSFnet, refer to Section 2 of Chapter 1 where the physical topology of NSFnet, the lightpaths configured on the shortest paths (in terms of the number of physical links), the routes between cities in the shortest path (in terms of the number of lightpaths), and the traffic matrix between cities are given. When constructing the auxiliary graph for the NSFnet, the weights of edges (representing the transit traffic between two lightpaths) are scaled to the range from 0 to 99.

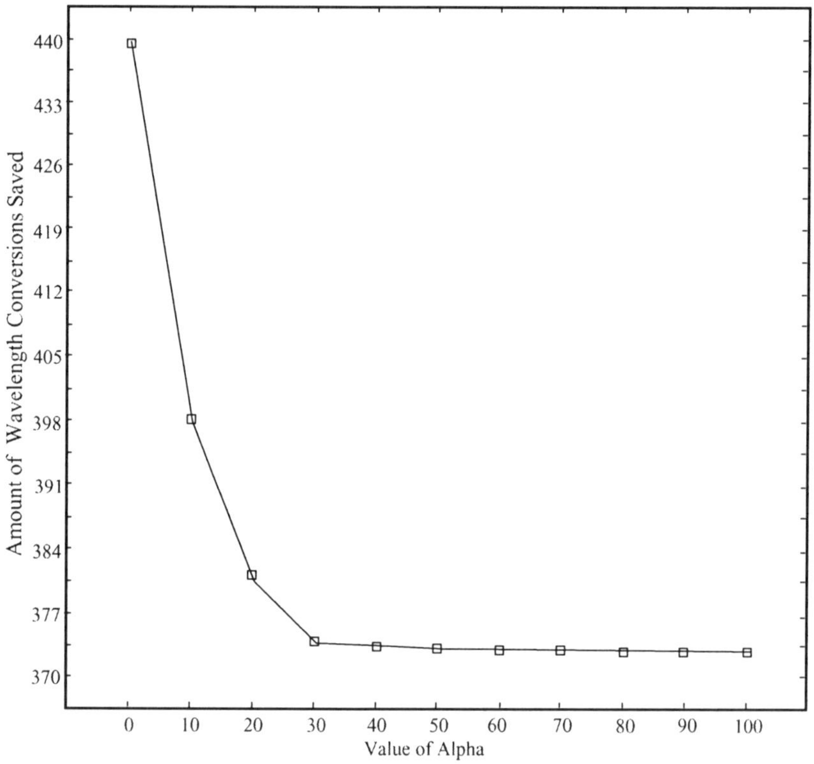

Figure 7.5. The amount of wavelength conversions saved against the value of α in NSFnet.

Fig. 7.5 shows the amount of wavelength conversions saved as the α value varies from 0 to 99. Due to the small network size, the number of wavelengths required in the system remains to be 5 regardless of the change of the α value. From Fig. 7.5, we can see that a significant saving of wavelength conversions can be achieved by using the proposed wavelength assignment method (i.e., the case when $\alpha = 0$). As the increase of the α value, Algorithm 7.6 emphasizes more on minimizing the number of wavelengths, with less consideration of wavelength conversions.

For the simulation on the NSFnet, a set P of 200 lightpaths is configured on $G(V, E)$. The end nodes of lightpaths are randomly selected from the network nodes and their routes are the shortest paths. The traffic matrix $\lambda(s, d)$ is randomly generated in a range between 0 - 1000. The edge-weighted auxiliary graphs $G_a(U_a, E_a)$

is constructed based on the configuration of lightpaths P. At each simulation point (i.e., for each α value), the simulation runs 100 times and the results presented below are the mean values. Each time, the simulation starts with the generation of a new physical graph $G(V, E)$. To make the mean values more sensible, the weights of edges in the auxiliary graph are scaled into the range between 0 and 99.

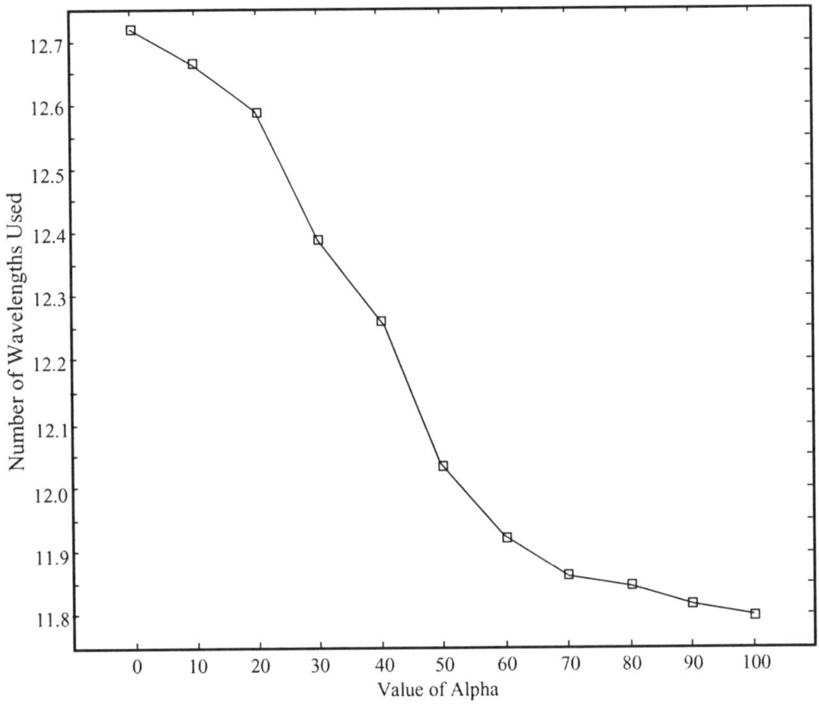

Figure 7.6. The number of wavelengths used against the value of α in general networks.

Fig. 7.6 shows the number of wavelengths required versus the α values. α varies in the range of the weight of edges, i.e., from 0 to 99. Although we can see the trend of using less wavelengths as the increase of α, the reduction of wavelengths is subtle. This suggests that minimizing wavelength conversions can be achieved at little cost of wavelengths in the system. Algorithm 7.5 for finding a clique also tries to maximize the clique size under the goal of maximizing the weight of the clique. Another reason for the insignificant reduction of wavelengths as the increase of α may be due to the approximations of Algorithm 7.3 and Algorithm 7.5 (for finding a clique-partition). Considering Algorithm 7.5, the change of α values will change the starting nodes in the search of maximum cliques in the graph. The selection of the starting node of a clique is supposed to be the dominant factor in determining the size of the clique. However, in a graph where vertex-degrees are relatively uniform, the size of a clique does not vary much no matter from which vertex it starts. It is anticipated that by using a better heuristic to compute the maximum clique, it might result in a significant reduction of wavelengths as the increase of α.

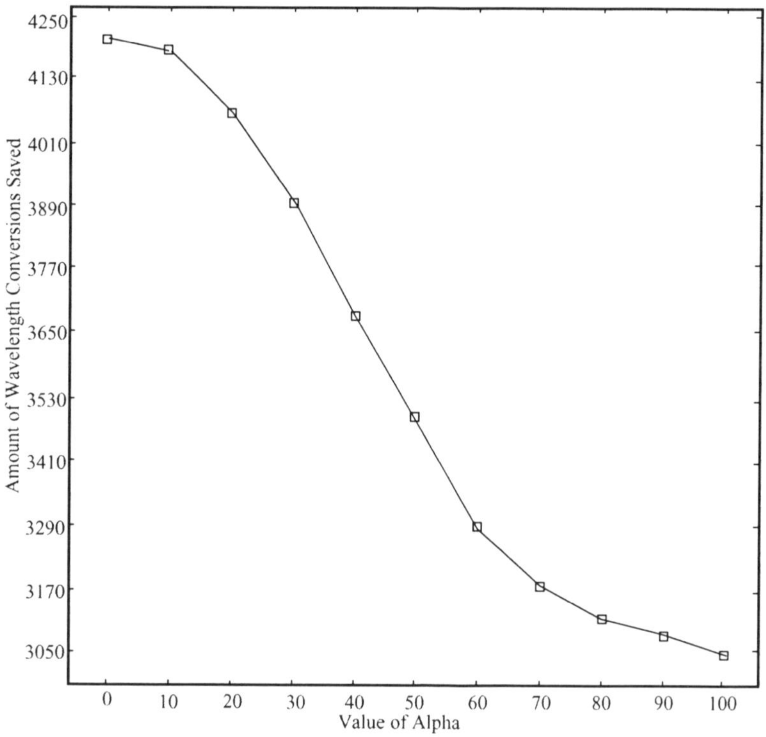

Figure 7.7. The amount of wavelength conversions saved against the value of α in general networks.

Fig. 7.7 shows the amount of wavelength conversions saved versus the α values. When $\alpha = 0$, Algorithm 7.6 tries to maximize the amount of wavelength conversions saved, because in this case the weight of the clique-partition represents the amount of wavelength conversions saved. When $\alpha = 99$, Algorithm 7.6 maximizes the size of each clique. This is exactly the case of traditional wavelength assignment aiming at minimization of the number of wavelengths used. The amount of wavelength conversions saved in this case can be regarded as the benchmark to measure the effectiveness of the proposed algorithm in the saving of wavelength conversions. From Fig. 7.7d, we can see that the saving of wavelength conversions decreases as the increase of the α value (which is consistent with Fig. 7.5). The difference between the values of saved conversions at $\alpha = 0$ and at $\alpha = 99$ is the maximal saving that is achieved by using the proposed wavelength assignment methods (i.e., Algorithms 7.4-6).

2. Wavelength Assignment for Broadcast Connections

In the preceding section we have studied the minimum wavelength conversion problem for unicast connections. In this section, we will study the same problem for broadcast connections. In WDM networks, broadcast connections are supported by light-trees (refer to the next two chapters for the detailed discussion on this topic).

In the following we will consider how to route and allocate wavelengths to a broadcast connection so that the total of wavelength conversions required is minimized. To tackle this problem, we will first introduce and study two related problems, the minimum wavelength-covering problem and the minimum vertex-wavelength-covering problem.

2.1 Optimization Formulation and Theory

We model a WDM networks as a connected graph $G(V, E, w)$. Here w is a mapping from edge-set E to a set W of wavelengths used in the networks. $w(e)$ represents the wavelengths available on edge e, and it is called *edge-wavelength-set* of edge e. Fig. 6 illustrates such a network. It consists of 11 vertices, and 5 wavelengths $\{w_1, w_2, w_3, w_4, w_5\}$ are used in the network. Note that not all of the wavelengths are available on an edge. For example, only two wavelengths w_2 and w_3 are available on edge between vertices v_4 and v_8, thus $w(v_4v_8) = \{w_2, w_3\}$. Similarly, $w(v_8v_9) = \{w_2\}$.

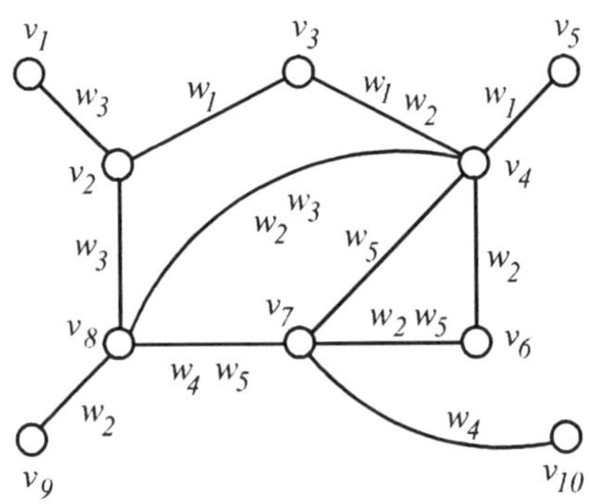

Figure 7.8. An example of WDM networks.

The minimum wavelength conversion problem for broadcast connections can be formulated as follows.

PROBLEM 7.2 *Minimal Wavelength Conversion Problem for Broadcast Connections*

Instance A graph $G(V, E, w)$.

Solution A spanning tree T of G with each edge in T assigned one of wavelengths available on the edge.

Objective Minimizing the number of vertices that are incident to the edges in T which are assigned with different wavelengths.

Fig. 7.8 shows an optimal solution to the Problem 7.2 in the example of Fig. 7.7. The tree contains two vertices v_4 and v_8 (in black) where wavelength conversions are required, because they are incident to four edges in the tree which are assigned with different wavelengths, respectively. As a contrast, vertices v_2 and v_7 are incident to two edges in the tree which are assigned the same wavelengths w_3 and w_4, respectively.

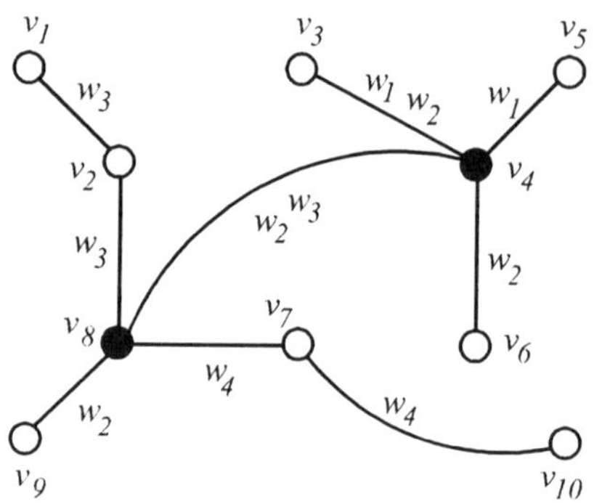

Figure 7.9. An optimal solution to the minimal wavelength conversion problem.

For a vertex $u \in V$, we denote by $w_v(u)$ the union of the edge wavelength sets of edges incident to u, and call it *vertex-wavelength-set* of vertex u. For example, in Fig. 7.8 vertex v_8 is incident to four edges, v_2v_8, v_9v_8, v_7v_8 and v_4v_8, thus $w_v(v_8) = \{w_3\} \cup \{w_2\} \cup \{w_4, w_5\} \cup \{w_2, w_3\} = \{w_2, w_3, w_4, w_5\}$.

A subset S of wavelength-set W is called a *wavelength-cover* if edges whose edge-wavelength-sets contain at least one wavelength in S form a connected spanning graph of V. In Fig. 7.8, $\{w_1, w_2, w_3, w_4\}$ is a wavelength-cover since edges with some wavelength(s) in $\{w_1, w_2, w_3, w_4\}$ form a spanning tree of network $G(V, E, w)$ (as shown in Fig. 7.10). In addition, it is easy to see that removing any wavelength from $\{w_1, w_2, w_3, w_4\}$ will make the resulting set not a wavelength-cover any more.

A subset S of vertex-set V is called a *vertex-wavelength-cover* if the union of vertex-wavelength-sets of vertices in S is a wavelength-cover. In Fig. 7.8, $\{v_4, v_8\}$ is a vertex-wavelength-cover since $w_v(v_4) \cup w_v(v_8) = \{w_1, w_2, w_3, w_5\} \cup$

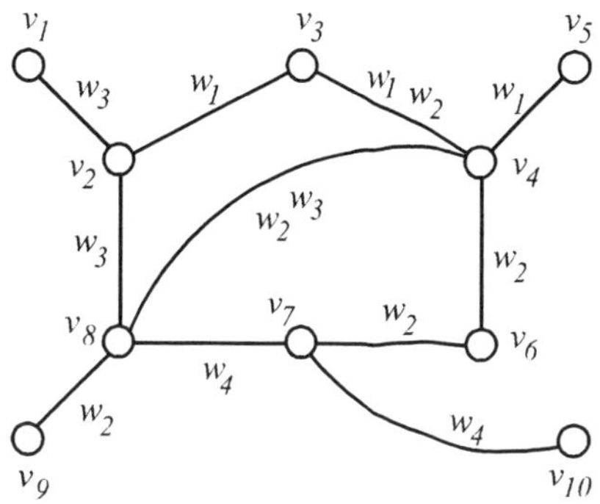

Figure 7.10. An example of wavelength-cover.

$\{w_2, w_3, w_4, w_5\} = \{w_1, w_2, w_3, w_4, w_5\}$. In addition, it is easy to see that neither $\{v_4\}$ nor $\{v_8\}$ is a vertex-wavelength-cover.

A subset S of vertex-set V is called *wavelength-connected* if connecting every pair of vertices in S that share at least one common wavelength produces a connected graph. Here the connected graph may not be a subgraph of graph G. For the example in Fig. 7.8, $\{v_1, v_2, v_4\}$ is wavelength-connected since $w_v(v_1) = \{w_3\}$, $w_v(v_2) = \{w_1, w_3\}$, $w_v(v_4) = \{w_1, w_2, w_3, w_5\}$ and each pair of the three vertices share at least one wavelength, connecting each of them produces a triangle, which is nota subgraph of given graph G. For the same example, $\{v_1, v_6\}$ is not wavelength-connected since $w_v(v_1) = \{w_3\}$, $w_v(v_6) = \{w_2, w_5\}$.

In order to see the difference between the concepts of vertex-wavelength-cover and wavelength-connectedness, we look at another simple example in Fig. 7.11. Observe that both sets $\{v_2, v_4, v_6\}$ and $\{v_1, v_3\}$ are vertex-wavelength-cover since both set have wavelengths w_1, w_2, w_3 and $\{w_1, w_2, w_3\}$ is clearly a wavelength-cover. However, set $\{v_2, v_4, v_6\}$ is not wavelength-connected since no pair of vertices in $\{v_2, v_4, v_6\}$ shares a common wavelength; Set $\{v_1, v_3\}$ is wavelength-connected since v_1 and v_3 share wavelength w_2. It is easy to see that the three components that consist of links carrying wavelengths w_1, w_2, and w_3, respectively, can not be joined together by vertices v_2, v_4, v_6, because they do not make a wavelength-connected set. On the other hand, these three components can be joined together by vertices v_1 and v_3 since they compose a wavelength-connected set. See Fig. 7.11(b). Note that there is no a spanning light-tree with wavelength conversions occurred at v_2, v_4, and v_6, but there exists a spanning light-tree with wavelength conversions occurred at vertices v_1 and v_3, respectively.

By the above definitions we can easily deduce the following two lemmas.

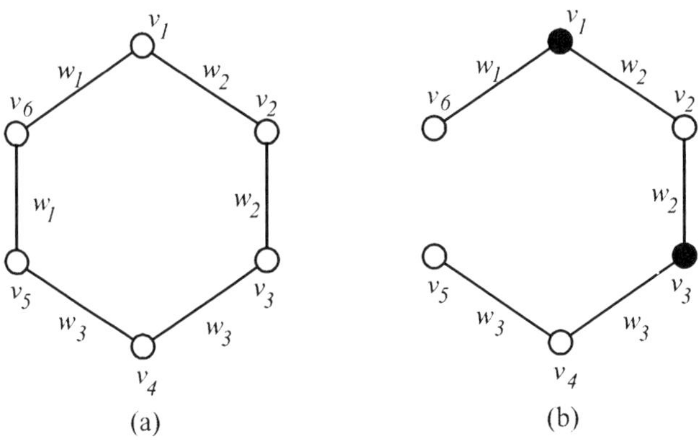

Figure 7.11. (a) An example network. (b) A spanning light-tree with wavelength conversions occurred at vertices v_1 and v_3.

LEMMA 7.3 (i) *Suppose that T is a solution to Problem 7.2. Let $W_T \subseteq W$ be the set of wavelengths that are assigned to the edges in T. Then W_T is a wavelength-cover of G.* (ii) *If subset $W' \subseteq W$ is a wavelength-cover of G, then Problem 7.2 has a solution T each of whose edges is assigned a wavelength in W'.*

LEMMA 7.4 *Suppose that T is a solution to Problem 7.2. Let $V_T \subseteq V$ be the set of vertices incident to edges in T that are assigned different wavelengths. Then V_T is a wavelength-connected vertex-wavelength-cover of G if $V_T \neq \emptyset$.*

In the Lemma 7.4, if $V_T = \emptyset$, then $\{v\}$ for any $v \in V$ is a wavelength-connected vertex-wavelength-cover of G. In this case, no wavelength conversion is required since there is one wavelength in W that is available on all edges in T.

LEMMA 7.5 *Suppose that $E(w) = \{e \in E \,|\, w \in w(e)\}$ constitutes a connected subgraph of $G(V, E, w)$ for any $w \in W$. If subset $V' \subseteq V$ is a wavelength-connected vertex-wavelength-cover of G with minimal number of vertices greater than one, then Problem 7.2 has a solution T such that the number of vertices incident to edges in T with different wavelengths is at least $|V'|$.*

Observe that if the condition in Lemma 7.5 is necessary. See Fig. 7.12(a) for a simple counterexample, where $E(w_1)$ is not a connected subgraph. It can be verified that $\{v_2, v_3\}$ is a wavelength-connected vertex-wavelength-cover and there does not exist a spanning tree such that the number of vertices incident to edges in the tree is at most two. However, we can easily modify the graph such that the condition is satisfied. Suppose that for wavelength w, all edges with w form $k > 1$ disjoint components $C_1, \cdots, C_k$. We can replace one wavelength w by k different wavelengths $w_1, \cdots, w_k$ in such a way that edges that make component C_i carry wavelength w_i for $1 \le i \le k$. See Fig. 7.12(b). Note that new wavelengths are not introduced physically in the network, they are only used for the simplicity of analysis and discussion.

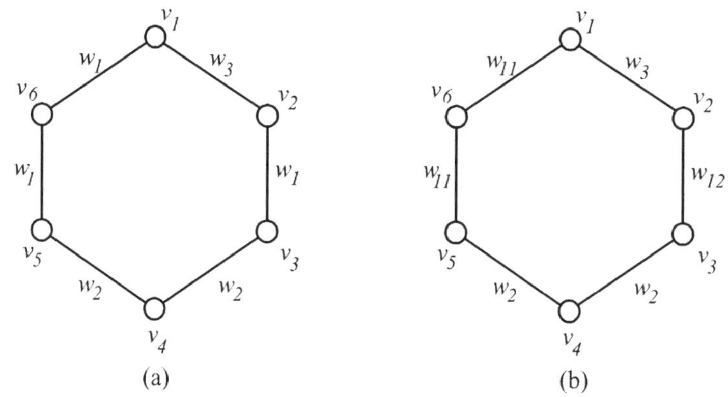

Figure 7.12. A counterexample: (a) Condition is not satisfied. (b) Condition is satisfied after modification.

Note that in the conclusion of Lemma 7.5 the number of vertices incident to the edges in T that are assigned different wavelengths may arbitrarily larger than $|V'|$. Fig. 7.13 gives such a simple example of a tree network. In Fig. 7.13(a) vertices u and v (in black) make a wavelength-connected vertex-wavelength-cover with minimal number of vertices. In Fig. 7.13(b) the number of vertices incident to the edges in the tree that are assigned different wavelengths is four.

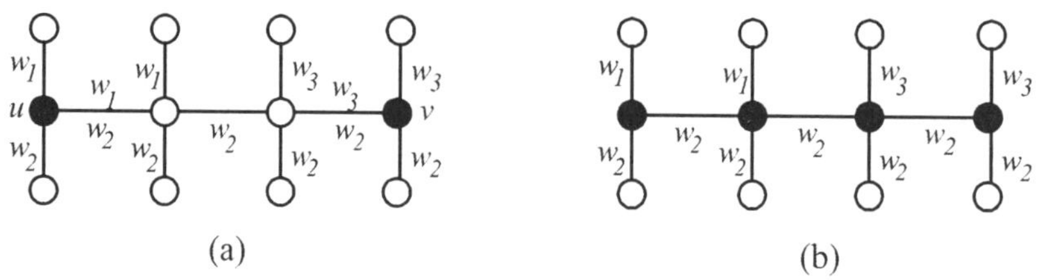

Figure 7.13. A counterexample: (a) The wavelength-connected vertex-wavelength-cover. (b) The solution to Problem 7.2.

By Lemmas 7.4-5 and the above analysis the solution to the following problem gives a lower bound on the solution to the minimum wavelength conversion problem (Problem 7.2).

PROBLEM 7.2' *Minimum Wavelength Conversion of Broadcast Connections*

Instance A connected graph $G(V, E, w)$ that for each wavelength $w \in W$, all edges that carry w form a connected subgraph of G.
Solution A vertex-wavelength-cover of G that is wavelength-connected.
Objective Minimizing the cardinality of the vertex-wavelength-cover.

In the following, we will focus on Problem 7.2'. First we study two closely related problems.

PROBLEM 7.3 *Minimum Wavelength-Covering Problem*

Instance A connected graph $G(V, E, w)$ that for each wavelength $w \in W$, all edges that carry w form a connected subgraph of G.
Solution A wavelength-cover of G.
Objective Minimizing the cardinality of wavelength-cover.

It can be verified that for the example in Fig. 7.8 $\{w_1, w_2, w_3, w_4\}$ is an optimal solution to Problem 7.3 (refer to Fig. 7.9).

PROBLEM 7.4 *Minimum Vertex-Wavelength-Covering Problem*

Instance A connected graph $G(V, E, w)$ that for each wavelength $w \in W$, all edges that carry w form a connected subgraph of G.
Solution A vertex-wavelength-cover of G.
Objective Minimizing the cardinality of the vertex-wavelength-cover.

It can be verified that for the example in Fig. 7.8 $\{v_4, v_8\}$ is an optimal solution to Problem 7.4 (see Fig. 7.9).

In the following subsections, we will study Problem 7.3 and Problem 7.4, respectively. We will show that both problems are $\mathcal{NP}$-hard and then we will derive both a lower bound and an upper bound of the approximation of these two problems. Using these results, we will give some theoretical results about the hardness of Problem 7.2' and propose a greedy algorithm to solve it in an efficient way.

2.2 Minimum Wavelength-Covering Problem

In this subsection, we study the minimum wavelength-covering problem (Problem 7.3). We first give a lower bound for the approximation of the problem and then propose a greedy algorithm that reaches nearly best possible approximation ratio. The following gives a negative result which implies that Problem 7.3 is unlikely to have a polynomial-time approximation with a constant performance ratio. The analysis is based on a result on the minimum set-covering problem obtained in [6] which improved an early result proved in [4]. Given a finite family $\mathcal{F}$ of subsets of a universal set U, a subfamily of $\mathcal{F}$ is called a *set-cover* if every element in U is in at least one subset of the subfamily. The *minimum set-covering problem* is to find the set-cover that has the minimal number of subsets. For the hardness of approximation of the minimum set-covering problem, we have the following negative result proved in [4].

LEMMA 7.6 *For any $0 < \rho < 1$, there is no polynomial-time approximation algorithm for the minimum set-covering problem with performance ratio $\rho \ln n$ unless $\mathcal{NP} \subset \mathcal{DTIME}(n^{poly \log n})$.*

In the above lemma $\mathcal{DTIME}(n^{poly \log n})$ denotes the class of combinatorial problems that can be solved deterministically in time $\mathcal{DTIME}(n^{poly \log n})$ where n is the input size of problems. It is believed (although not proved) that $\mathcal{NP} \subset \mathcal{DTIME}(n^{poly \log n})$ is not true. Thus Lemma 7.6 implies that it is impossible to find a good approximate solution to the minimum set-covering problem in polynomial-time. The following theorem shows that the same negative result is true for minimum wavelength-covering problem.

THEOREM 7.4 *Problem 7.3 has no polynomial-time approximation with performance ratio $\rho \ln |V|$ for $\rho < 1$ unless $\mathcal{NP} \subset \mathcal{DTIME}(|V|^{poly \log |V|})$.*

PROOF Suppose, by contradiction argument, that Problem 7.3 has a polynomial-time approximation with performance ratio $\rho \ln |V|$, where $\rho < 1$. Then based on this assumption, we will show that the minimum set-covering problem also has a polynomial-time approximation with the same performance ratio. This contradicts Lemma 7.6.

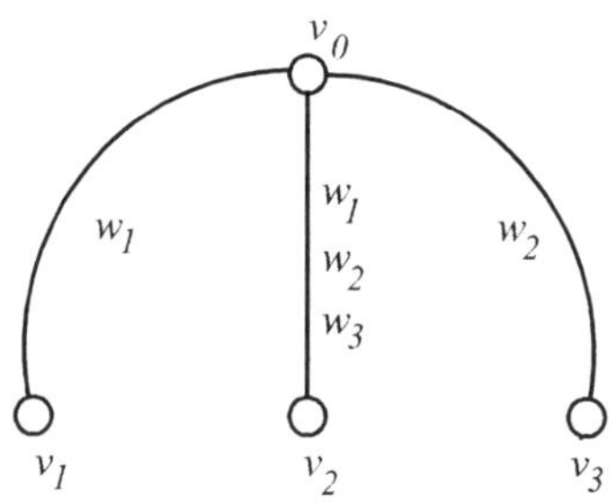

Figure 7.14. Reduction from the minimum set-covering problem to the minimum wavelength-covering problem.

For any family $\mathcal{F}$ of subsets $\{S_i\}$ of universal set $U = \{u_1, \cdots, u_n\}$, construct a graph $G(V, E, w)$ whose edges are assigned a set of wavelengths as follows: G has $n + 1$ vertices consisting of all elements in U and a special vertex v_0, i.e.,

$V = \{v_1 = u_1, \cdots, v_n = u_n\} \cup \{v_0\}$, and there is an edge between each element in U and special vertex v_0, i.e., $E = \{(v_0, v_1), \cdots, (v_0, v_n)\}$. For each $u \in S_i$, assign wavelength w_i to edge (v_0, u). For example, suppose $U = \{u_1, u_2, u_3\}$, $\mathcal{F} = \{S_1, S_2, S_3\}$, $S_1 = \{u_1, u_2\}$, $S_2 = \{u_2, u_3\}$ and $S_3 = \{u_2\}$. Then Fig. 7.14 shows the obtained graph $G(V, E, w)$. Edge (v_0, v_1) has wavelength w_1 since $u_1 \in S_1$, edge (v_0, v_2) has wavelengths w_1, w_2, and w_3 since u_2 belongs to S_1, S_2, and S_3, edge (v_0, v_3) has wavelength w_2 since $u_2 \in S_2$. Note that $\{S_1, S_2\}$ is a set-cover of U and $\{w_1, w_2\}$ is a wavelength-cover of the constructed G. In fact, it can be verified that $\{S_{i_1}, \cdots S_{i_k}\}$ is a set-cover of U if and only if $\{w_{i_1}, \cdots w_{i_k}\}$ is a wavelength-cover of G. This implies immediately that if $\{w_{i_1}, \cdots w_{i_k}\}$ is a wavelength-cover of G within a factor of $\rho \ln n$ from the optimal wavelength-cover, then $\{S_{i_1}, \cdots S_{i_k}\}$ is a set-cover of U within a factor of $\rho \ln n$ from the optimal set-cover. The proof is then finished. ■

We now propose an efficient greedy algorithm for the minimum wavelength-covering problem. The algorithm starts with an empty set W_c and repeatedly choose a wavelength to be put into W_c until it becomes a wavelength-cover. A wavelength is chosen as follows. Let H be a graph with vertex-set V. Initially, H has an empty edge-set. Let $C(H)$ be the number of components in H. Initially, $C(H) = |V|$. Let E_i denote the set of all edges with wavelength w_i, and let $C_r(E_i, H)$ be the number of components reduced by adding E_i to graph H. The greedy way is to choose the wavelength that reduces the most number of components in H. This means, the wavelength that maximizes $C_r(E_i, H)$ is chosen. After this wavelength is added into W_c, graph H is modified by putting E_i into it (thus the number of components in current H is less than that of precedent H). The algorithm terminates when $C(H)$ is reduced to 1, and it outputs W_c, which is a wavelength-cover since graph H is reduced to original graph G.

ALGORITHM 7.7 *Producing a Wavelength-Cover*

Input Graph $G(V, E, w)$.
Output Wavelength-cover W_c of G.

$W_c := \emptyset$,
Produce graph H with vertex-set $V(H) := V$ and edge-set $E(H) := \emptyset$,
$E_i := \{e \in E \mid w_i \in w(e)\}$.
while H is not connected **do**
  choose wavelength w_i that maximizes $C_r(E_i, H)$;
  $W_c := W_c \cup \{w_i\}$;
  $E(H) := E(H) \cup E_i$.
end-while
return W_c.

Fig. 7.15 illustrates the process of constructing a wavelength-cover by applying Algorithm 7.7 to the example in Fig. 7.8. Initially, $C(H) = 10$. We first choose w_2

in Fig. 7.15(a) since $C_r(E_2, H) = 5$ while $C_r(E_1, H)$, $C_r(E_3, H)$, $C_r(E_4, H)$, and $C_r(E_5, H)$ are all no greater than 3. In the same way, w_1, w_3, and w_4 were chosen one by one in Fig. 7.15(b-d).

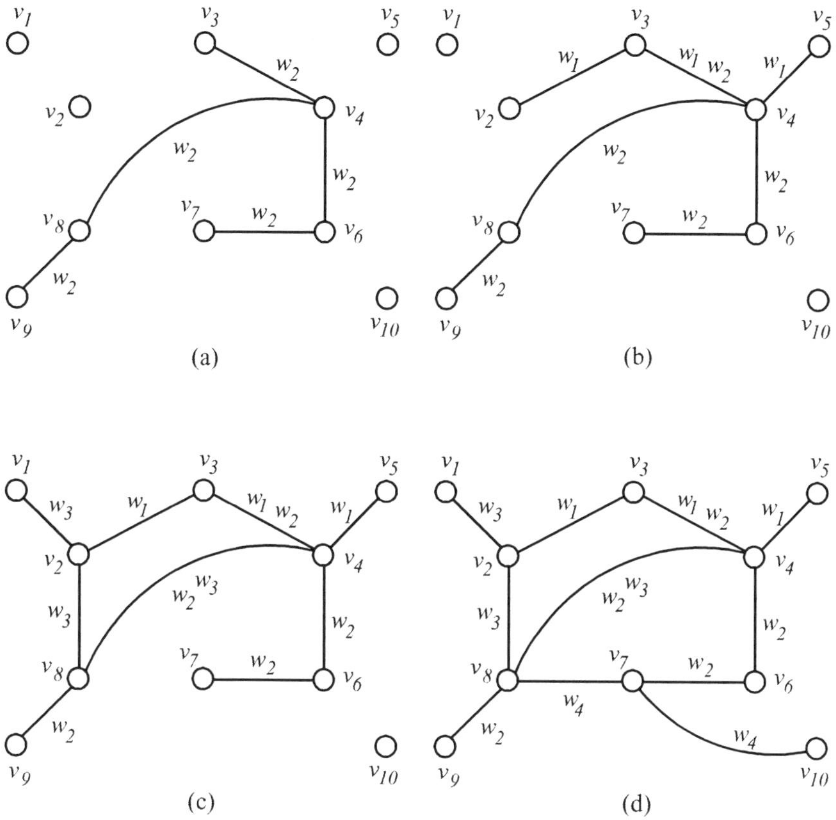

Figure 7.15. Producing a wavelength-cover.

THEOREM 7.5 *Given a graph $G(V, E, w)$, Algorithm 7.7 produces a wavelength-cover in time $O(|W||E| + |W|^2|V|)$, where W is the set of wavelengths used in the network.*

PROOF It is clear that Algorithm 7.7 outputs a wavelength-cover. Regarding its time-complexity, Algorithm 7.7 first takes time $O(|W||E|)$ to produce set E_i for $1 \leq i \leq |W|$. Then, the while-loop executes at most $|W|$ times. Each time we need to compute $C_r(E_i, H)$ for each wavelength $w_i \notin W_c$ and find the one which maximizes $C_r(E_i, H)$. This takes time $O(|W||V|)$. Thus in the total the while-loop takes time $O(|W|^2|V|)$. Therefore, the time-complexity of Algorithm 7.7 is $O(|W||E| + |W|^2|V|)$. ∎

The following theorem shows that Algorithm 7.7 is an approximation algorithm for Problem 7.3 with the best possible performance ratio.

THEOREM 7.6 *Algorithm 7.7 produces an approximation solution to Problem 7.3 within a factor of* $(\ln |V| + 1)$ *from the optimal solution.*

PROOF Suppose that W_c is the set (a wavelength-cover) output by Algorithm 7.7 and $|W_c| = k$. Let $w_1, \cdots, w_k$ be the wavelengths in W_c in the order of their appearance. Denote by H_i the graph with vertex-set V and edge-set $E_i \cup \cdots \cup E_i$ for $1 \leq i \leq k$ and H_0 is the graph H with empty edge-set. Let W_c^* be the set of an optimal wavelength-cover.

For each H_{i-1}, we have

$$\max_{w_j \in W_c^*} C_r(E_j, H_{i-1}) \geq \frac{C(H_{i-1}) - 1}{|W_c^*|}.$$

This is true since adding $\cup_{j \in W_c^*} E_j$ to H_{i-1} will produce a connected graph, thus reduce the number of components of H_{i-1} by $C(H_{i-1}) - 1$. Thus there must be a wavelength $w_j \in W_c^*$ such that when E_j is added into H_{i-1}, the number of components of H_{i-1} decreases at least $(C(H_{i-1}) - 1)/|W_c^*|$.

According to the rule of Algorithm 7.7, the chosen wavelength w_i maximizes $C_r(E_i, H)$, we have

$$C_r(E_i, H_{i-1}) \geq \max_{w_j \in W_c^*} C_r(E_j, H_{i-1}) \geq \frac{C(H_{i-1}) - 1}{|W_c^*|},$$

and

$$\begin{aligned} C(H_i) = C(H_{i-1}) - C_r(E_i, H) &\leq C(H_{i-1}) - \frac{C(H_{i-1}) - 1}{|W_c^*|} \\ &= \frac{C(H_{i-1})(|W_c^*| - 1) + 1}{|W_c^*|}. \end{aligned}$$

This implies

$$\begin{aligned} 1 \leq C(H_{|W_c|-1}) - 1 &\leq \left(\frac{|W_c^*| - 1}{|W_c^*|}\right)^{|W_c|-1} (C(H_0) - 1) \\ &= \left(\frac{|W_c^*| - 1}{|W_c^*|}\right)^{|W_c|-1} (|V| - 1). \end{aligned}$$

Therefore, we have

$$|W_c| \leq \log_{\frac{|W_c^*|-1}{|W_c^*|}} (|V| - 1) - 1.$$

Now dividing both sides of above inequality by $|W_c^*|$, we obtain

$$\begin{aligned}\frac{|W_c|}{|W_c^*|} &\leq \frac{1}{|W_c^*|}\log_{\frac{|W_c^*|-1}{|W_c^*|}}(|V|-1)+\frac{1}{|W_c^*|}\\ &\leq \frac{\ln|V|}{|W_c^*|\ln\frac{|W_c^*|}{|W_c^*|-1}}+1.\end{aligned}$$

Since $(i/(i-1))^i \geq e$, we get

$$\frac{|W_c|}{|W_c^*|} \leq \ln|V|+1.$$

This is the approximation performance ratio of Algorithm 7.7. The proof is then finished. ■

2.3 Minimum Vertex-Wavelength-Covering Problem

In this subsection we will study the minimum vertex-wavelength-covering problem (Problem 7.4). We will apply the same approach used in the preceding subsection for the minimum wavelength-covering problem (Problem 7.3). The following theorem shows that finding a good approximation solution to Problem 7.4 is as hard as finding a good approximation solution to Problem 7.3.

THEOREM 7.7 *Problem 7.4 has no polynomial-time approximation with performance ratio* $\rho \ln|V|$ *for* $\rho < 1$ *unless* $\mathcal{NP} \subset \mathcal{DTIME}(|V|^{poly\log|V|})$.

PROOF We prove the theorem again by reducing the minimum set-covering problem to Problem 7.4. For an input family $\mathcal{F}$ of subsets of a universal set $U = \{u_1, \cdots, u_n\}$, construct a graph $G(V, E, w)$ as follows: The vertex-set V consists of all elements in U, all subsets in $\mathcal{F}$, and a special vertex v_0. The edge-set E consists of (v_0, S) for all $S \in \mathcal{F}$ and (u, S) for all $u \in S \in \mathcal{F}$. Define a mapping w from E to wavelength set $W = \{w_1, \cdots, w_n, w_{n+1}\}$ by

$$\begin{aligned} w((v_0,S)) &= \{w_{n+1}\}, \quad \text{for all } S \in \mathcal{F},\\ w((u_i,S)) &= \{w_i\}, \quad \text{for all } u_i \in S \in \mathcal{F},\end{aligned}$$

To illustrate the reduction, we use the same instance of the minimum set-covering problem in the proof of Theorem 7.4. Recall that $U = \{u_1, u_2, u_3\}$, $\mathcal{F} = \{S_1, S_2, S_3\}$, $S_1 = \{u_1, u_2\}$, $S_2 = \{u_2, u_3\}$, and $S_3 = \{u_2\}$. The reduction produces a graph $G(V, E, w)$ as shown in Fig. 7.16. Note that $\{S_1, S_2\}$ is a vertex-wavelength-cover of $G(V, E, w)$ and a set-cover of U as well.

Let $V' \subset V$ be a vertex-wavelength-cover of $G(V, E, w)$. Without loss of generality, we may assume $V' \subseteq \mathcal{F}$ since otherwise we can easily find another vertex-wavelength-cover included in $\mathcal{F}$ with the same or smaller cardinality by replacing

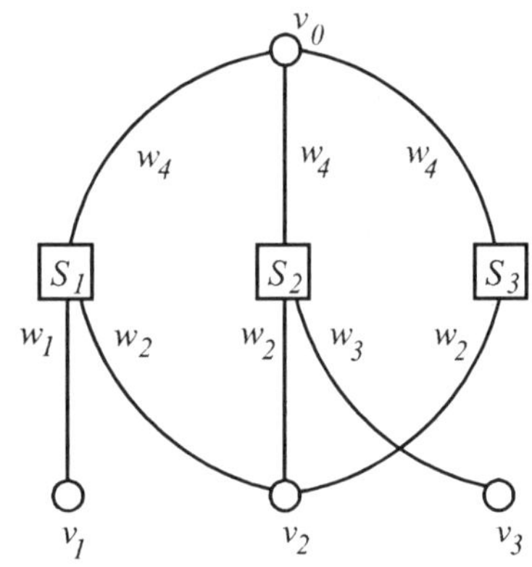

Figure 7.16. The reduction from the minimum set-covering problem to the minimum vertex-wavelength-covering problem.

each vertex not in $\mathcal{F}$ with one of its adjacent vertices in $\mathcal{F}$. For the example in Fig. 7.16, if v_0 is in V' then we can replace it by any one of S_1, S_2, and S_3. If u_3 is in V', we can replace it by S_2. We now show that V' is a vertex-wavelength-cover of $G(V, E, w)$ if and only if V' is a set-cover of U. First, suppose that V' is a vertex-wavelength-cover of G. Since $\{w_1, \cdots, w_n, w_{n+1}\}$ is the only wavelength-cover of G, the union of $w_v(u)$ for $u \in V'$ must be $\{w_1, \cdots, w_n, w_{n+1}\}$. Then V' must cover $u_1, \cdots, u_n$ because u_i is connected to some vertex in V' with an edge of wavelength w_i. Conversely, suppose that V' is a set-cover of U, then the union of $w_v(u)$ for $u \in V'$ must include $w_1, \cdots, w_n$, and w_{n+1}. Since these wavelengths form a wavelength-cover of G, V' is a vertex-wavelength-cover of G. Therefore, an optimal solution $V_{opt} \subseteq \mathcal{F}$ to the minimum set-covering problem is also an optimal solution to the minimum vertex-wavelength-covering problem.

Now note that $|V| = n + |\mathcal{F}| + 1$. Suppose that V' is a polynomial-time approximation of the optimal vertex-wavelength-cover with performance ratio $\epsilon \ln |V| = \epsilon \ln(n + |\mathcal{F}| + 1)$, where $\epsilon < 1$. When $|\mathcal{F}| \leq n$ and n is sufficiently large, we have

$$\epsilon \ln(n + |\mathcal{F}| + 1) \leq \epsilon' \ln n, \quad \text{for } \epsilon < \epsilon' < 1.$$

Therefore, V' is a polynomial-time approximation with performance ratio $\epsilon' \ln n$ for the minimum set-covering problem in the special case $|\mathcal{F}| \leq n$. From [6], we know that Lemma 7.6 still holds under the condition $|\mathcal{F}| \leq n$. Thus the minimum vertex-wavelength-cover has no polynomial-time approximation with performance ratio $\epsilon \ln |V|$ for $\epsilon < 1$ unless $\mathcal{NP} \subset \mathcal{DTIME}(|V|^{poly \log |V|})$. The proof is then finished. ■

The following theorem gives a positive result on the approximations of the minimum vertex-wavelength-covering problem.

THEOREM 7.8 *Problem 7.4 has a polynomial-time approximation with performance ratio* $(\ln|V|+1)$.

PROOF We prove the theorem by showing that an instance of Problem 7.4 can be transformed to an instance of the minimum wavelength-covering problem (Problem 7.3) in such a way that a solution to the latter problem is also a solution to the former problem.

Consider any instance $G(V,E,w)$ of Problem 7.4 with $V=\{v_1,\cdots,v_n\}$. We construct an instance of $G(V,E,w')$ of Problem 7.3. It has the same vertex-set V and edge-set E but a different mapping w' that is defined as follows:

$$w'(e) \equiv \{w_i \mid w \in w_v(v_i) \text{ for } w \in w(e)\}, \quad \text{for } e \in E.$$

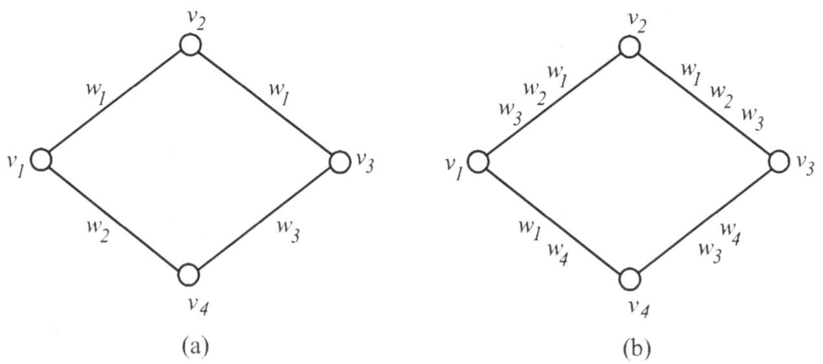

Figure 7.17. The transformation from (a) the minimum vertex-wavelength-covering problem to (b) the minimum wavelength-covering problem.

Fig. 7.17(a) shows an instance of Problem 7.4, and Fig. 7.17(b) shows the instance of Problem 7.3 constructed from the instance in Fig. 7.17(a). In Fig. 7.17(a), edge (v_1,v_2) has wavelength w_1, which is in the wavelength sets of v_1, v_2, and v_3. Thus in Fig. 7.17(b), edge (v_1,v_2) has wavelength set $w'((v_1,v_2))=\{w_1,w_2,w_3\}$. Similarly, edge (v_2,v_3) in Fig. 7.17(b) has the same wavelength set. In Fig. 7.17(a), edge (v_1,v_4) has wavelength w_2, which is in the wavelength sets of v_1 and v_4. Thus in Fig. 7.17(b), edge (v_1,v_4) has wavelength set $w'((v_1,v_4))=\{w_1,w_4\}$. Finally, edge (v_3,v_4) has wavelength w_3, which is in the wavelength sets of v_3 and v_4. Thus in Fig. 7.17(b), edge (v_3,v_4) has wavelength set $w'((v_3,v_4))=\{w_3,w_4\}$.

We now show that $\{v_{i_1},\cdots,v_{i_k}\}$ is a vertex-wavelength-cover of $G(V,E,w)$ if and only if $\{w_{i_1},\cdots,w_{i_k}\}$ is a wavelength-cover of $G(V,E,w')$. First, suppose that $\{v_{i_1},\cdots,v_{i_k}\}$ is a vertex-wavelength-cover of $G(V,E,w)$. Let U be the union of wavelength-sets of these vertices, then there is a subset of edges $E'\subset E$ such that E' contains a spanning tree of G and each edge in E' has a wavelength in U.

Now consider $G(V, E, w')$, according to the definition of w', each edge in E' has a wavelength $w_i \in \{w_{i_1}, \cdots, w_{i_k}\}$. Thus $\{w_{i_1}, \cdots, w_{i_k}\}$ is a wavelength-cover of $G(V, E, w')$. Conversely, suppose that $\{w_{i_1}, \cdots, w_{i_k}\}$ is a wavelength-cover of $G(V, E, w')$. Then there is a subset of edges $E' \subset E$ that contains a spanning tree of G and each edge in E' has a wavelength in the union of wavelength sets of vertices $\{v_{i_1}, \cdots, v_{i_k}\}$. Therefore, $\{v_{i_1}, \cdots, v_{i_k}\}$ is a vertex-wavelength-cover of $G(V, E, w)$.

By Theorem 7.6, the minimum wavelength-covering problem has a polynomial-time approximation with performance ratio $(\ln |V| + 1)$. Therefore, the minimum vertex-wavelength-covering problem also has a polynomial-time approximation with the same performance ratio. The proof is then finished. ■

2.4 Minimum Wavelength Conversion Problem

Using the results obtained in the preceding two subsections on the minimum wavelength-covering and minimum vertex-wavelength-covering problems, we can study the hardness of the minimum wavelength conversion problem (Problem 7.2') in this subsection and present an approximation algorithm to solve it. We first show the following theorem which gives a negative result on the approximation of Problem 7.2'.

THEOREM 7.9 *Problem 7.2' has no polynomial-time approximation algorithm with performance ratio* $\rho \ln |V|$ *for* $\rho < 1/2$ *unless* $\mathcal{NP} \subset \mathcal{DTIME}(|V|^{poly \log |V|})$.

PROOF Given $G(V, E, w)$, let set S be a solution to Problem 7.4. If S is wavelength-connected, then it is also a solution to Problem 7.2'; Otherwise, we can modify S to become wavelength-connected and thus a solution to Problem 7.2'. Given a subset $U \subseteq V$, we construct a graph $G(U)$ of vertex-set U in the following way. There is an edge between any two vertices $u_1, u_2 \in U$ if $w_v(u_1) \cap w_v(u_2) \neq \emptyset$. It can be verified that $G(U)$ is connected if and only if U is wavelength-connected. A subset $S' \subseteq S$ is called *wavelength-connected component* if S' is the vertex-set of a component in graph $G(S)$. Note that S has at most $|S|$ wavelength-connected components. Thus we can add a vertex $v \in V \setminus S$ to S such that it connects at least two wavelength-connected components and thus reduce the number of wavelength-connected components by at least one. This can be done since if we can not find such a vertex, then graph G is not connected, contradicting the assumption that it is connected.

As an example, let us consider the example in Fig. 7.6, $\{v_2, v_7\}$ is a vertex-wavelength-cover but is not wavelength-connected. It has two wavelength-connected components $\{v_2\}$ and $\{v_7\}$. We can add vertex v_8 to connect them since it shares wavelength w_3 with vertex v_2 and wavelengths w_2, w_4, and w_5 with v_7. Now $\{v_2, v_7, v_8\}$ is wavelength-connected and thus is a solution to Problem 7.2'.

At most $(|S| - 1)$ vertices are needed to connect all the wavelength-connected components in S. Therefore, there is a solution to Problem 7.2' with at most $2|S| - 1$ vertices.

Let A_{vwc} and A_{wc} be the sizes of the approximation solutions to Problem 7.4 and Problem 7.2', respectively. In addition, let O_{vwc} and O_{wc} be the sizes of the optimal solutions to Problem 7.4 and Problem 7.2', respectively. Following the above argument, we have $A_{wc} \leq 2A_{vwc}$ and $O_{wc} \leq O_{vwc}$. Moreover, it is clear that $O_{wc} \geq O_{vwc}$ and $A_{wc} \geq A_{vwc}$. Thus, $A_{wc}/O_{wc} \geq A_{vwc}/2O_{vwc}$. This implies $A_{vwc}/O_{vwc} \leq 2A_{wc}/O_{wc}$. If Problem 7.2' has an approximation with performance ratio $\rho \ln |V|$ where $\rho < 1/2$, then Problem 7.4 has an approximation with performance ratio $\rho \ln |V|$ where $\rho < 1$. The theorem then follows immediately from Theorem 7.7. The proof is then finished. ■

We now present a greedy approximation algorithm for Problem 7.2' with performance ratio $2(\ln |V| + 1)$. Given graph $G(V, E, w)$, it first finds a vertex-wavelength-cover of G. Let U be an initially empty set, we repeatedly choose a vertex to add into U until U becomes a vertex-wavelength-cover. In each step, we choose a vertex as follows: Let H be a graph of vertex-set V, initially the edge-set $E(H)$ is empty. Let $C(H)$ be the number of components of H. Then initially $C(H) = |V|$. Let E_i denote the set of all edges with wavelength w_i. Let $C_r(E_i, H)$ be the number of components reduced by adding E_i to graph H. When we choose a vertex u to add into U, we compute $\cup_{w_i \in w_v(u)} E_i$, which are all the edges in the vertex-wavelength-set of u. Adding these edges to graph H will reduce $C(H)$. At each step, the greedy strategy is to pick a vertex that reduces the most number of components in H, i.e., vertex u that maximizes $C_r(E_i, H)$. The goal is to reduce $C(H)$ to one. At this stage, graph H is connected, and therefore, U is a vertex-wavelength-cover. After a vertex-wavelength-cover U is found, we check if it is wavelength-connected. If it is, then U is a solution to Problem 7.2'; Otherwise, we modify U by putting vertices from $V \setminus U$ to it until U becomes wavelength-connected. The vertex we choose to put into U is the vertex that reduces the most number of wavelength-connected components in U.

Fig. 7.18 illustrates the above described process of constructing a wavelength-connected vertex-wavelength-cover of the example in Fig. 7.8. In Fig. 7.18(a) we choose vertex v_4 since it reduces the most number (i.e., nine) of components of graph H. In Fig. 7.18(b) we choose vertex v_7 for the same reason. It is clear that $\{v_4, v_7\}$ is a wavelength-connected vertex-wavelength-cover, thus no further operation is needed for this simple example.

ALGORITHM 7.8 *Constructing a Wavelength-Connected Vertex-Wavelength-Cover*

Input Graph $G(V, E, w)$

Output Wavelength-connected vertex-wavelength-cover U

Step 1 *Construct a vertex-wavelength-cover*

$U := \emptyset$,

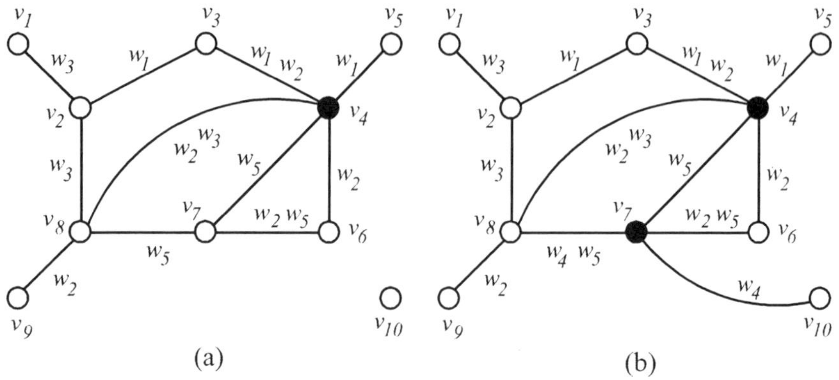

Figure 7.18. Producing a wavelength-connected vertex-wavelength-cover.

produce graph H with vertex-set $V(H) := V$ and edge-set $E(H) := \emptyset$,
$E_i := \{e \in E \mid w_i \in w(e)\}$.
while H is not connected **do**
choose a vertex u that maximizes $C_r(\cup_{w_i \in w_v(u)} E_i, H)$;
$U := U \cup \{u\}$;
$E(H) := E(H) \cup (\cup_{w_i \in w_v(u)} E_i)$.
end-while
Step 2 *Make U wavelength-connected*
while U is not wavelength-connected **do**
choose a vertex u such that $U \cup \{u\}$ has
the least number of wavelength-connected components,
$U := U \cup \{u\}$,
end-while
return U

THEOREM 7.10 *Given a graph $G(V, E, w)$, Algorithm 7.8 produces a wavelength-connected vertex-wavelength-cover in time $O(|W||V|^3)$, where W is the set of wavelengths used in the network.*

PROOF In Step 1, it takes time $O(|W||E|)$ to produce E_i. The while-loop executes at most $|V|$ times. In the loop for each $u \notin U$ it takes time $O(|W||V|)$ to compute $C_r(\cup_{w_i \in w_v(u)} E_i, H)$ and there are at most $|V|$ vertices that we can choose from. Thus the loop finishes in time $O(|W||V|^3)$. In Step 2, the while-loop also executes at most $|V|$ times. In the loop for each vertex $u \notin U$ we need to compute the number of wavelength-connected components in $U \cup \{u\}$ and choose the vertex that produces the least number of wavelength-connected components. The computation takes time $O(|W||V|)$ and there at at most $|V|$ vertices that we can choose from. Thus the loop also finishes in time $O(|W||V|^3)$. Furthermore, in the total the running time is at most $O(|W||V|^3)$ since $|E|$ is at most $O(|V|^2)$. ■

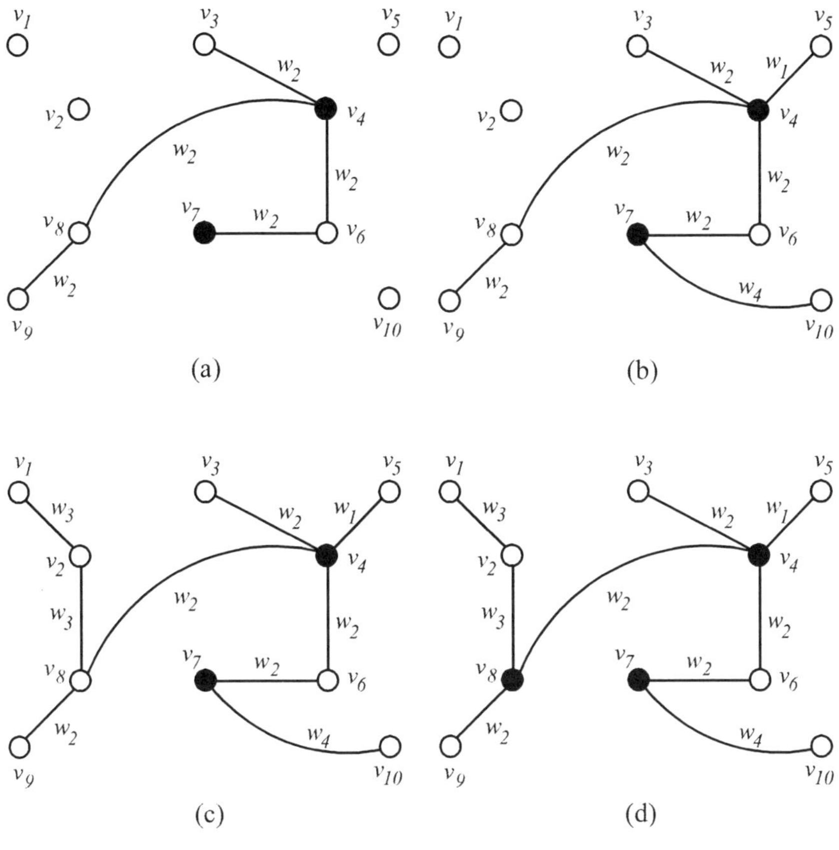

Figure 7.19. Producing a spanning tree of $G(V, E, w)$.

Now we turn back to the original version of the minimum wavelength conversion problem (Problem 7.2). After obtaining a wavelength-connected vertex-wavelength-cover U of $G(V, E, w)$, we need to construct a spanning tree from it and assign a wavelength to each of the edges in the tree. Again we can do in a greedy way as follows (The basic idea is very similar to Algorithm 7.8). First, find a wavelength that is in at least two vertex-wavelength-sets of the vertex-wavelength-cover and spans the most number of vertices (edges) in current G, and then remove these edges from current G and put them into tree T being under construction; Repeat this process until either there does not exist any wavelength which is in two vertex-wavelength-sets, or T is a spanning tree of G (in this case the process stops since the solution is produced). Secondly, find a wavelength that is in one of the vertex-wavelength-sets of the vertex-wavelength-cover and spans the most number of vertices (edges) in current G, and then remove these edges and put them into tree T being under construction; This process is repeated until T is a spanning tree of G (since U is a vertex-wavelength-cover).

Fig. 7.19 illustrates the above described process of constructing a solution to Problem 7.2 in the example of Fig. 7.8 (continued from Fig. 7.18). In Fig. 7.19(a) we choose wavelength w_2 since it is in both $w_v(v_4)$ and $w_v(v_7)$ and spans the most number (i.e., six) of vertices of G. In Fig. 7.19(b-c) for the same reason we choose wavelengths w_1, w_4 and w_3, respectively. Here we consider w_1 and w_4 first since there are edges in current G that is incident to v_4 and v_7, while w_3 does not satisfy this condition. The purpose of do so is to reduce the number of wavelength conversions that will occur, because we expect that the wavelength conversion is required at the vertices in the given vertex-wavelength-cover. In Fig. 7.19(d) a spanning tree is produced. However, a wavelength conversion is required at vertex v_8 which is not in the given vertex-wavelength-cover. Recall the optimal solution in Fig. 7.9 for the same example needs only two wavelength conversions.

ALGORITHM 7.9 *Constructing a Spanning Tree with Assigned Wavelengths*

Input A graph $G(V, E, w)$ and a vertex-wavelength-cover U
Output A spanning tree T of G (each edge in T is assigned a wavelength)

  $E_i := \{e \in E \,|\, w_i \in w(e)\}$.
  $T := \emptyset$,
  while T is not a spanning tree of G **do**
    while there are $u, v \in U$ and $w_i \in w_v(u) \cap w_v(v)$ **do**
      if $\exists\, u, v \in U$ and $w_i \in W$ with $w_i \in w_v(u) \cap w_v(v)$ such that
        at least one vertex in U is incident to an edge in E_i,
      then choose such w_i that maximizes the number of vertices
          incident to edges in E_i but not in T and U,
          find a spanning tree in E_i,
          put them into T.
      else choose $w_i \in w_v(u) \cap w_v(v)$ that maximizes the number
          of vertices incident to edges in E_i but not in T and U,
          find a spanning tree in E_i,
          put them into T.
    end-while
    if there is w_i not used such that
      at least one vertex in U is incident to an edge in E_i,
    then choose such w_i that maximizes the number of vertices
        incident to edges in E_i but not in T and U,
        find a spanning tree in E_i,
        put them into T.

else choose $w_i \in w_v(u) \cap w_v(v)$ that maximizes the number
of vertices incident to edges in E_i but not in T and U,
find a spanning tree in E_i,
put them into T.
end-while
return T

THEOREM 7.11 *Given a graph $G(V, E, w)$ and a wavelength-connected vertex-wavelength-cover of G, Algorithm 7.9 produces a spanning tree of G and assigns a wavelength to each of the edges in the tree in time $O(|W||V|^2)$.*

PROOF Note that there are two key operations in Algorithm 7.9. One is to find a wavelength w_i such that the number of vertices that are not in U and T and incident to edges in E_i is the maximal; The other is to produce a spanning tree in E_i. Since there at most $|W|$ wavelengths in $\{w_i \mid w_i \in w_v(u), u \in U\}$ and it takes at most time $O(|E_i|)$ to construct a spanning tree, the running-time of Algorithm 7.0 is $O(|W||V|^2)$. ■

2.5 Simulation Study

To evaluate the performance of the proposed greedy method (Algorithm 7.8) for the minimum wavelength conversion problem (Problem 7.2'), we use a brute force algorithm to compute the optimal solutions to Problem 7.2' and then run Algorithm 7.8 to produce the approximate solutions to Problem 7.2', and then we obtain the approximation performance ratio of Algorithm 7.8 in general case. Given a network $G(V, E, w)$, the brute force algorithm computes an optimal solution as follows: It generates subsets $V' \subseteq V$ with size varying from 1 to $|V|$. Whenever a subset V' is generated, it is checked if V' is a solution to Problem 7.2'. If it is a solution, the algorithm stops and outputs the set as the solution; Otherwise, the algorithm continues to generate the next subset and repeats this process. Clearly, this simple-minded method guarantees to produce an optimal solution. However, its running time may be exponential to $|V|$ and therefore is not useful in practice.

We simulate the greedy algorithm and the above brute force algorithm on 50 randomly generated graphs. The number of vertices ranges from 10 to 500 while the number of wavelengths ranges from 4 to 64. The results showed that out of the 50 tested instances, Algorithm 7.8 produced the optimal solutions in 26 of them. A summary of the test results is shown in Table 7.4. The first column is the number of vertices in the tested instances. The second instance is the number of simulated instances that the produced solutions were optimal. The third column is the number of simulated instances that the produced solutions were nonoptimum. The fourth column is the average performance ratio of the simulated instances.

Note that the average approximation performance ratio over these 50 simulated instances is 1.169, which is much better than the theoretical guaranteed approxi-

Table 7.4. Simulation results.

Number of vertices	Number of times optimum produced	Number of times nonoptimum produced	Average of performance ratio
10	2	3	1.2
15	4	1	1.050
20	6	1	1.071
25	1	6	1.297
30	3	5	1.191
35	3	3	1.222
40	2	3	1.133
50	1	1	1.375
100	1	1	1.167
200	1	0	1.000
300	1	0	1.000
500	1	0	1.000

mation performance (under the worst-case analysis) claimed in Theorem. In other words, the proposed greedy algorithm is both efficient and effective.

3. Discussion

In Section 1 of this chapter, we have discussed the wavelength assignment problem (Problem 7.1) aiming at minimization of wavelength conversions occurred in the network. This is different from the previously discussed works on the wavelength assignment problem, that aims at minimizing either the usage of wavelength resources (as in Chapter 3) or the system blocking probabilities (as in Chapter 6). The significance of this work [3] includes two aspects. First, the overall network speed can be greatly increased by reducing the number of wavelength conversions experienced by the connections during data transmission. Second, it takes less time for the connection setup and less overhead for the connection maintenance when there are less wavelength conversions throughout a connection.

In Section 2 of this chapter, we have studied the problem of establishing a broadcast connection with minimal number of wavelength conversions occurred at the routing nodes. The importance of this work [2, 7, 8] is as follows. If we can find a spanning tree such the same wavelength is available on every edge of the tree, then broadcasting data does not need any wavelength conversion, In such the most ideal case, the broadcast connection can be implemented through a light-tree [9] which is based on light-splitting technique. Otherwise, the broadcast connection has to experience a wavelength conversion at every routing node to which some edges incident does not have a common available wavelength.

References

[1] J. Håstad, Clique is hard to approximate within $n^{1-\epsilon}$, *Acta Mathematica*, 182 (1999), 105-142.

[2] D.-Y. Li, X.-F. Du, X.-D. Hu, L. Ruan, and X.-H. Jia, Minimizing number of wavelengths in multicast routing trees in WDM networks, *Networks*, 35 (4) (2000), 260-265.

[3] X.-H. Jia, D.-Z. Du, X.-D. Hu, H.-J. Huang, and D.-Y. Li, A new wavelength assignment method for minimal wavelength conversions in WDM networks, *Proceedings of the 9-th IEEE International Conference on Computer Communication and Networks* (ICCCN'2000),(2000), 621-624.

[4] C. Lund and M. Yannakakis, On the hardness of approximating minimization problems, *Journal of ACM*, 41 (5) (1994), 960-981.

[5] B. Ramamurthy and B. Mukherjee, Wavelength conversion in WDM networking, *IEEE Journal on Selected Areas in Communications*, 16 (7) (1998), 1061-1073.

[6] R. Raz and S. Safra, A sub-constant error-probability low-degree test, and a sub-constant error-probability PCP characterization of NP, *Proceedings of the 29-th Symposium Theory of Computing* (STOC), (1997), 475-484.

[7] L. Ruan, D.-Z. Du, X.-D. Hu, X.-H. Jia, D.-Y. Li, and Z. Sun, Converter placement supporting broadcast in WDM networks, *IEEE Transaction on Computers*, 50 (7) (2001), 750-758.

[8] L. Ruan, X.-F. Du, X.-D. Hu, X.-H. Jia, and D.-Y. Li, Approximations for color-covering problems, *AMS/IP Study in Advanced Mathematics*, American Mathematics Society, Providence, RI, USA, 20 (2001), 503-507.

[9] L. H. Sahasrabuddhe and B. Mukherjee, Light-trees: optical multicasting for improved performance in wavelength-routed networks, *IEEE Communications Magazine*, 37 (2) (1999), 67-73.

Chapter 8

QoS GUARANTEED MULTICAST

Multicast is a point to multipoint communication, by which a source node sends messages to multiple destination nodes. Multicast uses a tree structure as the route to transmit messages to multiple destinations. Unicast, a point-to-point communication, and a point-to-all communication (broadcast), are the two special cases of multicast. End-to-end delay is an important QoS (Quality of Service) parameter in data communications. QoS multicast requires that the delay of messages from the source to any destination should be within a bound. There are many network applications nowadays which need the support of QoS multicast, such as multi-media conferencing systems, video on demand systems, real-time control systems, and so on.

In WDM networks a light-tree concept was proposed in [12] to support multicast. A *light-tree*, as a point to multi-point extension of a lightpath, is a tree in the physical topology and occupies the same wavelength in all fiber links in the tree. Each fork node of the tree is a multicast-capable optical switch, where a power splitter is used to split an input optical signal into multiple signals which are then forwarded to output ports without electrical conversions. Hence, light-trees facilitate single-hop multicast.

There are two (not necessarily separate) steps for a multicast connection establishment in all optical WDM networks: routing and wavelength assignment. *Routing* is to construct a tree in the given network which is rooted from the source node and connects all the destination nodes; *Wavelength assignment* is to assign a wavelength to the generated routing tree in such a way that when two trees share a common link they must be assigned different wavelengths. In this chapter we will discuss how to establish QoS guaranteed multicast in WDM networks.

The rest of this chapter is organized as follows. Section 1 presents a well known algorithm proposed in [2, 8] for producing a routing tree with minimal cost. Section 2 presents a simple and elegant algorithm proposed in [7] for producing a routing tree with low cost and short delay. Section 3 and Section 4 discusses the establishment of QoS guaranteed multicast in WDM networks under the static model and dynamic model, respectively. Since multicast routing is substantially different from the unicast routing, some new approaches are introduced to integrate routing with wavelength assignment by using re-routing and wavelength re-assignment. Section 5 concludes the chapter.

1. Multicast Routing of Minimal Cost

There are basically two ways to implement routing to multiple destinations [13]. One is based on point-to-point connections (see Fig. 8.1(a)). Suppose that a piece of data needs to be transmitted from a source node to 100 destinations (recipients), 100 separate point-to-point connections (multiple lightpaths) from the source to each of 100 destinations are employed. As a result, 100 copies of the data may have to be sent over a single link, which consumes large amount of bandwidth. Moreover, this kind of connection model requires a large number of wavelengths, because two lightpaths that share a common link must be assigned two distinct wavelengths. Another one is based on *multicast*, which is a point-to-multipoint communication that enables a node to send or forward data to multiple destinations (see Fig. 8.1(b)). It makes better use of the available bandwidth by transmitting at most one copy of the data on each link. Moreover, as a light-tree [12] can be used, this requires one wavelength occupies the same wavelength in all fiber links in the light-tree. Hence, light-trees greatly facilitate single-hop point connection to multipoint communications.

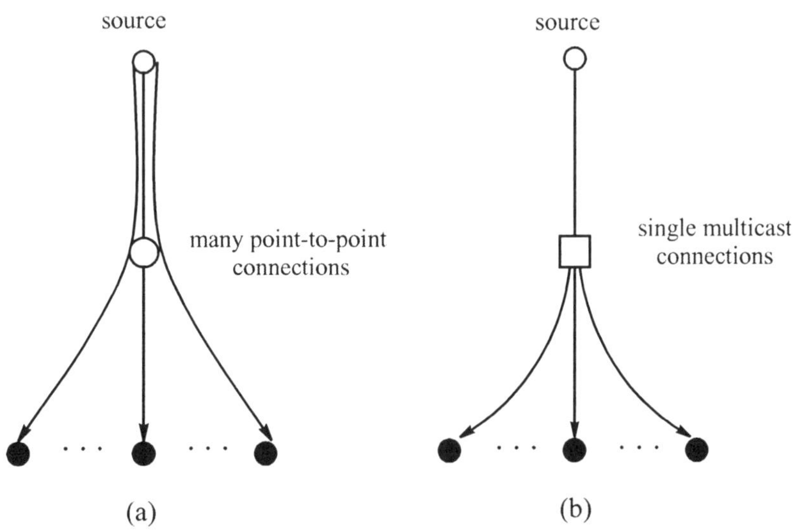

Figure 8.1. (a) Unicast versus (b) multicast.

We consider k bidirectional multicast connections in the system with *Quality of Service* (QoS) requirement, denoted by $\{ r(s, D, \Delta) \mid i = 1, 2, \cdots, k \}$. Each connection, $r(s, D, \Delta)$, is for setting up a QoS multicast connection from source node s to a group of destination nodes D ($s \notin D$) and the data transmission delay from s to any node in D should be within bound Δ. The actual value of Δ depends on the application. Multicast becomes *unicast* if $|D| = 1$, and *nonselective broadcast* if $|D| = |V| - 1$.

A *multicast routing* for $r(s, D, \Delta)$ is to find a tree T on G rooted from s and connecting all of the nodes in D, where an arbitrary subset of nodes in $V \setminus D$ can be included on T, but the leaf-set of T is a subset of D. We are to find the QoS routing trees (one for each of the QoS multicast connections) and assign wavelengths to them.

Let $T(s, D, \Delta)$ be the routing tree for connection $r(s, D, \Delta)$. When multicasting a message from source s to D along tree $T(s, D, \Delta)$, the light signal is split at source s and forwarded to the output ports leading to its children, which then transmit the signal to their children until all nodes in the tree (thus, all nodes in D) receive it. Based on the splitting mechanism, optical multicasting has some improved characteristics over electronic multicast, since "splitting light" is conceptually easier than copying a packet in an electronic buffer. We assume an optical signal can be split into an arbitrary number of optical signals at a switch. Thus, there is no restriction on node degree in a routing tree.

Since every node in a routing tree is multicast-capable, a message flows through each branch of the tree once and only once to reach all the destinations. Therefore, the cost of multicasting a message, called the cost of the tree, is the sum of the cost of all links in the tree. It can be formally defined as:

$$c(T(s, D, \Delta)) \equiv \sum_{e \in T(s,D,\Delta)} c(e). \tag{8.1}$$

In this section we focus on one of the important requirements for constructing a routing tree, low network cost. The *network cost* of a routing tree is defined as the sum of the costs of all links in the tree. The gain of network cost savings is especially significant in multimedia applications where communication occurs more frequently and messages are usually very large in size due to the transmission of audio or video files. Given a multicast connection, finding a routing tree in the given network with the minimum network cost can be reduced to the well-known *Minimum Steiner Tree Problem* (MSTP).

PROBLEM 8.1 *Minimum Steiner Tree Problem in Networks*

Instance An edge-weighted graph $G(V, E)$ and a subset $S \subset V$.
Solution A Steiner tree interconnecting all vertices in S.
Objective Minimizing the total weights of edges in the tree.

In the above, when $S = V$ the MSTP is reduced to the minimum spanning tree problem, that can be optimally solved in polynomial time by Kruskal algorithm [9]. When $|S| = 2$ the MSTP is reduced to the shortest path problem that can also be solved in polynomial time by Dijkstra algorithm [3].

In general, minimum Steiner tree problem is $\mathcal{NP}$-hard [6]. In the following we present a well-known approximation algorithm for MSTP that is based on minimum spanning tree [2, 8]. It works as follows. Firstly, find the shortest path in G between each pair of vertices in S and computes the distance between them. This can be done in polynomial time by Dijkstra algorithm [3]. Secondly, construct an auxiliary graph that is complete graph of S and the weight of the edge between two vertices is the distance between them in G. Thirdly, produce a minimum spanning tree of the auxiliary graph. This can also be finished in polynomial time by Kruskal algorithm [9]. Fourthly (and lastly), modify the obtained tree into a Steiner tree in original graph by substituting each edge in the tree with the corresponding shortest path in the original graph. This operation may cause cycles, so some redundant edges must be removed.

ALGORITHM 8.1 *Constructing a Steiner Tree.*

Input An edge-weighted graph $G(V, E)$ and a subset $S \subset V$.
Output A Steiner tree T interconnecting all vertices in S.
Step 1 *Construct an auxiliary graph*
 for all vertex pairs in S **do**
  find the shortest path $p(v_i, v_j)$ in G between vertices v_i and v_j;
  $d(v_i, v_j) :=$ the total weight of edges in $p(v_i, v_j)$.
  construct an edge-weighted complete graph $\overline{G}$ of S
  such that its edge (v_i, v_j) has weight $d(v_i, v_j)$.
 end-for
Step 2 *Produce a Steiner tree*
 produce a minimum spanning tree T of $\overline{G}$.
 modify T into a Steiner tree T_S in G as follows
 replace each edge in T by its corresponding path in G;
 remove redundant edges from T to keep it a tree structure in G.
 return T_S.

Fig. 8.2 illustrates the above algorithm applied to a simple instance. Fig. 8.2(a) shows a graph $G(V, E)$ of six vertices and S containing five of them (in black). The shortest path between vertices v_1 and v_4 (in dashed line) is via vertex v_0 and has distance 2. Fig. 8.2(b) shows the auxiliary graph and Fig. 8.2(c) gives a spanning tree of it. Continued from Fig. 8.2(c), Fig. 8.2(d) shows the resultant subgraph of G by replacing each edge in the spanning tree with its corresponding path. Observe that it is not a tree. Fig. 8.2(e) shows the obtained Steiner tree after removing edge (v_3, v_4). It has weight 6 while the minimum Steiner tree (in Fig. 8.2(f)) has weight 5.

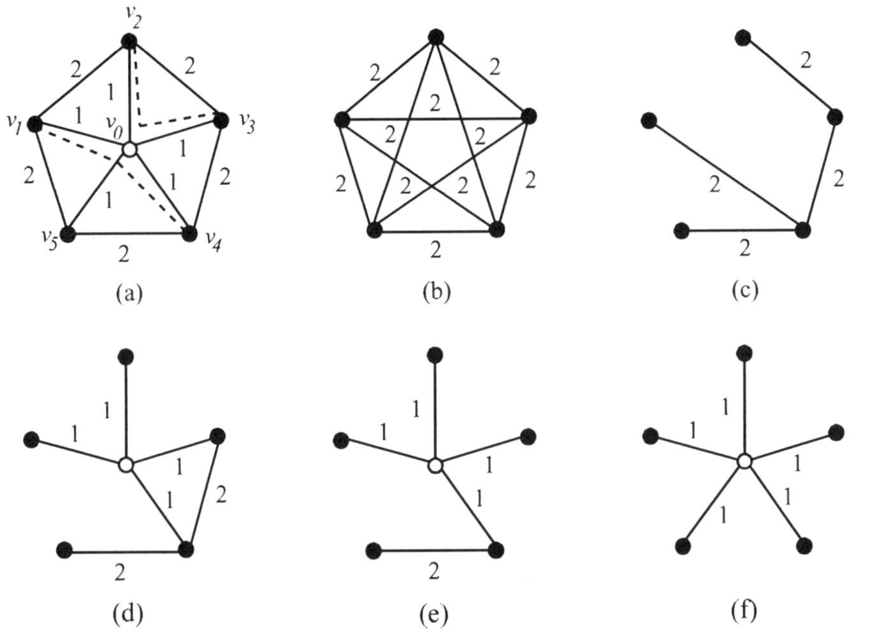

Figure 8.2. Algorithm 8.1: (a,b,c) Step 1 and (d,e,f) Step 2.

The following theorem [2] shows that Algorithm 8.1 has a guaranteed performance ratio less than 2.

THEOREM 8.1 *Given a graph $G(V, E)$ and any subset $S \subset V$, Algorithm 8.1 produces a Steiner tree of S whose weight is at most $2(|S|-1)/|S|$ times that of the minimum Steiner tree of S.*

PROOF Let T_{opt} be the minimum Steiner tree of S. Then we can obtain a (directed) closed circuit C of T_{opt} in G such that each edge is traversed exactly twice. See Fig. 8.3(a). Clearly, the weight of C is twice of that of T_{opt}. Now we denote by C' the simple cycle in $\overline{G}$ obtained from C by visiting each vertex of S in order that they appear in C. See Fig. 8.3(b). Since each edge in C' corresponds a shortest path between its endpoints in G, the weight of C' is not greater than that of C. Next we remove the most heavy edge in C' and produce a spanning tree T' in auxiliary graph $\overline{G}$. Since the most heavy edge has at least $1/|S|$ times that of the weight of C'.

Thus, we have

$$weight(T') \leq \left(1 - \frac{1}{|S|}\right) weight(C')$$

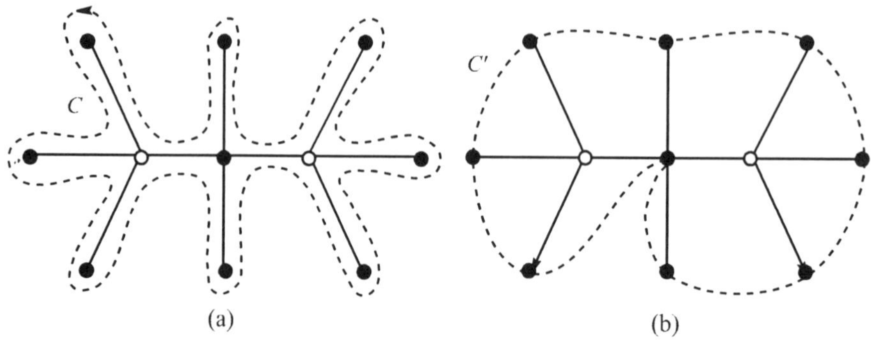

Figure 8.3. For the proof of Theorem 8.1.

$$\leq \frac{|S|-1}{|S|} weight(C) \leq \frac{2(|S|-1)}{|S|} weight(T_{opt})$$

Moreover, as T is a minimum spanning tree in $\overline{G}$, its weight is not greater than that of T'. Therefore, we have

$$weight(T_S) \leq weight(T) \leq weight(T').$$

The proof is then finished. ■

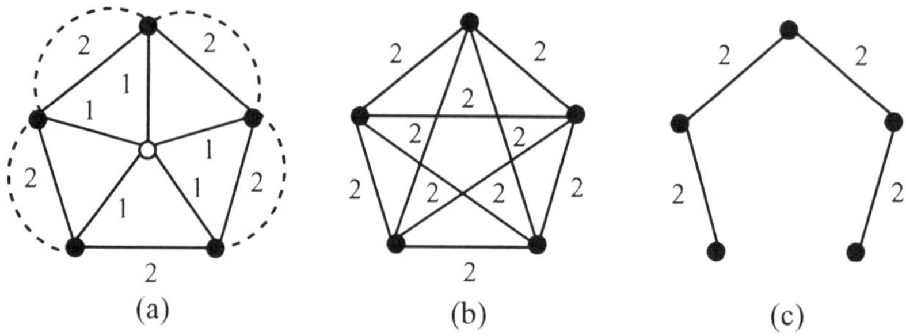

Figure 8.4. The approximation performance ratio is tight.

The ratio $2|S|/(|S|+1)$ is tight. Consider again the example shown in Fig. 8.2(a). When the shortest paths between vertices v_5 and v_1, v_1 and v_2, v_2 and v_3, v_3 and v_4 use the edges between these vertex pairs, respectively, as shown in the dashed lines in Fig. 8.4(a), the same auxiliary graph (Fig. 8.4(b)) is obtained. However, Algorithm 8.1 produces a minimum spanning tree (Fig. 8.4(c)) at Step 1 which is different from the one as shown in Fig. 8.2(c). Note that Algorithm 8.1 outputs the Steiner tree at Step 2 which has the same structure as Fig. 8.4(c), it has weight 8 while the minimum Steiner tree has weight 5. This example can be easily generalized to the case $|S| \geq 5$ such that the minimum Steiner tree has weight $|S|$ while the Steiner tree produced by Algorithm 8.1 has weight $2(|S|-1)$.

2. QoS Guaranteed Multicast Routing

In this section, we study another important requirement for constructing a routing tree, bounded delay transmission. The delay from the source to any destination along a routing tree shall not exceed a prespecified bound. Clearly, the *Shortest Path Tree* (SPT), which consists of the shortest paths from the source to every destination, has the shortest delay to any destination. However, it usually has high network cost. The requirement of minimizing network cost often conflicts with the bounded delay requirement in multicast routing.

In this section, the WDM network under consideration is modelled as weighted connected graph $G(V, E, c, d)$, where each edge $e \in E$ is associated with two weight functions $c(e)$ and $d(e)$, where $c(e)$ represents the communication cost of edge e and $d(e)$ the delay of edge e. We assume that both function c and function d are additive:

$$c(E') \equiv \sum_{e \in E'} c(e), \ \ \forall E' \subseteq E \ \text{ and } \ d(E') \equiv \sum_{e \in E'} d(e), \ \ \forall E' \subseteq E. \tag{8.2}$$

In order to guarantee that audio or video signals can be effectively used in interactive multimedia communications, the delay of transmitting a message from source node s to any destination in D is required to be upper bounded. As messages from source s to destination $u \in D$ are transmitted along the path in a routing tree T, so the delay from source s to destination u is the total delays of every link in the path. For any subgraph $G' \subseteq G$, let $P_{G'}(u, v)$ be the shortest path on G' from u to v, then the delay from s to u on T is defined as:

$$d(P_T(s,u)) \equiv \sum_{e \in P_T(s,u)} d(e). \tag{8.3}$$

The Δ-bounded delay requirement for routing tree T is as follows,

$$d(P_T(s,u)) \leq \Delta, \ \ \forall u \in D. \tag{8.4}$$

In our theoretical analysis and simulation study, the bounded delay requirement (8.4) is transformed into the following α-short delay requirement,

$$d(P_T(s,u)) \leq \alpha \cdot d(P_G(s,u)), \ \ \forall u \in D, \tag{8.5}$$

where $\alpha \equiv \min\{\frac{\Delta}{d(P_G(s,u))} \mid u \in D\}$. Thus from now on we do not distinguish these two delay requirements.

In this section we will study how to construct a routing tree that satisfies the delay requirement and has cost as minimal as possible. As we have appointed out that constructing a routing tree of minimal cost is $\mathcal{NP}$-hard (regardless of the delay requirement), we can only expect to find a routing tree with low (may not be minimal) cost; Or more formally, to construct a routing tree T satisfying β-approximation requirement:

$$c(T) \leq \beta \cdot c(T_{opt}), \quad \text{for any connection,} \tag{8.6}$$

where T_{opt} is the optimal routing tree and $\beta \geq 1$. Theorem 8.1 proves that Algorithm 8.1 produces a routing tree T_S satisfying requirement (8.6) with $\beta = 2$. As $c(T_{opt})$ can not be computed in polynomial-time, in practice the above requirement (8.6) is relaxed as the following β-low cost requirement:

$$c(T) \leq \beta \cdot c(T_S), \quad \text{for any connection,} \tag{8.7}$$

In the following a routing tree is called *feasible* if it satisfies both cost and delay requirements (8.7) and (8.5).

Notice that the network cost of a routing tree is additive over the whole tree under the metric of cost function c, while delay requirement is additive over individual path from the source node to each destination node on the tree under the metric of delay function d. Sometimes, a feasible routing tree may not exist for the following two reasons:

1) Values of β and α may be too stringent (close to one). It is $\mathcal{NP}$-hard to decide whether a given graph contains a routing tree that satisfies both requirements (8.5) and (8.7) when $(\beta - 1)(\alpha - 1) < 2$ (refer to [7]). Hence we assume that β and α satisfy the following cost-delay ratio-relation:

$$(\beta - 1)(\alpha - 1) \geq 2. \tag{8.8}$$

This relation means that we can ask for reducing the network cost, but only at the expense of increasing transmission delay, and likewise the other way round. Hence the tradeoff between cost and delay has to be made.

2) Cost function c and delay function d may be totally irrelevant. In such a case, a routing tree with low cost may have long delay, and a routing tree with short delay may have high cost. To guarantee that a feasible routing tree exits and can be constructed, we assume that functions c and d are correlated, that is, there exists a constant ρ such that

$$d(e) = \rho \cdot c(e), \ \forall \, e \in E. \tag{8.9}$$

Note that relation (8.9) means that delay function d is proportional to cost function c, which is the case in most data communication networks where the cost parameter is measured by the number of hops, because a path with less number of hops usually has shorter network delay. The multicast routing problem with QoS requirement can be formalized as follows.

PROBLEM 8.2 *QoS Guaranteed Multicast Routing Problem*

Instance A graph $G(V, E, c, d)$ and a multicast connection $r(s, D, \Delta)$.
Solution A Steiner tree T of G that interconnects all vertices in $D \cup \{s\}$ and satisfy the delay requirement (8.4).
Objective Minimizing the cost of T.

We now present the algorithm proposed in [7] for solving Problem 8.2. It first generates a low cost routing tree by solving the minimum Steiner tree problem (i.e., Problem 8.1) heuristic for the Steiner tree problem. It then modifies this tree into the one which meets the QoS requirement by checking delay requirement (8.5) for all destinations. If the delay requirement is not met for a destination, the shortest path (in given graph) in terms of delay function d from the source to the destination will replace the corresponding path (in the tree) linking the destination.

ALGORITHM 8.2 *Constructing a QoS Guaranteed Routing Tree*

Input A multicast connection $r(s, D, \Delta)$ and a graph $G(V, E, c, d)$
Output A QoS routing tree T_{QS} if $T_{QS} \neq \emptyset$ (otherwise $T_{QS} = \emptyset$)
Step 1 *Construct a low cost routing tree*
  produce a tree T_S by employing Algorithm 8.1 with cost function c.
Step 2 *Modify T_S into a QoS routing tree T_{QS}*
  while T_{QS} does not meet delay requirement (8.5) for some $u \in D$ **do**
    add $P_G(s, u)$ (under delay function d) into T_{QS},
    remove redundant edges in T_{QS} to keep it a tree structure.
    if T_{QS} still does not meet delay requirement (8.5) **then**
      return $T_{QS} = \emptyset$.
  end-while
  return T_{QS}.

In Step 2, we can apply the depth-first-search [7] on T_{QS} as follows. Traverse an edge-by-edge walk from source s (as the root) through the nodes of T_{QS}. Each edge is traversed twice, once in each direction. If node u in D is visited the first time and the delay requirement is not met, the shortest path from s to u on G in terms of function d is added into t_A. If the delay requirement still cannot be met for node u after the modification, then Algorithm 8.2 returns $T_{QS} = \emptyset$. In this case, there does not exist a routing tree that satisfies the delay requirement (8.5). To ensure that Algorithm 8.2 returns $T_{QS} = \emptyset$, Δ must be no less than $d(P_G(x, y))$ for $x, y \in V$.

Fig. 8.5 illustrates Algorithm 8.2 applied to the simple example [7] as shown in Fig. 8.5(a), where the cost and delay functions are equal. We consider the case of broadcast with source $s = v_1$ and $\alpha = 2$. A minimum spanning tree and a shortest-path tree are shown in Fig. 8.5(b) and Fig. 8.5(c), respectively. Fig. 8.5(d) shows the state of Algorithm 8.2 just after vertex v_5 is visited. Note that vertex v_5 is the first vertex that its shortest path to the root is added into the tree, since its current distance was more than twice the shortest distance in the graph. After this modification, its current distance is reduced from 40 to 15 and its parent is now changed from vertex v_4 to the root. More than that, the current distance from vertex v_4 to the root has been also reduced from 30 to 25 and its parent from vertex v_3 to vertex v_5. The numbers besides the vertices give the distance from the vertices to the root in current tree. Similarly, Fig. 8.5(e) shows the state after vertex v_7 has been visited that is the next vertex to have its shortest path added. Fig. 8.5(f) shows the final state of Algorithm 8.2. The produced tree consists of edges with arrows, which shows the paths in the tree from vertices towards the root. The following theorem shows that Algorithm 8.2 is efficient.

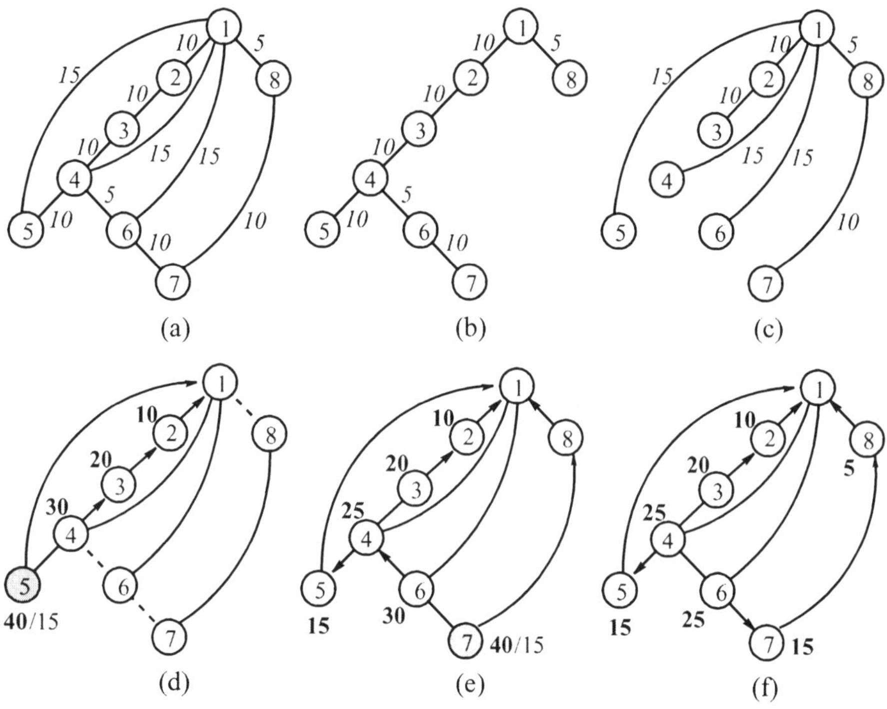

Figure 8.5. The illustration of Algorithm 8.2.

THEOREM 8.2 *Given a multicast connection $r(s, D, \Delta)$ in graph $G(V, E, c, d)$, Algorithm 8.2 returns a $Q \circ S$ routing in time $O(|D||V|^2)$.*

PROOF Since delay requirement (8.5) is enforced in the execution of Algorithm 8.2, a $Q \circ S$ routing tree can be found (if there exists one). In Step 1, it takes time

$O((|D|+1)|V|^2)$ to generate a shortest path in graph G between each pair of vertices in $\{s\}\cup D$, and time $O((|D|+1)^2)$ to construct a minimum spanning tree in auxiliary graph $\overline{G}$. In Step 2, the depth-first-search can be done in time $O(|V|)$. Therefore, the running time of Algorithm 8.2 is $O(|D||V|^2)$ since $|D| < |V|$. ■

In the above example, observe that the routing tree in Fig. 8.5(f) produced by Algorithm 8.2 has cost 70 while the minimum spanning tree in Fig. 8.5(b) has cost 60. The following theorem [7] shows that in general the delay requirement (8.5) could be satisfied by adding a comparatively small cost into T_S.

THEOREM 8.3 *Given a graph $G(V, E, c, d)$ whose delay function d is proportional to its cost function c, if Algorithm 8.2 produces a routing tree for multicast connection $r(s, D, \Delta)$, then for $\beta \geq 1 + 2/(\alpha - 1)$,*

$$c(T_{QS}) \leq \beta \cdot c(T_S), \text{ where } \alpha \equiv \min\left\{\frac{\Delta}{d(P_G(s,u))}\,\Big|\, u \in D\right\} > 1. \qquad (8.10)$$

PROOF The theorem can be proved by applying the same argument used in [7], which considers the cast of broadcast, to the complete graph G on $\{s\} \cup D$.

Let $v_1, v_2, \cdots, v_k$ be the vertices that caused the shortest paths to be added during the depth first search in the order that they were encountered, and let $v_0 = s$. When the shortest path $P_G(s, v_j)$ from s to v_j ($j \geq 1$) was added into the current tree T_{QS}, the total cost of the added edges is at most $c(P_G(s, v_j))$. In addition, the edges on the path to v_j consisting of the shortest path (in terms of function d) to v_{j-1} followed by the path in T_S from v_{j-1} to v_j are modified in order, so that

$$d(P_{T_{QS}}(s, v_j)) \leq d(P_G(s, v_{j-1})) + d(P_{T_S}(v_{j-1}, v_j)). \qquad (8.11)$$

The shortest path $P_G(s, v_j)$ from s to v_j in G (in terms of function d) was added because

$$d(P_{T_{QS}}(s, v_j)) > \Delta \geq \alpha \cdot d(P_G(s, v_j)). \qquad (8.12)$$

Combining inequalities (8.11) and (8.12), we get

$$\alpha \cdot d(P_G(s, v_j)) < d(P_G(s, v_{j-1})) + d(P_{T_S}(v_{j-1}, v_j)). \qquad (8.13)$$

To sum over j in inequality (8.13), we obtain

$$\alpha \sum_{j=1}^{k} d(P_G(s, v_j)) < \sum_{j=1}^{k} (d(P_G(s, v_{j-1})) + d(P_{T_S}(v_{j-1}, v_j)), \qquad (8.14)$$

which implies

$$(\alpha - 1) \sum_{j=1}^{k} d(P_G(s, v_j)) < \sum_{j=1}^{k} d(P_{T_S}(v_{j-1}, v_j)). \qquad (8.15)$$

Under the assumption that delay function d is proportional to cost function c, we deduce from inequality (8.15) that the total cost of the added paths is at most

$$(\alpha - 1)\sum_{j=1}^{k} c(P_G(s, v_j)) < \sum_{j=1}^{k} c(P_{T_S}(v_{j-1}, v_j)). \tag{8.16}$$

Notice that the depth first search traverses each edge exactly twice, and hence the sum on the right-hand side of inequality (8.16) is at most twice the cost of T_S. This yields, the total cost of the added paths is less than $2 \cdot c(T_S)/(\alpha - 1)$. Therefore, we have

$$c(T_{QS}) \leq \left(1 + \frac{2}{\alpha - 1}\right) c(T_S) \leq \beta \cdot c(T_S).$$

In addition, for $v \in D \backslash \{v_1, \cdots, v_k\}$, requirement (8.5) is met. For $v \in \{v_1, \cdots, v_k\}$, requirement (8.5) is met after the shortest path $P_G(s, v)$ from s to v is added into T_{QS}. Therefore, T_{QS} satisfies requirement (8.5). The proof is then finished. ■

Theorem 8.2 shows that for each destination node u, the delay from source s to u along T_{QS} is at most α times the delay of the shortest path from s and u in graph G, that is,

$$d(T_{QS}) \leq \alpha \cdot d(P_G(s, u)), \quad \forall\, u \in D, \tag{8.17}$$

and the cost of T_{QS} is at most β times the cost of T_S. The relationship between ratios α and β indicates that reducing the network cost is at the expense of increasing the delay, and likewise for reducing the delay. There is always a tradeoff between the cost and the delay of routing trees.

3. Static Multicast Routing and Wavelength Assignment

In this section, we will study the problem of multicast routing and wavelength assignment under static model. This problem can be formulated as follows.

PROBLEM 8.3 *Static QoS Routing and Wavelength Assignment Problem*

Instance A graph $G(V, E, c, d)$ and a set of channels $\{r_i(s_i, D_i, \Delta) \,|\, i = 1, \cdots, k\}$.
Solution A feasible routing tree and a wavelength assigned to each r_i such that two trees must be assigned different wavelengths if they share a common link.
Objective Minimizing the number of wavelengths used.

This is a challenging problem since it involves not only QoS multicast routing (as we have studied in the preceding section), but also wavelength assignment. We will present four algorithms for rerouting and wavelength (re)assignment, which can be integrated into one effective method for solving Problem 8.3.

3.1 Algorithm Design and Analysis

In this subsection, we will first present an algorithm for wavelength assignment for multicast connections. And the we present two rerouting methods that aim at reducing the number of wavelengths used by the algorithm. The first method reroutes some of routing trees to reduce the maximal link load by avoiding use of the links whose load is the maximum, because the number of wavelengths required is greater than or equal to the maximal link load. The second method reroutes routing trees whose wavelengths are the least used in order to reduce the number of wavelengths in wavelength assignment, that is to free out the least used wavelengths.

To assign wavelengths to a set of routing trees without causing wavelength conflict, we introduce an auxiliary graph G_a, where each vertex in G_a represents a routing tree and there is an edge between two vertices in G_a if and only if the two routing trees share a common link in G. Assigning wavelengths to the trees is reduced to the problem of coloring all vertices in G_a such that no two adjacent vertices receive the same color (because two adjacent vertices in G_a indicate that their corresponding routing trees share a common link in G). This is the exactly same approach used for solving Problem 3.1 (refer to Section 1 of Chapter 3).

ALGORITHM 8.3 *Assigning Wavelengths to Routing Trees*

Input A set of routing trees $\{ T(r_i) \mid i = 1, 2, \cdots, k \}$ in graph $G(V, E)$
Output A set W of wavelengths assigned to $\{ T(r_i) \mid i = 1, 2, \cdots, k \}$
Step 1 *Create an auxiliary graph* $G_a(V_a, E_a)$

$V_a := \{ T(r_i) \mid i = 1, 2, \cdots, k \}$,
$E_a := \{ (T(r_i), T(r_j)) \mid T(r_i) \cap T(r_j) \neq \emptyset, i \neq j \}$.

Step 2 *Assign wavelengths to vertices in* V_a *sequentially*

$V' := V_a$.
$W := \emptyset$.
while $V' \neq \emptyset$ **do**
choose $T(r_j) \in V'$ which has the least degree,
$V_j := \{ T(r_i) \in V' \mid (T(r_j), T(r_i)) \notin E_a \}$,
find a maximal set $V_j^{max} \subseteq V_j$ such that
no edge between any pair of vertices in V_j^{max},
assign all vertices in V_j^{max} a wavelength w,
$V' := V' \setminus V_j^{max}$.
$W := W \cup \{w\}$.
end-while
return $\{ T(r_i) \mid i = 1, 2, \cdots, k \}$ with W.

THEOREM 8.4 *Given a set of routing trees* $\{ T(r_i) \mid i = 1, 2, \cdots, k \}$ *in graph* $G(V, E)$*, Algorithm 8.3 assigns wavelengths to them in time* $O(k^2 |V|^2)$.

PROOF At Step 1, $G_a(V_a, E_a)$ can be produced in time $O(k^2 |V|^2)$, because each $T(r_i)$ has $(|V| - 1)$ edges for $i = 1, \cdots, k$ and there are at most k^2 edges in $G_a(V_a, E_a)$. At Step 2, every given routing tree is assigned a wavelength, and two trees are assigned the same wavelength if and only if they do not share any edge. In addition, choosing a vertex of the least degree and finding a maximal set both can be done by checking adjacency among vertices in V_a. Thus the wavelength assignment can be finished in $O(k^2|V|^2)$. ■

3.1.1 Optimization through Load Balancing

In the following we study how to reroute a given set of routing trees so that they requires less number of wavelengths. The basic idea is as follows. First, calculate the load on each link. Second, choose a tree which contains the links having the maximum load. Third, reroute the tree by employing Algorithm 8.2 on the subgraph of $G(V, E, c, d)$ obtained by removing the links having the maximum load. This rerouting process is repeated until the maximal link load cannot be reduced any further.

ALGORITHM 8.4 *Balancing Network Load*

Input A set of routing trees $\{ t(r_i) \mid i = 1, 2, \cdots, k \}$
Output A set of re-routed trees $\{ t(r_i) \mid i = 1, 2, \cdots, k \}$ with balanced load
Step 1 *Calculate the load on each link*
    $L(e) := \{ Tt(r_i) \mid e \in T(r_i) \}$, for $e \in E$,
    $l_{max} := \max\{ |L(e)| \mid e \in E \}$.
Step 2 *Reroute the trees to balance the load*
    $E_{max} := \{ e \in E \mid l(e) = l_{max} \}$.
    while $E_{max} \neq \emptyset$ **do**
        choose an edge $e_{max} \in E_{max}$.
        $R := \{ r_i \mid e_{max} \in T(r_i) \}$.
        // rerouting any tree in R successfully will reduce load on e_{max} by 1
        while $R \neq \emptyset$ **do begin**
            choose an $r_i \in R$,
            $j := l_{max} - 1$.
            while $j \geq 1$ **do** reroute r_i on the most lightly loaded links
                $E_j := E \setminus \{ e \mid |L(e)| \geq j \}$,
                // avoid using links whose load $\geq j$
                run Algorithm 8.2 for r_i on $G(V, E_j, c, d)$.
                if $T_{QS} \neq \emptyset$ **then** // rerouting r_i on E_j succeeds
                    $T(r_i) := T_{QS}$,
                    $j := j - 1$. // try to reroute r_i on E_{j-1}
                else exit **while**-j. // cannot reroute r_i on lighter links
            end-while-j
            if $j = l_{max} - 1$ **then** // rerouting r_i fails
                $R := R \setminus \{r_i\}$, // try to reroute another tree in R
            else $E_{max} := E_{max} \setminus \{e_{max}\}$, // load on e_{max} is reduced by 1
            exit **while**-R. // work on another edge in E_{max}
        end-while-R
        if $R = \emptyset$ **then** // none of trees in R can be rerouted

stop. // the maximal link load cannot be reduced any further
end-while-E_{max}
if $E_{max} = \emptyset$ **then** // the maximal link load was reduced by 1
go to Step 1. // reduce the maximal link load further

At Step 2, in the while-loop of j we search for the smallest index j such that the tree of $r_i \in R$ can be rerouted on the most lightly loaded edges. This can be considered as a greedy strategy for the multicast version of the load balancing problem (Problem 2.5). The following theorem shows the correctness and efficiency of Algorithm 8.5.

THEOREM 8.5 *Given a set of routing trees for multicast channels* $\{r_i(s_i, D_i, \Delta) | i = 1, \cdots, k\}$ *in graph* $G(V, E, c, d)$, *Algorithm 8.4 finishes rerouting them in time* $O((K|V|^3 L_{max}^2 + k|E|)L_{max})$, *where* $K = \max\{\,|D_i|\,|\,i = 1, 2, \cdots, k\,\}$ *and* $l_{max} := \max\{\,|L(e)|\,|\,e \in E\,\}$.

PROOF Step 2 is to find out all links having the maximal link load, and then for each such link, try to reroute one tree occupying it. It has three layers of loops. The outside loop is on each link having the maximal load. The second layer loop tries to successfully reroute one tree occupying the concerned link. The inner loop reroutes the tree to the most lightly loaded links. In this loop, we set a load threshold (i.e., value of j) for links to be removed from graph G before rerouting the chosen tree, say r_i. The initial value of j is set to (l_{max} - 1), so that links whose load is greater than or equal to (l_{max} - 1) will be removed. This is to ensure that after rerouting r_i, the maximal link load in the network will still be less than l_{max} (thus the maximal load is reduced). Inside this loop, each time the load threshold (i.e., j's value) is decremented by 1. That is, r_i will be rerouted to more lightly loaded links upon each interaction of the loop. When Algorithm 8.2 returns $T_{QS} = \emptyset$, r_i cannot be rerouted at the level with this value of j. Now, we consider the following two cases.

1) $j = l_{max} - 1$. In this case, it is impossible to reroute r_i without using links with load l_{max} or $(l_{max} - 1)$. It exits the while-loop of j and tries to reroute another tree in R, because successfully rerouting any tree in R would reduce the load on e_{max} by 1. When R becomes empty, that means all trees in R have been tried and none of them can be rerouted. In this case, the load on e_{max} cannot be reduced. The algorithm stops.

2) $j < l_{max} - 1$. In this case, r_i is re-routed successfully on $G(V, E_{j+1}, c, d)$ whose link loads are less than or equal to j. This means that the load on link e_{max} can be reduced by 1 (the one was occupied by r_i). The next link in E_{max} will be chosen for load reduction. The maximal link load in the system, l_{max}, is reduced by 1 when E_{max} becomes empty.

At Step 1, it takes time $O(k|E|)$ to calculate the load on each link. At Step 2, the while-loop of E_{max} would run at most $(|V| - 1)$ times, because each routing tree has at most $(|V| - 1)$ edges. The while-loop of R would run at most L_{max} times (for each tree in R). Each while-loop of j would be executed at most $(L_{max} - 1)$ times (for $j = L_{max} - 1, \cdots, 2, 1$) and each time Algorithm 8.2 (whose time complexity is $O(|D_i||V|^2)$) will be called. Note that once the maximal load is reduced by 1, Algorithm 8.4 will go back to Step 1. This process can be repeated at most $(L_{max} - 1)$ times, because the maximal load can be decreased (at most) from L_{max} to 1. Thus, Algorithm 8.4 can terminate in time $O((K|V|^3 L_{max}^2 + k|E|)L_{max})$. The proof is then finished. ■

3.1.2 Optimization through Wavelength Reassignment

In the following we present a different way to reroute a give set of routing trees so that the number of wavelengths used can be reduced by reassigning some of them. The basic idea is as follows. First, calculate the set of routing trees which a wavelength is assigned to. Second, reroute the routing trees which are assigned with the least used wavelength, so that they can be assigned with other wavelengths. The rerouting process is repeated until the number of wavelengths used cannot be reduced any further.

ALGORITHM 8.5 *Reassigning Wavelengths to Routing Trees*

Input A set W of wavelengths assigned to routing trees $\{\, T(r_i) \mid i = 1, 2, \cdots, k \,\}$
Output Rerouted trees $\{\, T(r_i) \mid i = 1, 2, \cdots, k \,\}$ with reassigned wavelengths
Step 1 *Calculate the distribution of currently used wavelengths*
  $\mathcal{T}(w) := \{\, T(r_i) \mid T(r_i)$ is assigned wavelength $w \,\}$, for $w \in W$,
  sort $|\mathcal{T}(w)|$ in ascending order, i.e., $|\mathcal{T}(w_1)| \le |\mathcal{T}(w_2)| \le \cdots \le |\mathcal{T}(w_{|W|})|$.
Step 2 *Reroute the trees that are assigned the least used wavelength*
  while $\mathcal{T}(w_1) \ne \emptyset$ **do** // reroute all trees in $\mathcal{T}(w_1)$ to save w_1
    choose $T(r_i) \in \mathcal{T}(w_1)$,
    $j := |W|$.
    while $j > 1$ **do begin** // reroute $T(r_i)$ to use wavelength w_j
      $E_j := E \setminus \{\mathcal{T}(w_j)\}$, // remove $\mathcal{T}(w_j)$ from network graph G
      run Algorithm 8.2 for r_i on $G(V, E_j, c, d)$.
      if $T_{QS} \ne \emptyset$ **then** // rerouting r_i succeeds and r_i can use w_j
       $T(r_i) := T_{QS}$,
       assign r_i wavelength w_j, // because $T(r_i)$ disjoint with $\mathcal{T}(w_j)$
       $\mathcal{T}(w_1) := \mathcal{T}(w_1) \setminus \{T(r_i)\}$;
       exit **while**-j. // reroute another tree in $\mathcal{T}(w_1)$
       else $j := j - 1$. // try to use another wavelength w_{j-1}
    end-while-j
    if $j = 1$ **then** // r_i cannot use any other wavelength but w_1
     stop.
  end-while-$\mathcal{T}(w_1)$
  if $\mathcal{T}(w_1) = \emptyset$ **then** // wavelength w_1 can be saved
   $W := W \setminus \{w_1\}$,
  go to Step 1. // try to save another wavelength in W

In Algorithm 8.5, the input can be obtained by employing Algorithm 8.3. The following theorem shows the correctness and efficiency of Algorithm 8.5.

THEOREM 8.6 *Given a set of routing trees for multicast channels* $\{r_i(s_i, D_i, \Delta)|i = 1, \cdots, k\}$ *in graph* $G(V, E, c, d)$, *Algorithm 8.5 reassigns wavelengths to them in time* $O(K|V|^2|W|^3/k)$, *where* $K = \max\{\,|D_i|\,|\,i = 1, 2, \cdots, k\,\}$.

PROOF In Step 2, when exiting the inner while-loop-j, there are two possible cases.

1) $j = 1$. In this case, there exists a routing tree $T(r_i)$ which cannot be rerouted and reassigned to any other wavelength. That is, the number of wavelengths cannot be reduced. The algorithm stops.

2) $j > 1$. In this case, routing tree $T(r_i)$ can be rerouted and reassigned with another wavelength w_j. The next tree in $\mathcal{T}(w_1)$ will be chosen for rerouting. When $\mathcal{T}(w_1)$ becomes empty, this means that w_1 can be freed.

At Step 1, it takes time $O(k)$ to determine the distribution of currently used wavelengths. At Step 2, the inner while-loop runs at most $(|W| - 1)$ times (for $j = |W|, \cdots, 2$) and each run of Algorithm 8.2 needs time $O(|D_i||V|^2)$. The outer while-loop would run at most $|\mathcal{T}(w_1)|$ times (for each tree in $\mathcal{T}(w_1)$), where $|\mathcal{T}(w_1)| \leq |W|/k$. When w_1 is saved and removed from W, Algorithm 8.5 will go back to Step 1. This process is repeated at most $(|W| - 1)$ times. Therefore, Algorithm 8.5 can terminate in time $O(K|V|^2|W|^3/k)$. ■

3.2 Simulation Study

In the preceding subsections, we have proposed two algorithms, Algorithm 8.4 and Algorithm 8.5, for optimizing wavelength assignment. In order to evaluate the performance of these two algorithms, we simulate four different combinations of Algorithm 8.2, Algorithm 8.3, Algorithm 8.4, and Algorithm 8.5. In the following, for the simplicity of notations we represent these four algorithms by *A*, *B*, *C* and *D*, respectively. *AB* denotes non-optimization (that runs A first and then B), *ACB* denotes load balancing optimization (that runs A first and then C, in the end B), *ABD* denotes wavelength assignment optimization, and *ACBD* denotes integrated optimization.

The network is generated by using the method described in Section 2 of Chapter 1. The QoS multicast channels are generated randomly. For channel $r_i(s_i, D_i, \Delta_i)$, nodes s_i and D_i are randomly picked up from the nodes in generated network graph. The delay bound Δ_i is set as: $\Delta_i \equiv \alpha \cdot \max\left\{d(P_G(s_i, u))\,\middle|\, u \in D_i\right\}$. It is common in real-time communications to set the delay bound to be a constant times the worst case delay.

The objective of simulation work is to find the effectiveness of proposed optimization algorithms. That is, to determine the saving in wavelengths under various network environments. Throughout the simulations, we introduce a lower bound of number of wavelengths as a performance benchmark. The lower bound is defined as

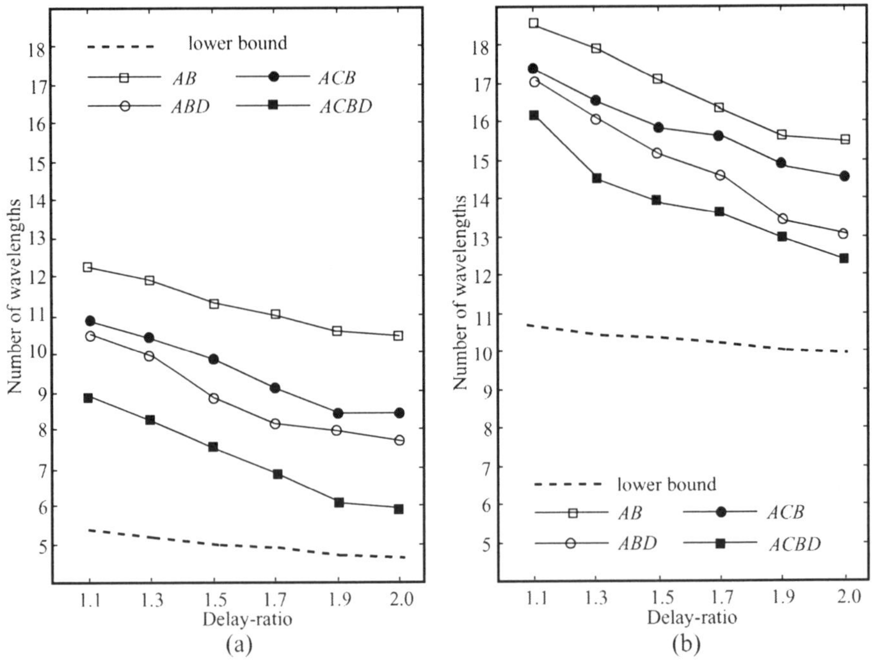

Figure 8.6. The number of wavelengths used against the delay ratio for twenty multicast channels: (a) five destinations and (b) ten destinations.

the maximal link load in the system, which is obtained by using method *AC* (without considering wavelength assignment). At each simulation point, the simulation runs 50 times. Each time a different set of QoS multicast channels is generated, and methods *AB*, *ACB*, *ABD* and *ACBD* are applied, respectively. The number of wavelengths presented in the figures below are the mean values of 50 simulation runs.

In the simulations, we simulate the number of wavelengths against three parameters: delay ratio α, number of multicast destinations, and the number of multicast channels (i.e., the number of channels to be established).

Fig. 8.6-8 show the number of wavelengths versus delay ratio α varying from 1.1 to 2. The numbers of channels are set at 20, 10, and 5, respectively. Fig. 8.6-8(a) display the cases where the number of destinations is 5, and Fig. 8.6-8(b) are the cases where the number of destinations is 10. From Fig. 8.6-8, the following observations can be made.

1) Algorithm 8.5 reduces the number of wavelengths more effectively than Algorithm 8.4. Comparing the performance of method *AB* with that of method *ACB*, we can see that it only reduces the number of wavelengths slightly by using load balance rerouting Algorithm 8.4 alone. Furthermore, by looking at the curves of

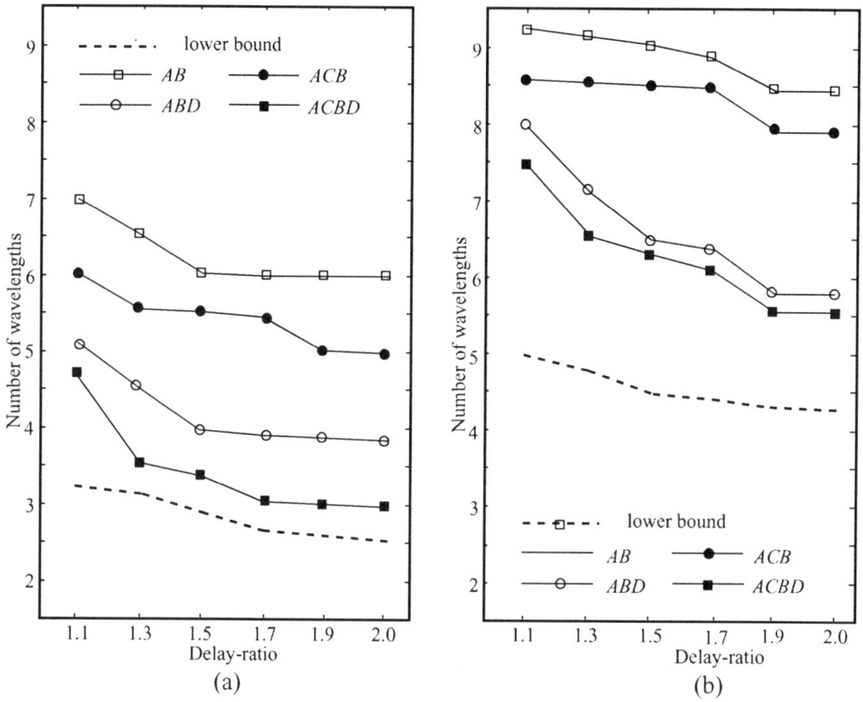

Figure 8.7. The number of wavelengths used against the delay ratio for ten multicast channels: (a) five destinations and (b) ten destinations.

method *ABD* and method *ACBD*, we can find that method *ACBD* does not improve the performance significantly over method *ABD*. The reason is that Algorithm 8.5 tries to free out the least used wavelengths during rerouting, which saves the number of wavelengths directly; while Algorithm 8.4 aims at reducing the maximal link load during rerouting. Reducing the maximal link load in the system does not directly reduce the number of wavelengths (the wavelength assignment phase may negate some of the benefits gained at the load balancing phase).

2) The number of wavelengths monotonically decreases as the delay ratio increases. There are two reasons for this phenomenon: a) with a larger delay bound, Algorithm 8.4 and Algorithm 8.5 can have more chance to succeed in rerouting QoS routing trees, which results in more wavelength savings. b) according to Theorem 1, QoS routing trees of larger delay ratios have smaller cost. Usually, routing trees having smaller cost have less number of links. They, thus, have less chance to share links with each other. Therefore, Algorithm 8.5 is able to assign the same wavelength to more trees without causing wavelength conflict when α is larger.

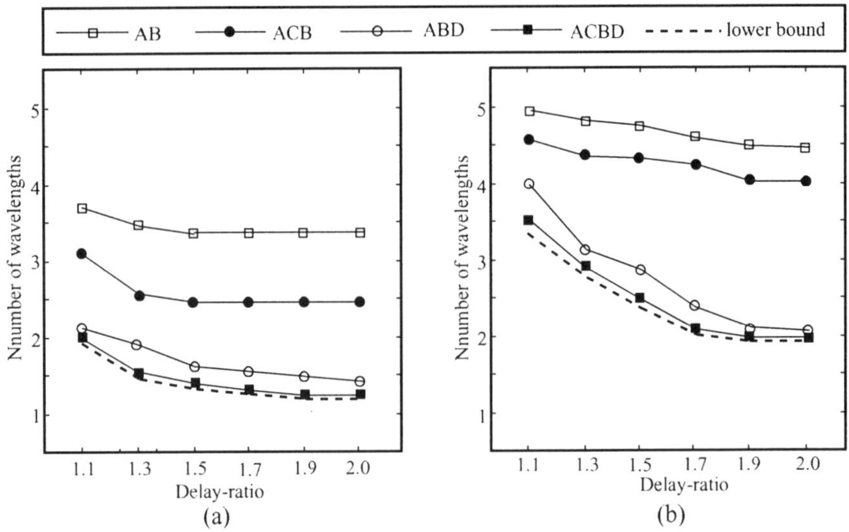

Figure 8.8. The number of wavelengths used against the delay ratio for five multicast channels: (a) five destinations and (b) ten destinations.

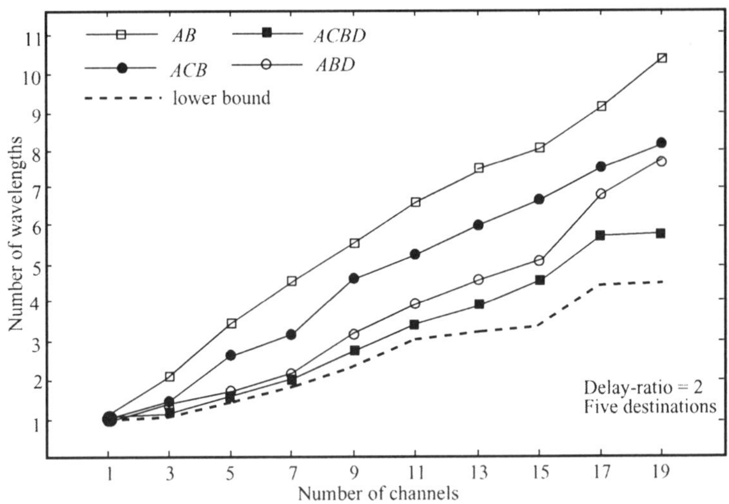

Figure 8.9. The number of wavelengths used against the number of multicast channels having five destinations.

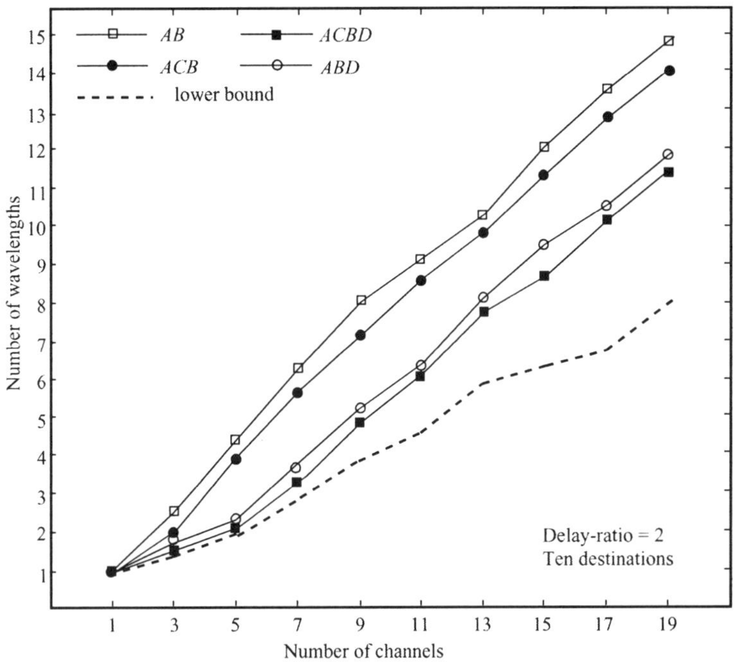

Figure 8.10. The number of wavelengths used against the number of multicast channels having ten destinations.

3) The optimization algorithms work more effectively in sparse networks. Comparing Fig. 8.6-8(a) with Fig. 8.6-8(b), we can see when the number of multicast destinations is smaller, the saving of wavelengths is more significant and the performance is closer to the defined low bound (the curves in dashed lines). A similar trend can be observed with the decrease of the number of channels in the system. This is because when the system has multicast connections with fewer destinations (or has fewer channels), the link loads are lighter (i.e., fewer trees share links with each other). The rerouting operations in the proposed optimization algorithms have more chance to succeed. It can also be anticipated that they would work more effectively in sparse networks, where there is more room for rerouting.

The second group of simulations concerns the number of wavelengths versus the number of channels. In this group of simulations, the delay-ratio α is fixed at 2.0. The number of channels varies from 1 to 20. Fig. 8.9 shows the case where the number of destinations is set to 5, and in Fig. 8.10, it is set to 10. From Fig. 8.9-10, we see a continuous change of wavelengths as the increase of channels. The curves in Fig. 8.9-10 are consistent with the above observation 3). The proposed optimization algorithms work better in the cases where routing trees are smaller (i.e., fewer multicast destinations) or there are fewer routing trees (i.e., fewer channels).

4. Dynamic Multicast Routing and Wavelength Assignment

In this section, we will study the problem of multicast routing and wavelength assignment under dynamic model. This problem can be formulated as follows.

PROBLEM 8.4 *Dynamic QoS Routing and Wavelength Assignment Problem*

Instance A graph $G(V, E, c, d)$ and a set of requests $\{r_i(s_i, D_i)|i = 1, \cdots, k\}$ that that arrive sequentially.

Solution A sequentially produced feasible routing tree $T(r_i)$ with an assigned wavelengths such that two trees must be assigned different wavelengths if they share a common link.

Objective Minimizing the number of wavelengths used.

Note that the above problem is the on-line version of Problem 8.3 and can be considered as the on-line version of routing and wavelength assignment problem (RWAP, Problem 3.2) since we are asked to do routing and make wavelength assignment for coming multicast requests one by one (with no knowledge what kind of requests will come later), that is, we need to design on-line algorithms for routing and wavelength assignment. In this sense Problem 8.4 is harder than Problem 8.3, especially for wavelength assignment. In the following subsections, we will first study Problem 8.4 in special networks and then in general networks.

4.1 Ring Networks

In this subsection for the simplicity of presentation we will only study Problem 8.4 in ring networks. The proposed approach and analysis is applicable for dealing with general tree of rings networks. A *tree of rings* can be generated by substituting some of nodes in a tree with some rings (see Fig. 8.11). The tree of rings can be considered as a general case of star, tree, ring and ring networks, which are popular types of interconnecting networks.

In the following for a multicast request $r(s, D)$, we denote by $MST(r)$ the *Minimum Steiner Tree* (MST) and by $SPT(r)$ the *Shortest Path Tree* (SPT) of request r. The network cost of routing tree $T(r)$ is defined as the number of edges it includes and denoted simply by $|T(r)|$. The delay from source s to a destination $d \in D$ on $T(r)$ is measured by the cost of the path $P_T(s, d)$ from s to d on $T(r)$, and denoted by $|P_T(s, d)|$. Such definitions reflect the situation in most data communication networks where the cost parameter is measured by the number of hops.

Let $R(n)$ be the ring network of n nodes. For the simplicity of presentation, we assume that n can be divided by 4. We label these n nodes clockwise as $0, 1, \cdots, n-2, n-1$, where the node label takes module of n. Each link between a pair of nodes carries two oppositely-directed fibers for data transmissions in the two directions of

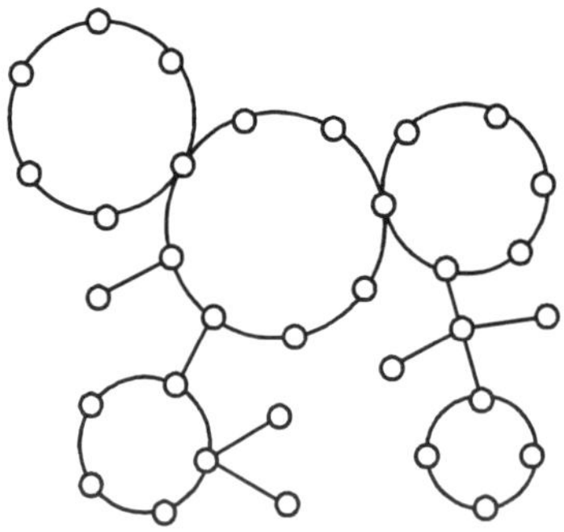

Figure 8.11. An example of tree of rings.

the link. Here we assume bidirectional connections. Given a request $r(s, D)$ on $R(n)$, there are $(|D| + 1)$ possible routing trees. For a simple example, consider request $r(0, \{n/2-2, n/2+1\})$, there are three routing trees as shown in Fig. 8.12, where $n = 12$. The routing tree shown in Fig. 8.12(c) has greater cost than $MST(r)$ shown in Fig. 8.12(b) and longer delay than $SPT(r)$ shown in Fig. 8.12(a). The following lemma will be used to prove some of the subsequent results.

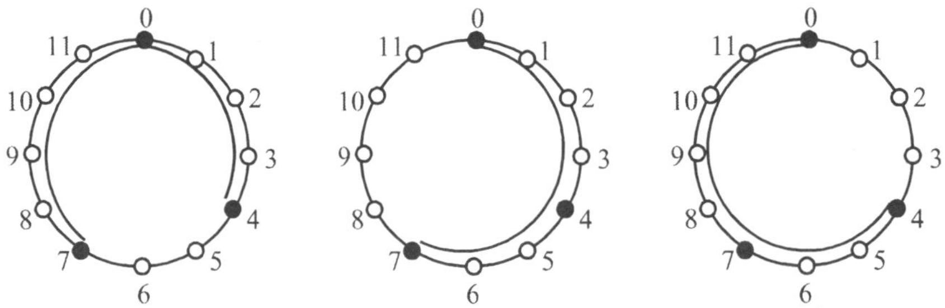

Figure 8.12. Multicast routing in a ring: (a) $SPT(r)$, (b) $MST(r)$, and (c) a routing tree.

LEMMA 8.1 *For any multicast request $r(s, D)$ on ring $R(n)$.*
(i) *If path $(s \to s+1 \to \cdots \to d-1 \to d)$ contains link $(i \to i+1)$, then path $(s \to s-1 \to \cdots \to d+1 \to d)$ contains link $(n/2+i+1 \to n/2+i)$.*
(ii) *If $|P_{MST}(s, d')| > n/2$, for some $d' \in D$, then there exists a destination node $d'' \in D$ on $P_{MST}(s, d')$ such that the distance between d'' and d' is less than $n/2$.*

PROOF Observe that path $(s \to s+1 \to \cdots \to d-1 \to d)$ and path $(s \to s-1 \to \cdots \to d+1 \to d)$ are edge-disjoint and they include all nodes in the ring. See Fig. 8.13(a). Thus it is easy to verify (i).

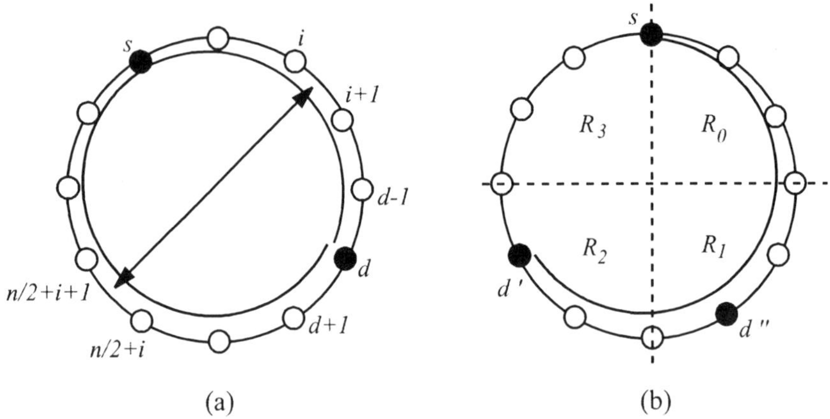

Figure 8.13. For the proof of Lemma 8.1.

We now consider (ii) and show the case of $s = 0$. Let us divide ring $R(n)$ into four parts $R_i = \{i\frac{n}{4} + j | j = 0, 1, \cdots, \frac{n}{4} - 1\}$, for $i = 0, 1, 2, 3$. See Fig. 8.13(b). Clearly, $MST(r) = SPT(r)$, if $D \subseteq R_0 \cup R_3$, or $D \subseteq R_0 \cup R_1$, or $D \subseteq R_1 \cup R_3$. Without loss of generality, we assume that $d' > n/2$. Notice that there exists a node $d'' \in D$ with $d' > d''$, otherwise $MST(r)$ consists of a path having distance $n - d' < n/2$, this contradicts $|P_{MST}(s, d')| > n/2$. Let $d'' \in D$ be the closest node to d' in $\{d | d \in D, d < d'\}$.

Case 1. For all $d \in D, d \leq d'$. In this case, $MST(r)$ consists of a single path, which implies $d' \leq n - d' + d''$. (Otherwise $MST(r)$ consists of two paths.) Thus $d' - d'' \leq n/2$.

Case 2. For some $d_0 \in D, d_0 > d'$. We assume that d_0 is the closest node to d' in $\{d | d \in D, d > d'\}$. If d_0 is not on $P_{MST}(s, d')$, then $|MST(r)| = n - d_0 + d' \leq n - d' + d''$, this implies $d' - d'' \leq d_0 - d' < n/2$. If d_0 is on $P_{MST}(s, d')$, then $d_0 \leq |MST(r)| \leq n - d' + d''$. This leads to $d' - d'' \leq n - d_0 < n/2$. The proof is then finished. ■

The following theorem shows that in rings the cost of SPT is less than two times that of MST while the delay of MST is less than two times that of SPT. This is not true in general.

THEOREM 8.7 *For any multicast request $r(s, D)$ on $R(n)$.*
(i) $1 \leq |SPT(r)|/|MST(r)| \leq 2(n-1)/(n+2)$.
(ii) $1 \leq |P_{MST}(s, d)|/|P_{SPT}(s, d)| \leq 2(n-2)/n$, *for any* $d \in D$.

PROOF It suffices to show the case of $s = 0$. (i) It is easy to verify that the ratio achieves the maximum of $2(n-1)/(n+2)$ when $\{n/2-1, n/2, n/2+1\} \subseteq D$ and $d \leq n/2+1$, for all $d \in D$, or $d \geq n/2-1$, for all $d \in D$. (ii) It is easy to verify that the ratio achieves the maximum of $2(n-2)/n$ when $D = \{1, 2, \cdots, n-3, n-2\}$ or $D = \{2, 3, \cdots, n-2, n-1\}$. ■

In the following we will use the competitive analysis [14] to study the performances of on-line algorithms. An on-line algorithm A is called $\alpha - competitive$ if for any input sequence σ, $C_A(\sigma) \leq \alpha C_{opt}(\sigma)$, where $\alpha \geq 1$ is a constant independent of σ, $C_A(\sigma)$ and $C_{opt}(\sigma)$ are the cost of on-line algorithm A and the cost of optimal off-line algorithm for σ, respectively. In our problems, the cost of algorithm A, $C_A(\sigma)$, is the network load for Problem 2.5 and the number of wavelengths used for Problem 8.4, respectively. However, we notice that his standard approach does not exactly fit our case. This is because, given a sequence of requests $\{r_i\}$, some previous requests may be torn down when a new request is called, thus an on-line algorithm A only need to consider those active requests, and $C_A(\{r_i\})$ be strictly less than $C_{opt}(\{r_i\})$ if the optimal off-line algorithm needs to process all requests in $\{r_i\}$. Therefore, in the following discussion we will evaluate an on-line algorithm by comparing its performance with that the optimal off-line algorithm on the set which consists of only active requests. Accordingly, we denote by $act(r_i)$ the set of requests which are active when request r_i comes.

We first study the performances of the optimal on-line algorithms for the on-line versions of load balancing problem (LBP, Problem 2.5) and Problem 8.4, respectively.

THEOREM 8.8 *The optimal algorithm for the on-line version of Problem 2.5 has competitive ratio at least 2.*

PROOF Let A be an optimal on-line algorithm, we then consider how algorithm A routes a sequence of requests $\{r_i(s_i, D_i)\}$ on $R(n)$, which is adversely delivered in the following way: First, deliver $r_1(0, \{n/2\})$. Now suppose, without lose of generality, that algorithm A routes r_1 clockwise, then deliver $r_2(1, \{2\})$. If algorithm A routes r_2 clockwise, link $(1 \rightarrow 2)$ has load of two, thus it can not be α-competitive, for $\alpha < 2$, since the optimal off-line algorithm will route these two requests one clockwise while the other anticlockwise, and every link has load at most one. If algorithm A routes r_2 anticlockwise, then deliver $r_3(3, \{4\})$. Therefore, it is easy to see that no matter how algorithm A routes r_3, anticlockwise or clockwise, there always exists a directed link with load of two. However, the optimal off-line algorithm will route r_1 anti-clockwise while r_2 and r_3 clockwise, in this way no link has load greater than one. This implies that algorithm A has competitive ratio at least 2. The proof is then finished. ■

By using a similar but more complicated argument, we can prove the following theorem.

THEOREM 8.9 *The optimal algorithm for Problem 8.4 in $R(n)$ has competitive ratio at least $n/2$.*

PROOF Let A be an optimal on-line algorithm, we then consider how algorithm A routes a sequence of requests on $R(n)$, which is adversely delivered in the following way: First, deliver $2n$ identical requests of $r(i, \{i+1\})$, for each $i = 0, 1, \cdots, n-1$. If algorithm A routes all $2n^2$ requests anticlockwise, then it must use $2n^2$ different wavelengths. However, the optimal off-line algorithm will route them all clockwise so that it just needs $2n$ wavelengths. Thus algorithm A has competitive ratio at least n, and we are done. Suppose that algorithm A routes at least one request clockwise. Notice that at least n out of $2n$ identical requests of $r(i, \{i+1\})$ are assigned with n different wavelengths, and there exist n requests $\{r(i, \{i+1\}) | i = 0, 1, \cdots, n-1\}$ being assigned with n different wavelengths and at least one of them is routed clockwise. Now keep these n requests active and release the rest of requests. Second, deliver $2n$ identical requests of $r(i, \{i-1\})$, for each $i = n, n-1, \cdots, 1$. If algorithm A routes all $2n^2$ new requests clockwise, then it must use at least $2n^2 + 1$ different wavelengths for $2n^2$ new and n old requests. However, the optimal off-line algorithm will route $2n^2$ new requests all anticlockwise while n old requests clockwise, so that it just needs $(2n + 1)$ wavelengths. Thus algorithm A has competitive ratio at least $n/2$, and we are done. Suppose that algorithm A routes at least one of new requests anti-clockwise. Notice that at least n out of $2n$ identical requests of $r(i, \{i-1\})$ are assigned with n different wavelengths, and there exist n requests $\{r(i, \{i-1\}) | i = n, n-1, \cdots, 1\}$ being assigned with n different wavelengths and at least one of them is routed clockwise. Now keep these n requests active and release the rest of requests. Clearly, at least n wavelength are being used for $2n$ different requests. In the end, deliver request $r(0, \{1, 2, \cdots, n-1\})$. Algorithm A must introduced a new wavelength, since one active request is routed clockwise while another one routed anticlockwise. However, the optimal off-line algorithm will route requests $\{r(i, \{i+1\}) | i = 0, 1, \cdots, n-1\}$ clockwise while requests $\{r(i, \{i-1\}) | i = n, n-1, \cdots, 1\}$ anticlockwise, and the last one in any way. Routing in such a way, it only needs 2 wavelengths. This means that algorithm A has competitive ratio at least $(n+1)/2$. The proof is then finished. ■

Now, we study the performances of MST and SPT Algorithms for the on-line LBP in rings.

THEOREM 8.10 *SPT algorithm for the on-line version of Problem 2.5 has 2-competitive on ring networks.*

PROOF Given a sequence of requests $\{r_i(s_i, D_i)\}$ on $R(n)$, let $l_i = C_{SPT}(act(r_i))$, for i. Then we can assume, without loss of generality, that when r_i is coming, there exist l_i directed paths which are routed by SPT algorithm for l_i active requests and traverse link $(0 \to 1)$. Notice that each of these l_i directed paths is routed clockwise and has distance no more than $n/2$ to its source node, as SPT algorithm is applied.

Now consider any one of these l_i paths, say path p which is a branch of SPT for request $r(s, D)$. Let d' and d'' be two nodes on path p such that $d' \in D$ is the closet node to node 1 while $d' \in D \cup \{s\}$ is the closest node to node s. If an off-line algorithm does not use clockwise directed path ($d \to \cdots \to d''$) that traverses link $(0 \to 1)$, then it must use anti-clockwise directed path ($s \to \cdots \to d'$) that traverses link $(n/2 + 1 \to n/2)$ due to Lemma 8.1(i). Therefore, to route these l_i active requests by any off-line algorithm will produce load at least $\lceil l_i/2 \rceil$ on either link $(0 \to 1)$ or link $(n/2 + 1 \to n/2)$, this implies $C_{opt}(act(r_i)) \geq \lceil l_i/2 \rceil$. Therefore, SPT algorithm has competitive ratio at most 2. This, along with Theorem 8.82, proves the theorem. ∎

THEOREM 8.11 *MST algorithm for the on-line version of Problem 2.5 has competitive at most 4 and at least 3 on ring networks.*

PROOF Given a sequence of requests $\{r_i\}$ on $R(n)$, let $l_i = C_{MST}(act(r_i))$, for i. Then we can assume, without loss of generality, that when r_i is coming, there exist l_i directed paths which are routed by MST for l_i active requests and traverse link $(0 \to 1)$. Now consider any one of these l_i paths, say path p, which is a branch of MST for request $r(s, D)$. Let d' and d'' be two nodes on path p such that $d'' \in D$ is the closet node to node 1 while $d' \in D \cup \{s\}$ is the closest node to node 0. Notice that by Lemma 8.1(ii) the distance between node d and d'' is less than $n/2$. If an off-line algorithm does not use clockwise directed path ($d' \to \cdots \to d''$) that traverses link $(0 \to 1)$, then it must use anticlockwise directed path ($s \to \cdots \to d''$) by Lemma 8.1(i). Moreover, by case study it is easy to verify that routing in that way the optimal algorithm will use at least one of three directed links in $\{(1 \to 0), (n/2 \to n/2 + 1), (n/2 + 1 \to n/2)\}$. Thus no matter how the optimal off-line algorithm routes these l_i active requests, at least one directed link in $\{(0 \to 1), (1 \to 0), (n/2 \to n/2 + 1), (n/2 + 1 \to n/2)\}$ has load at least $\lceil l_i/4 \rceil$. Therefore, MST algorithm has competitive ratio at most 4.

To show the lower bound, we consider three requests $r(0, \{1\})$, $(11, \{4, 15\})$, $r(4, \{1, 6, 8, 10, 11, 13, 15\})$ on $R(16)$. MST algorithm will route them like this: $(0 \to 1), (11 \to 12 \to \cdots \to 15 \to 0 \to 1 \to \cdots \to 3 \to 4)$, and $(4 \to 5 \to \cdots \to 15 \to 0 \to 1)$. Link $(0 \to 1)$ has load of three. The optimal off-line algorithm will route them like this: $(0 \to 1), (11 \to 12 \to 13 \to 14 \to 15) \cup (11 \to 10 \to \cdots 5 \to 4)$, and $(4 \to 5 \to \cdots \to 9 \to 10) \cup (4 \to 3 \to \cdots \to 1 \to 0 \to \cdots \to 12 \to 11)$. Each link has load at most one. Thus the lower bound is three. The proof is then finished. ∎

We now study the performance of *SPT-based* and *MST-based* algorithms for the on-line RWAP. They first route incoming request by SPT or MST algorithms, and then assign the request a currently used wavelength if possible.

THEOREM 8.12 *SPT-based algorithm for the on-line version of Problem 3.2 has competitive ratio at most n on $R(n)$.*

PROOF Let $r = \{r_i(s_i, D_i)\}$ be a sequence of requests. Suppose, without loss of generality, that $SPT(r_{j+1})$ includes a clockwise directed path $(0 \rightarrow 1 \rightarrow 2 \rightarrow \cdots \rightarrow d-1 \rightarrow d)$, that is, it starts from source node 0 and ends at a destination node d, where $d \leq n/2$. Thus there are $l_j - 1$ active requests which are assigned with $l_j - 1$ different wavelengths and whose routing trees include at least one directed link in clockwise directed path $(0 \rightarrow 1 \rightarrow 2 \rightarrow \cdots \rightarrow d-1 \rightarrow d)$. This implies that there exists a clockwise directed link, say link $(i \rightarrow i+1)$, for some $i, 0 \leq i \leq d-1$, which has load at least

$$\lceil \frac{l_j - 1}{d} \rceil \geq \lceil \frac{l_j - 1}{n/2} \rceil.$$

Now applying the same argument used in the proof of Theorem 8.10, we can deduce that no matter how the optimal off-line algorithm to route (and assign) those active requests whose routing trees traverse link $(i \rightarrow i+1)$, there exists a directed link having load at least

$$\left\lceil \frac{1}{2} \left(\left\lceil \frac{l_j - 1}{n/2} \right\rceil + 1 \right) \right\rceil \geq \frac{l_j}{n}.$$

Hence the optimal off-line algorithm needs at least l_j/n wavelengths. The proof is then finished. ∎

By using the same argument we can obtain the parallel results of MST-based algorithm.

THEOREM 8.13 *MST-based algorithm for the on-line version of Problem 3.2 has competitive ratio at most* $4n$ *on* $R(n)$.

PROOF Let $r = \{r_i(s_i, D_i)\}$ be a sequence of requests, and $C_{MST}(act(r_{j+1})) = l_j$, for $j \in J_{MST}(r)$, then $C_{MST}(act(r_j)) = l_j - 1$. Now suppose, without loss of generality, that $MST(r_{j+1})$ includes a clockwise directed path $(0 \rightarrow 1 \rightarrow \cdots \rightarrow d)$, that is, it starts from source node 0 and ends at a destination node d, where $d \leq n-2$. Thus there are $l_j - 1$ active requests which are assigned with $l_j - 1$ different wavelengths and whose routing trees include at least one link in clockwise directed path $(0 \rightarrow 1 \rightarrow \cdots \rightarrow d)$. This implies that there exists a clockwise directed link, say link $(i \rightarrow i+1)$, for some $i, 0 \leq i \leq d-1$, which has load at least

$$\lceil \frac{l_j - 1}{d} \rceil \geq \lceil \frac{l_j - 1}{n-2} \rceil.$$

Now applying the same argument used in the proof of Theorem 8.11, we can deduce that no matter how the optimal off-line algorithm to route (and assign) those active requests whose routing trees traverse link $(i \rightarrow i+1)$, there exists a directed link having load at least

$$\left\lceil \frac{1}{4} \left(\left\lceil \frac{l_j - 1}{n-2} \right\rceil + 1 \right) \right\rceil \geq \frac{l_j}{4(n-2)}.$$

Hence the optimal off-line algorithm needs at least $l_j/4(n-2)$ wavelengths. The proof is then finished. ∎

The *greedy algorithm* for the on-line LBP is to route each incoming request without using those most heavily loaded links. The *greedy algorithm* for the on-line RWAP is to route each coming request in such a way that a currently used wavelength can be assigned to the request (a new wavelength will not be introduced unless it has to). By using the same technique, we can prove the following two theorems. They show that for ring networks greedy algorithms for the on-line LBP and the on-line RWAP not only may have larger network cost and longer delay but also have larger competitive ratios than SPT and MST algorithms.

THEOREM 8.14 *The greedy algorithm for the on-line version of Problem 2.5 in $R(n)$ has competitive ratio at most n and at least $n/2$.*

PROOF Let $\{r_i\}$ be a sequence of requests on $R(n)$, and denote by G the greedy strategy for the on-line LBP. To prove the upper bound, let $C_G(act(r_{j+1})) = l_j$. Then there exist one clockwise directed link $(u \to u+1)$ and one anticlockwise directed link $(v+1 \to v)$ which have load of $l_j - 1$. This implies that there are $l_j - 1$ clockwise directed paths passing through $(u \to u+1)$ and $l_j - 1$ anticlockwise directed paths passing through $(v+1 \to v)$. Now consider how the optimal off-line algorithm routes those $2(l_j-1)+1$ directed paths (as incoming request r_{j+1} also needs to be routed). Suppose that it route l' (l'', respectively) clockwise (anticlockwise, respectively) directed paths anticlockwise (clockwise, respectively) while keeping $l_j - 1 - l'$ ($l_j - 1 - l''$, respectively) clockwise (anti-clockwise, respectively) directed paths unchanged. In addition, suppose, without loss of generality, that the optimal off-line algorithm routes r_{j+1} clockwise. Hence in $R(n)$ there are at least $l_j - 1 - l' + l'' + 1$ ($l_j - 1 - l'' + l'$, respectively) clockwise (anticlockwise, respectively) directed paths. As there are n directed links in each direction, there exists a clockwise (anticlockwise, respectively) directed link which has load at least $\lceil (l_j - 1 - l' + l'' + 1)/n \rceil$ ($\lceil (l_j - 1 - l'' + l')/n \rceil$, respectively). Therefore, at least one directed link has load at least

$$\frac{1}{2}\left(\left\lceil \frac{l_j - 1 - l' + l'' + 1}{n} \right\rceil + \left\lceil \frac{l_j - 1 - l'' + l'}{n} \right\rceil\right) \geq \frac{l_j}{n}.$$

Hence the upper bound of competitive ratio is n.

To prove the lower bound, consider a sequence of requests on $R(n)$ as follows: First, deliver $(i+1)$ identical requests of $r(i, \{i+1\})$ in the order of $i = 0, 1, \cdots, n-1$. Without loss of generality, we assume that request $r(0, \{1\})$ is routed clockwise. Then it can be easily verified that for each $i = 1, 2, \cdots, n-1$, the greedy strategy first routes i identical requests $r(i, \{i+1\})$ clockwise and then one request $r(i, \{i+1\})$ anticlockwise. Now keep these n anticlockwise routed requests active and release rest of the requests. Note that link $(i \to i-1)$ carries $(n-1)$ requests, for each $i = n-1, n-2, \cdots, 1$. Second, deliver a sequence of requests $\{r(i, \{i-1\}), i = n, n-1, \cdots, 1\}$ It is easy to see that greedy strategy will route them all anticlockwise.

Now for each directed link on $R(n)$, it carries $(n-1)$ requests. In the end, deliver request $r(0,\{2\})$. Notice that no matter how greedy strategy routes this request. The load of network will increase from $(n-1)$ to n. However, the optimal off-line algorithm for LBP will route n requests of $\{r(i,\{i+1\})|\, i=0,1,\cdots,n-1\}$ clockwise while n requests of $\{r(i,\{i-1\})|\, i=n,n-1,\cdots,1\}$ anticlockwise, and the last request $r(0,\{2\})$ either clockwise or anticlockwise. In such a way every link on $R(n)$ carries at most two requests. This shows that the lower bound of competitive ratio is $n/2$. The proof is then finished. ■

THEOREM 8.15 *The greedy algorithm for the on-line version of Problem 3.2 has competitive ratio at most $n^2/2$ on $R(n)$.*

PROOF Let $\{r_i\}$ be a sequence of requests on $R(n)$, and denote by G the greedy strategy for the on-line RWAP. Let $C_G(act(r_{j+1})) = l_j$. Suppose, without loss of generality, that routing tree for r_{j+1} includes a clockwise directed path $(0 \to 1 \to \cdots \to d)$, that is, it starts from source node 0 and ends at destination node d. Note that $C_G(act(r_j)) = l_j - 1$. So there are $l_j - 1$ active requests which are assigned $l_j - 1$ different wavelengths and whose routing trees include at least one clockwise and one anticlockwise directed link in directed path $(0 \to 1 \to \cdots \to d)$ and $(0 \to n-1 \to \cdots \to d)$, respectively. This implies that there exists a clockwise (anti-clockwise, respectively) directed link, say link $(u \to u+1)$ $((v+1 \to v)$, respectively), for some $u, 0 \le u \le d-1$ ($v, n \ge v \ge d+1$, respectively), which has load at least $\lceil (l_j-1)/d \rceil$ $(\lceil (l_j-1)/(n-d) \rceil$, respectively). Now applying the same argument used in the proof of Theorem 8.14, we can deduce that no matter how optimal off-line algorithm routes (and assign) r_{j+1} and those active requests which traverse links $(u \to u+1)$ and $(v+1 \to v)$, there exists a directed link which has load at least

$$\frac{1}{2}\left\lceil \frac{(l_j-1)\frac{4}{n}+1}{n} \right\rceil \ge 2\frac{l_j}{n^2}.$$

Hence the upper bound of competitive ratio is $2l_j/n^2$. The proof is then finished. ■

Table 8.1. The competitive analysis of algorithms for the on-line LBP and the on-line RWAP.

	Load Balancing		Wavelength Assignment		Cost	Delay
	Lower Bound	Upper Bound	Lower Bound	Upper Bound		
Optimal	2	2	$n/2$	n	1	1
SPT	2	2	$n/2$	n	2	1
MST	3	4	$n/2$	$4n$	1	2
Greedy	$n/2$	n	$n/2$	$n^2/2$	$n-1$	$n-1$

In the above table we summarize the obtained results on lower/upper bounds of competitive ratios of optimal, SPT, MST and greedy algorithms for the on-line LBP and the on-line RWAP. The table also gives the worst-case ratios of network cost and delay over the optimal values of the routing trees generated by those methods. From the table we can see that SPT and MST algorithms can produce multicast routing trees that have good guaranteed performance for both LBP and WAP in the tree of ring networks.

4.2 General Networks

In this subsection we will study the dynamic QoS Routing and wavelength assignment problem (Problem 8.4) in general networks.

4.2.1 Single-Phase Algorithm

We adopt the greedy strategy for wavelength assignment again that routes a request in such a way that a currently used wavelength can be assigned to it. This simple idea gives birth to the following single-phase algorithm for Problem 8.4. Let W be the set of currently being used wavelengths in the network, and $E(w)$ be the set of active routing trees whose corresponding requests are assigned with wavelength w, for $w \in W$.

ALGORITHM 8.6 *Single-Phase Algorithm*

Input A multicast request $r(s, D)$, a graph $G(V, E, c, d)$, W and $E(W)$
Output A feasible routing tree T of $r(s, D)$ with an assigned wavelength
$W_1 := W$.
while $W_1 \neq \emptyset$ **do begin** // try to use a currently used wavelength in W
    choose $w \in W_1$,
    $E' := E \setminus E(w)$, // remove $E(w)$ from network graph
    generate T_{QS} by Algorithm 8.2 for $r(s, D)$ on $G(V, E', c, d)$,
    if $T_{QS} \neq \emptyset$ **then**
      $T := T_{QS}$,
      assign wavelength w to T, // because T is disjoint with $E(w)$
      return T.
    else $W_1 := W_1 \setminus \{w\}$. // try to use another wavelength in W
end-while
generate T_{QS} by Algorithm 8.2 for $r(s, D)$ on $G(V, E, c, d)$,
$T := T_{QS}$,
assign a new wavelength $w \notin W$ to T,
return T.

In the above algorithm as soon as we find a currently used wavelength that can be properly assigned to the current request, we assign it this wavelength. There are some simple policies for searching such a wavelength. (a) The earliest introduced wavelength (or the latest introduced wavelength) is checked first. (b) The most heavily used wavelength (or the most lightly used wavelength) is checked first.

THEOREM 8.16 *Given a request $r(s, D)$ and a connected graph $G(V, E, c, d)$, Algorithm 8.6 constructs a feasible routing tree and assigns it a wavelength in time $O(|W||D||V|^2)$.*

PROOF It follows from Theorem 8.2 that Algorithm 8.6 constructs a feasible routing tree for a given request either on $E \setminus E(w)$, for some $w \in W$, or on E; In the former case this currently being used wavelength w is assigned to the request, and in the latter case a new wavelength is introduced and assigned to the request. Clearly, to assign a wavelength in such a way will not cause wavelength conflict.

The time-complexity follows from Theorem 8.2 and two facts: 1) Algorithm 8.2 is called at most ($|W| + 1$) times, and 2) It requires only a constant time to assign $r(s, D)$ a wavelength. ■

4.2.2 Two-Phase Algorithm

Given a set of requests routed on the network, the number of wavelengths required is at least the network load. Based on this fact, we propose another greedy algorithm for Problem 8.4. This algorithm consists of two phases: routing and wavelength assignment. At the routing phase, we try to generate a routing tree in such a way that the network load is as low as possible. At the second phase, apply the greedy strategy for wavelength assignment. This is exactly the same method that we described in Section 4 of Chapter 3.

Let W be the set of wavelengths currently being used, and $E(w)$ the set of routing trees whose corresponding requests are assigned with wavelength w, for $w \in W$. In addition, let L_{max} be the current network load of G, and $L(j)$ be the set of links that there are j routing trees traversing it, for $1 \le j \le L_{max}$, and let $L(L_{max} + 1) = \emptyset$.

ALGORITHM 8.7 *Two-Phase Algorithm*

Input A multicast request $r(s; D)$, a graph $G(V, E, c, d)$, W and $E(W)$
Output A feasible routing tree T of $r(s; D)$ with an assigned wavelength
Step 1 *Construct a feasible routing tree*
  $k := L_{max} + 1$.
  while $k \ge 1$ **do** // route $r(s, D)$ on most lightly loaded edges as much as possible
    $E_k := E \setminus L(k) \cup L(k+1) \cup \cdots \cup L(L_{max} + 1)$,
    // remove edges with load $\ge k$
    run Algorithm 8.2 with $r(s, D)$ and $G(V, E_k, c, d)$ and
    generate a routing tree T_{QS}.
    if $T_{QS} \ne \emptyset$ **then** // routing $r(s, D)$ on E_k succeeds
    $T := T_{QS}$,
    $k := k - 1$; // try to route $r(s, D)$ on E_{k-1}
    else go to Step 2.
  end-while
Step 2 *Assign a wavelength to T*

$W_2 := W$.
while $W_2 \neq \emptyset$ **do begin** // use a currently used wavelength
choose $w \in W_2$.
if $T \cap E(w) = \emptyset$ **then** // T can use w
assign T wavelength w,
return T;
else $W_2 := W_2 \setminus \{w\}$. // try to use another wavelength in W
end-while
assign T a new wavelength $w \notin W$,
return T.

In the above algorithm, at Step 1 we search for the smallest k such that a feasible routing tree can be generated in subnetwork $G(V, E_k)$ of $G(V, E)$, which excludes those links whose load are greater than k. Thus it can be considered as a greedy strategy for LBP. To speed up locating k among $\{1, 2, \cdots, L_{max}, L_{max} + 1\}$, we can use binary search so that k can be determined after running Algorithm 8.2 at most $\lceil \log L_{max} \rceil$ times. There are two possible results of Step 1: (1) $k = L_{max} + 1$. This implies that the network load is increased by one; (2) $k \leq L_{max}$. This implies that the network load does not increase.

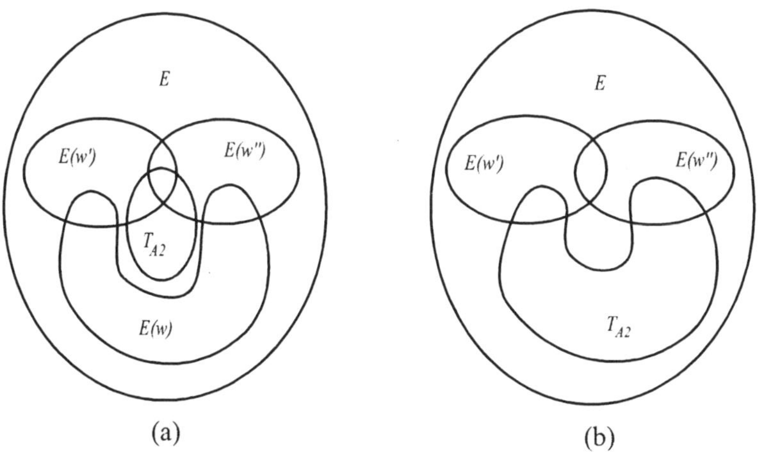

Figure 8.14. (a) The network load increases, but a new wavelength needs not to be introduced. (b) The network load does not increase, but a new wavelength has to be introduced.

At Step 2 searching for a currently used wavelength that can be properly assigned to the request is nontrivial in both cases, because the relationship between the network load and the number of wavelengths required is that the increase in the network load does not necessarily demand a new wavelength, and keeping the network load unchanged does not necessarily mean no need to introduce a new wavelength. A simple example is demonstrated in Fig. 14(a), where no feasible routing tree exists in $G(V, E_2, c, d)$, so a feasible routing tree T is constructed in original $G(V, E, c, d)$. As T includes some edges in $E(w') \cap E(w'')$, the network load is increased from two to three. However the request can be assigned a currently used wavelength w, since $E(w) \cap T = \emptyset$. In addition, a simple example is demonstrated in Fig. 14(b), where a feasible routing tree T is constructed in $G(V, E_2, c, d)$ and the network load is not increased. However neither of two currently used wavelengths w' and w'' can be assigned to the request, as T includes some edges in $E(w')$ and $E(w'')$, respectively. Thus a new wavelength has to be introduced.

THEOREM 8.17 *Given a request $r(s, D)$ and a connected graph $G(V, E, c, d)$, Algorithm 8.7 constructs a feasible routing tree and assigns it a wavelength in time $O(|V|^2|D| \log L_{max})$, where L_{max} is the network load.*

PROOF It follows from Theorem 8.2 that routing tree T constructed in Step 1 is a feasible routing tree for $r(s, D)$. A currently used wavelength $w \in W$ is assigned to $r(s, D)$ on the condition of $T \cap E(w) = \emptyset$. Clearly, such an assignment is proper.

The time-complexity follows from Theorem 8.2 and two facts: (1) Algorithm 8.2 is invoked at most $\lceil \log(L_{max} + 1) \rceil$ times; (2) The searching process in Step 2 can be well integrated with the routing process in Step 1. This can be done by keeping track of the active requests that traverse the link and the wavelengths that they use for each link. Accordingly, when adding a link into the feasible routing tree under the construction, we know which of the currently used wavelengths can be properly assigned to it. In such a way, when completing the construction of whole routing tree, the wavelength assignment can also be done at the same time. Hence assigning the request a wavelength will not cause extra time. ■

4.2.3 Discussions of Algorithms

As we see in the previous two subsections that Algorithm 8.6 and Algorithm 8.7 apply the same greedy strategy for the purpose of saving wavelengths. The difference between them is that Algorithm 8.6 achieves this goal through a direct way, that is to construct a feasible routing tree on $G(V, E \setminus E(w))$, for some $w \in W$, while Algorithm 8.7 achieves this goal through an indirect way, that is to construct a feasible routing tree on $G(V, E_k)$. Such a difference in implementing the greedy strategy produces two outcomes which are demonstrated by Fig. 8.15.

(1) Algorithm 8.7 has more freedom to construct a feasible routing tree than Algorithm 8.6. They try to construct a feasible routing tree in $G(V, E_k)$ and $G(V, E \setminus E(w))$, respectively. When k is close to L_{max}, set E_k, which just excludes links with loads greater than $(k-1)$, contains more links than set $E \setminus E(w)$, which excludes links assigned wavelength w (some of them may not heavily loaded). In other words, Algorithm 8.6 has more chances to succeed than Algorithm 8.7.

(2) Algorithm 8.6 can be expected to use fewer number of wavelengths than Algorithm 8.7. Because the latter returns a feasible routing tree that may include

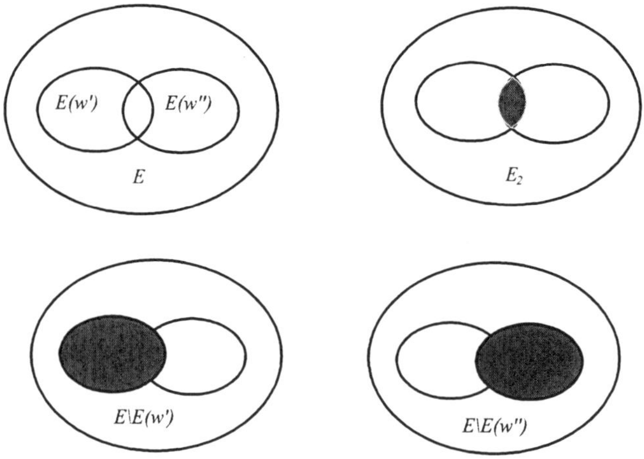

Figure 8.15. Algorithm 8.7 has more freedom to construct a routing tree than Algorithm 8.6.

some edges in $E(w)$ for a number of different $w \in W$, while the former returns a feasible routing tree that may exclude $E(w)$ for a certain $w \in W$.

4.2.4 Simulation Study

In the preceding three subsections, we have proposed two algorithms for Problem 8.4 that integrate routing and wavelength assignment on general WDM networks, Algorithm 8.6 and Algorithm 8.7. In order to evaluate their performances, we use routing Algorithm 8.2 as the bench mark in the following way: route a multicast request without considering load balancing and wavelength assignment, and then assign it a wavelength by employing the greedy strategy for wavelength assignment.

As we focus on how to integrate routing and wavelength assignment. And the purpose of simulation study is to see how well the proposed algorithms works. For the simplicity, we only concentrate on some typical instances.

The network used in the simulation is generated by using the method described in Section 2 of Chapter 1. The cost function c on link (u, v) in the generated graph is the distance between nodes u and v on the rectangular coordinated grid, and the delay function d is set equal to cost function c in the simulations. In the simulation study, β-low cost requirement (8.7) and α-short delay requirement (8.5) are enforced for α varying from 1.1 to 2 while β being fixed at 2 and β varying from 1.1 to 2 while α being fixed at 2, respectively. Note that in all these cases, the cost-delay ratio-relation (8.8) is not satisfied, which means that Algorithm 8.2 may fail to produce a feasible routing tree for some requests. However, in the simulation study such failure occurred twice only when $\beta = 2$ and $\alpha = 1.1$.

When applying the greedy strategy for wavelength assignment, Algorithm 8.2, Algorithm 8.6 and Algorithm 8.7 all adopt the same searching rule: check if one of currently used wavelengths can be assigned to the current request in the order that they were introduced.

The sizes of destination sets for all multicast requests are all fixed at 10. For each request, we randomly select a node in V as the source and then randomly select ten other nodes in V as the destinations. Algorithm 8.2, Algorithm 8.6 and Algorithm 8.7 are run for the same randomly generated sequence of 20 requests coming one by one and no of them is released after it arrives, respectively. This process is repeated 15 times. In the end for $i = 1, 2, \ldots, 20$, we obtain the average number of wavelengths used for first i requests when these three algorithm are applied, respectively.

For the simplicity of presentation, we denote Algorithm 8.2, Algorithm 8.6 and Algorithm 8.7 by A, $A1$ and $A2$, respectively, and denote the average number of wavelengths that they used $W_A(\beta; \alpha; i)$, $W_{A1}(\beta; \alpha; i)$ and $W_{A2}(\beta; \alpha; i)$.

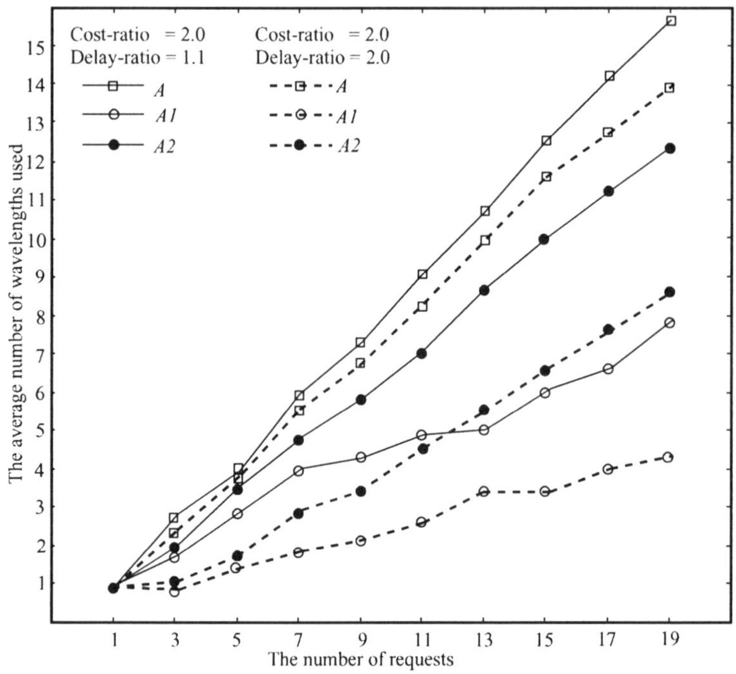

Figure 8.16. The average number of wavelengths used against the number of requests.

Fig. 8.16-17 display $W_{\{A,A1,A2\}}(2; 1.1, 2; i)$ and $W_{\{A,A1,A2\}}(1.1, 2; 2; i)$, respectively. They show how the average number of wavelengths used increases as 20 requests are routed and assigned a wavelength one after another. Fig. 8.18(a) and Fig.

8.18(b) display $W_{\{A,A1,A2\}}(2; 1.1, \cdots, 2; 10)$ and $W_{\{A,A1,A2\}}(1.1, \cdots, 2; 2; 20)$, respectively. They show how the changes of cost and delay ratios affect the average number of wavelengths used for first ten requests and twenty requests, respectively.

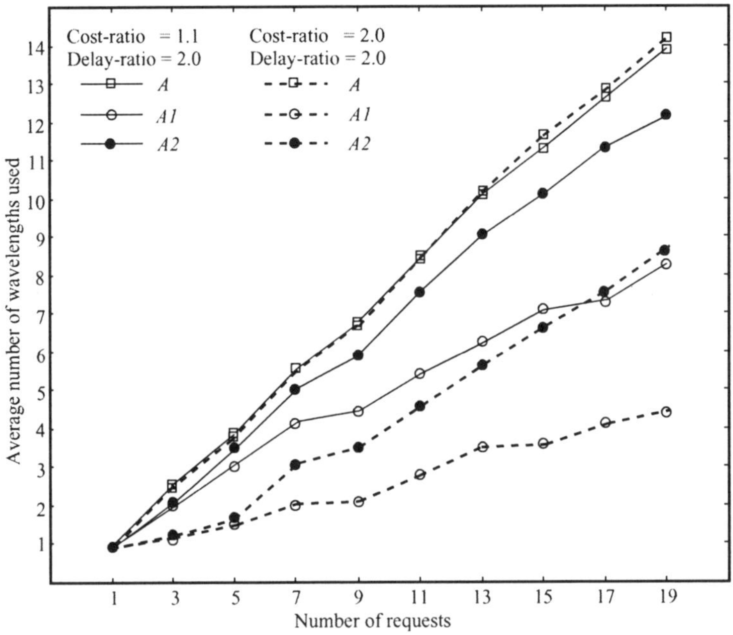

Figure 8.17. The average number of wavelengths used against the number of requests.

From these three figures, we can draw the following conclusions.

(1) Compared with non-integrated Algorithm 8.2, the proposed integrated Algorithm 8.6 and Algorithm 8.7 require considerably less number of wavelengths, although they use the same routing method and adopt the same greedy strategy for wavelength assignment. Still more, Algorithm 8.6 is much more effective than Algorithm 8.7, this is just as we expect in the preceding subsection.

(2) For Algorithm 8.2 and Algorithm 8.7, the average number of wavelengths used increases at a constant rate, while for Algorithm 8.6 this value almost keeps unchanged for three or four requests after it increases considerably. This difference can be more apparently observed in Fig. 8.16 and Fig. 8.17 by comparing $W_{\{A,A2\}}(2; 2; i)$ with $W_{A1}(2; 2; i)$. This implies that Algorithm 8.6, on average, is able to route those twenty requests in such a way that approximately four requests are assigned with the same wavelength.

(3) For Algorithm 8.2, the variation of the delay-ratio causes stronger affect on the average number of wavelengths to be used than that of the cost-ratio. This

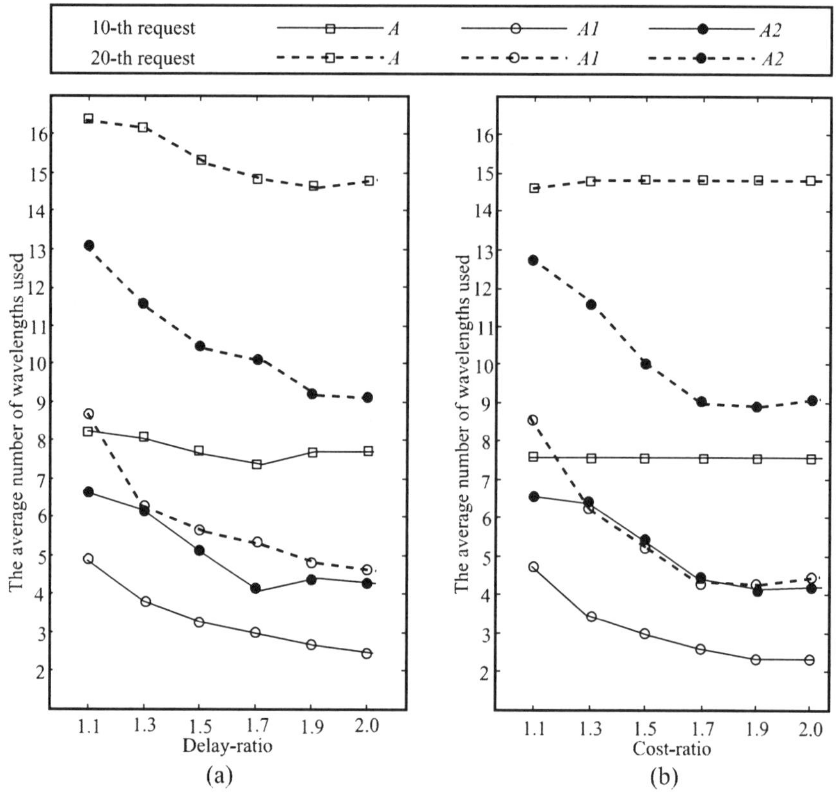

Figure 8.18. (a) The average number of wavelengths used for first ten (and twenty, respectively) requests against the delay-ratio. (b) The average number of wavelengths used for first ten (and twenty, respectively) requests against the cost-ratio.

can be easily observed by comparing $W_A(2; 1.1, \cdots, 2; 10, 20)$ in Fig. 8.18(a) with $W_A(1.1, \cdots, 2; 2; 10, 20)$ in Fig. 8.18(b) and $W_A(2; 1.1, 2; i)$ in Fig. 8.16 with $W_A(1.1, 2; 2; i)$ in Fig. 8.17, respectively. There is little difference between $W_A(1.1; 2; i)$ and $W_A(2; 2; i)$ (and among $W_A(1.1, \cdots, 2; 2; 10, 20)$). As a contrast, there is a considerable difference between $W_A(2; 1.1; i)$ and $W_A(2; 2; i)$ (and among $W_A(2; 1.1, \cdots, 2; 10, 20)$). The reason is that a feasible routing tree is generated by starting with an MST and then modifying it into a feasible one, thus β-low cost requirement can be more easily satisfied than α-short delay requirement. Hence with the delay-ratio being fixed at 2, when a routing tree of a request generated by Algorithm 8.2 satisfies 1.3-low cost requirement, the same routing tree also satisfies β-low cost requirement for $\beta \geq 1.3$ and accordingly is used for the same request.

As a result, the average number of wavelengths used by Algorithm 8.2 for first ten and twenty requests remains unchanged for $\beta \geq 1.3$.

4) For algorithms A and $A2$, the average number of wavelengths used does not monotonously decrease as the delay-ratio or the cost-ratio increases. This can be easily observed in Fig. 5 (a). The reason is that although a feasible routing tree of a request that satisfies cost or delay requirements with small ratios can certainly serve as a feasible routing tree satisfying cost or delay requirements with big ratios, the routing method may generate different feasible routing trees for small ratios and big ratios, respectively. Given a request, after an initial routing tree is generated, there exist more destinations that do not satisfy delay requirement with a small delay ratio than that with a big delay ratio. Thus the process of substituting the shortest path in the current tree with the shortest path in the network will be done more times in the former case than in the latter case. However it can be observed in Fig. 8.17, by comparing $W_A(1.1;2;9)$ with $W_A(2;2;9)$ and $W_A(1.1;2;13)$ with $W_A(2;2;13)$, that for the first 9 and 13 requests the average numbers of wavelengths used by Algorithm 8.2 increase as the cost-ratio increases from 1.1 to 2. The reason behind this is that there do not exist any feasible routing trees for one randomly generated 8-th request and 12-th request when cost-ratio is 1.1 and delay-ratio is 2, while feasible routing trees for all requests can be constructed when the cost-ratio is increased to 2. Thus there are more requests that need to be assigned a wavelength in the latter case than in the former case, this makes the average number of wavelengths used to increase.

5. Discussion

In Section 1 of this chapter, we have presented a well-known 2-approximation algorithm (Problem 8.1) for Steiner tree problem in networks. There is an algorithm due to Zelikovsky [15], which has approximation ratio 11/6. There are two reasons for not using this theoretically better algorithm. One is that Algorithm 8.2 is simple, and the other is that it has an important application in multicast routing under multi-drop lightpath and light-tree models which will be discussed in the next chapter.

In Section 2 of this chapter, we have presented a simple algorithm (Algorithm 8.2) due to Khuller et al [7] for balancing the cost and delay of a spanning/Steiner tree. It is assumed that the cost and delay functions on edges in given network are the same or proportional to each other. However, this may not be true in some applications where the cost is typically a measure of the amount of buffer space or channel bandwidth used while the delay is a combination costs of the propagation, transmission, and queueing delays. Marathe et al [10] studied a more general class of bicriteria network design problems as follows. Given an undirected graph and two minimization objectives (under different cost functions), with a budget specified on the first objective, find a subgraph from a given subgraph-class that minimizes the second objective subject to the budget on the first objective. They discussed three different criteria, the total edge cost, the diameter, and the maximum degree of the network. They developed a framework for bicriteria problems and their approximations. When the two criteria are the same they presented a "black box" parametric search technique. This black box takes in as input an (approximation) algorithm for the unicriterion situation and generates an approximation algorithm for the bicriteria case with only a constant factor loss in the performance guarantee. In particular, when the two criteria are the diameter and the total edge costs, that is the case we discussed in Section 2, they use a cluster-based approach to devise an approximation algorithms whose the output solutions violate both the criteria by a logarithmic factor.

Besides these works, Bharath-Kumar and Jaffe [2] proposed some routing algorithms for the tradeoff between the network cost and delay cost, where the delay cost of a routing tree is defined as the sum of the cost of all paths from the source to destinations divided by the number of destinations. This can be considered as the average delay of a multicast connection. In addition, all of these algorithms are centralized algorithms that require the complete information about the network topology. Jia [4] proposed some distributed algorithms that generate delay-bounded sub-optimal routing trees. The proposed methods can be integrated with the operation of multicast connection configurations.

In Section 3 of this chapter we have discussed the problem of establishing multicast connections, given QoS requirements, under static traffic model in WDM networks. The result of this research can be used as a guideline for configuration of a WDM network, such as the configuration of lightpaths or light-trees. The traffic type discussed in this paper (i.e., QoS multicast traffic) is a general one. It allows us to describe and consider all different types of traffic, such as unicast traffic, multicast traffic, QoS traffic and non-QoS traffic, in a uniform way, because they are special cases of QoS multicast traffic. An "optimal" solution to the proposed problem is applicable to other single-hop or light-tree based systems. Another significant aspect of this work is the discovery of the effectiveness of wavelength optimization methods in multicast domain. In particular, the optimization via reassigning the least used wavelength works more effectively than the optimization via traffic load-balancing.

In Section 4 of this chapter we have discussed the problem of establishing multicast connections, given QoS requirements, under dynamic traffic model in WDM networks. Similar to the static traffic model, we have found that rerouting aiming at wavelength re-assignment demonstrates a considerably better performance than rerouting aiming at load balancing. This means that rerouting to free out the least used wavelengths is more effective than rerouting for load balancing. Besides the simple dynamic model that once connection request arrives it must specify all its destinations (this is the model adopted in this section), Pankaj [11] considered a more complicated dynamic situation where a new request can arrive with only one destination specified and more destinations can be added or existing destinations can quit from an ongoing connection. For both cases, he constructed different network topologies based on expander graphs and showed that the number of wavelengths needed to support multicasting is $O(\log^2 |V|)$. Unfortunately, the constant in front of $\log^2 |V|$ is too large and the network topologies that achieve this bound are too complex to be very practical.

In the work referred above, it is assumed that each link can carry as many as necessary connections and there are plenty of wavelengths to support coming connections. Under such assumptions all connection requests can be satisfied and the goal is to minimize the number of wavelengths used. This can be considered as a dual of *call* or *admission control* version of the wavelength assignment problem. Under this popular formulation, the network load is upper bounded and the goal to maximize the throughput. A typical algorithm for such problems on general networks was proposed by Awerbuch et al [1]. The proposed algorithm was proved to be $O(\log |V|)$-competitive throughput if the network load is $\Omega(\log |V|)$, and when the network load network is unbounded, an $\Omega(|V|)$ lower bound on the competitive ratio of deterministic on-line algorithms was given.

References

[1] B. Awerbuch, Y. Azar, A. Fiat, S. Leonardi, and A. Rosén, On-line competitive algorithms for call admission in optical networks, *Algorithmica*, 31 (2001), 29-34.

[2] K. Bharath-Kumar and J. M. Jaffe, Routing to multiple destinations in computer networks, *IEEE Transactions on Communications*, 31 (1983), 343-351.

[3] E. W. Dijkstra, A note on two problems in connexion with graphs, *Numerical Mathematics*, 1 (1959), 269-271.

[4] X.-H. Jia, A Distributed algorithm of delay-bounded multicast routing for multimedia applications in wide area networks, *IEEE/ACM Transaction on Networking*, 6 (6) (1998), 828-837.

[5] X.-H. Jia, X.-D. Hu, L. Ruan, and J. H. Sun, Multicast routing, load balancing and wavelength assignment on Tree of Rings, *IEEE Communication Letters*, 6 (2) (2002), 79-81.

[6] R. M. Karp, Reducibility among combinatorial problems, in R. E. Miller and J. W. Thatcher (eds.), *Complexity of Computer Computations*, Plenum Press, New York, 1972, 85-103.

[7] S. Khuller, B. Raghavachari, and N. Young, Balancing minimum spanning trees and shortest-path trees, *Algorithmica*, 14 (1995), 305-321.

[8] L. Kou, G. Markowsky, and L. Berman, A fast algorithm for Steiner trees, *Acta Informatica*, 15 (2) (1981), 141-45.

[9] J. B. Kruskal, On the shortest spanning subtree of a graph and the travelling salesman problem, *Proceedings of American Mathematics Society*, 7 (1956), 48-50.

[10] M. V. Marathe, R. Ravi, R. Sundaram, S. S. Ravi, D. J. Rosenkrantz, and H. B. Hunt III, Bicriteria network design problems, *Journal of Algorithms*, 28 (1998), 142-171.

[11] R. K. Pankaj, Wavelength requirements for multicasting in all-optical networks, *IEEE/ACM Transactions on Networking*, 7 (3) (1999), 414-424.

[12] L. H. Sahasrabuddhe and B. Mukherjee, Light-Trees: optical multicasting for improved performance in wavelength-routed networks, *IEEE Communication Magazine*, 37 (2) (1999), 67-73.

[13] L. H. Sahasrabuddhe and B. Mukherjee, Multicast routing algorithms and protocols: a tutorial, *IEEE Network*, 14 (1) (2000), 90-102.

[14] D. Sleator and R. Tarjan, Amortized efficiency of list update and paging rules, *Communications of the ACM*, 28 (1985), 202-208.

[15] Z. Z. Zelikovsky, The 11/6-approximation algorithm for the Steiner problem on networks, *Algorithmica*, 9 (1993), 463-470.

Chapter 9

MULTICAST UNDER MULTI-DROP MODELS

In the preceding chapter we have discussed how to establish QoS guaranteed multicast connections in WDM networks. It was assumed that all nodes in the network were equipped with multicast-capable optical switches and an input optical signal can be split into arbitrary many output optical signals. In practice, however, this is not the case. In fact, since today's networks were designed to mainly support unicast (point-to-point) communication, routing nodes in most of networks do not have multicast capability [16]. In some networks, nodes just have a limited multicast capability, that is, an input optical signal can be split into a limited number of output optical signals. This is because an optical signal transmitted at a source node has fixed amount of power, each time an optical signal is split at a router onto multiple output ports a splitting loss is incurred which reduces the power of the signal at each of the outputs. Thus the application of routing light-trees based multicast is not easy.

In this chapter we study two multi-drop models for multicast routing in WDM networks. One is called *multi-drop lightpath* model, which can be considered as a generalization of point-to-point connection based lightpath model. The other is called *multi-drop light-tree* model, which can be considered as a generalization of point-to-multipoint connection based light-tree model. Under multi-drop lightpath model, data is sent from its source to a destination in a lightpath. During the data transmission along the lightpath, if an intermediate node itself is a destination node, then by splitting the optical signal, the data can be dropped at the destination node while it is forwarded to its adjacent neighbor in the lightpath. Each lightpath includes at most a fixed number of destination nodes where the data can be dropped. Accordingly, multicast routing is to find a set of such paths so that every destination is designated in a path to receive the data. Under multi-drop light-tree model, data is sent from its source to a destination in a light-tree. Each light-tree includes at most a fixed number of destination nodes where the data can be dropped. Accordingly, multicast routing is to find a set of such trees so that every destination node is designated in a tree to receive the data.

Compared with one-light-tree model, multi-drop models make multicast easier and more efficient to implement, but at the expense of demanding more network resources. The reason is that since the network cost of a light-tree is generally less than that of a set of lightpaths or light-trees rooted at the source, and a light-tree just needs one wavelength while a set of lightpaths or light-trees may require more than one wavelengths.

In this chapter we will study two routing problems under multi-drop models. One is the multicast routing and wavelength assignment problem under multi-drop model, that is how to establish a multicast connection under multi-drop model so that the number of wavelengths required is minimized. The other is the minimum cost of multicast routing problem under multi-drop model, that is how to establish a multicast connection under multi-drop model so that the network cost is minimized.

The rest of this chapter is organized as follows. Section 1 studies on how to determine the number of wavelengths required by a multicast connection, and how to find the optimal route for the connection to achieve the minimum number of wavelengths. Section 2 discusses how to produce a multicast routing such that it has minimal cost and requires minimal number of wavelengths, for a given source and a set of destinations. Section 3 concludes the chapter.

1. Routing for Minimization of Wavelength Usage

In this section, we study two routing problems under both multi-drop lightpath and multi-drop light-tree models. One is, given the limited number of drops allowed in a lightpath or light-tree, how to determine the minimal number of wavelengths required for establishing a multicast connection. The other is how to find an optimal routing and wavelength assignment for a given multicast connection, such that the number of wavelengths required is minimized.

We will first present an upper bound of the minimal number of wavelengths required for a multicast connection in general networks, which is expressed by the edge-connectivity of a given network. And then we will develop a polynomial time algorithm to find a sub-optimal routing and wavelength assignment for a given multicast connection request. In the end, we obtain the minimal number of wavelengths required in several widely used network topologies, and give efficient algorithms to find the optimal routing and wavelength assignment for given multicast connection.

1.1 Multi-Drop Lightpath Model

In a network where routing nodes have no multicast support (i.e., an input optical signal cannot be split on multiple output ports), the route of a multicast connection is a path, which links all the destinations. However, due to the power loss when an optical signal drops off at a destination along the path, there is a restriction on the maximal number of destinations that can be dropped on a path. In this section, we assume that the data can be dropped at most m destinations in a lightpath, where k is dependent on the power of light transmission or some other restriction of hardware.

The physical topology of WDM networks under consideration is modelled as an arc-weighted digraph $\vec{G}(V, A)$, where vertex-set V is the set of nodes (switches/routers) in the network and arc-set A is the set of links (fibers) between nodes. We assume that $\vec{G}(V, A)$ is *symmetric*, that is, there is an arc $(u, v) \in A$ from u to v if and only if there is an arc $(v, u) \in A$ from v to u. For the simplicity of presentation, we denote the underlying graph $\vec{G}(V, A)$ by $G(V, A)$ that is obtained by replacing two arcs between a pair of nodes with one edge between them. A multicast connection request is represented by $r(s, D)$, where s is the source node from which data is sent to a set of destination nodes D. We assume *unidirectional transmission* that data can be transmitted only in one direction from the source to the destination nodes, this means, data is allowed to be transmitted over the same link twice but in the the opposite directions.

We define a k*-path* as directed trail in digraph $\vec{G}$ that at most k nodes in the trail are allowed to receive the data (all other nodes in the trail, including destination nodes, can only forward the data to their neighbors in the trail), where an arc can appear in the trail at most once (but a node may appear more than once). The *multicast* k*-routing* under multi-drop lightpath model, denoted by $R_p(s, D, k)$, is to find a set of k-paths such every destination node in D is designated in exactly one lightpath to receive the data. The constrained multicast routing and wavelength assignment problem under multi-drop lightpath model can be formulated as follows.

PROBLEM 9.1 *Multicast Routing and Wavelength Assignment Problem*

Instance A digraph $\vec{G}(V, A)$, a multicast connection request $r(s, D)$, and an integer $k \geq 1$.
Solution A k-routing of $R_p(s, D, k)$ for $r(s, D)$ with assigned wavelengths to k-paths in $R_p(s, D, k)$.
Objective Minimizing the number of wavelengths used.

The following theorem shows that the above problem is $\mathcal{NP}$-hard.

THEOREM 9.1 *Problem 9.1 is* $\mathcal{NP}$*-hard.*

PROOF In the following we will show that the WAP in directed trees (Problem 3.1), which was proved to be $\mathcal{NP}$-hard (refer to Theorem 3.3), can be reduced in polynomial-time to Problem 9.1. Given a set of ordered node pairs in a tree $G_t(V, E)$, $(s_1, t_1), \cdots, (s_m, t_m)$, we construct a graph $G(V \cup \{v_0\}, E')$ with $E' = E \cup \{(v_0, s_i) \mid i = 1, 2, \cdots, m\}$. Now set $s := v_0$, $D := \{t_i \mid i = 1, 2, \cdots, m\}$, and $k = 1$. It is easy to verify that lightpaths from s_i to t_i for $i = 1, 2, \cdots, m$ require the same number of wavelengths that are needed to establish multicast connection $r(s, D)$ by k-routing, and vice versa. ■

1.1.1 General Networks

Let $W_p(G, s, D, k)$ denote the minimal number of wavelengths required for multicast request $r(s, D)$ in network $\vec{G}$ under multi-drop lightpath model. In the following discussion, we will give some upper bounds and lower bounds on $W_p(G, s, D, k)$ in general networks.

THEOREM 9.2 *Let $G(V, A)$ be a $2m$-edge connected graph. Then for any multicast request $r(s, D)$ and positive integer k,*

$$\lceil \frac{|D|}{\delta(s)k} \rceil \leq W_p(G, s, D, k) \leq \lceil \frac{|D|}{km} \rceil,$$

where $\delta(s)$ is the degree of source node s.

PROOF By the assumption, $G(V, A)$ has at least m edge-disjoint undirected spanning trees (refer to [8, 10]). Thus, the data can be set out from source s to k destinations by traversing arcs of edges in every spanning tree. This can produce a k-path. Since these m spanning trees are edge-disjoint, m k-paths can share one wavelength. Thus one wavelength can support the data transmission to km destinations. This implies the upper bound

$$W_p(G, s, D, k) \leq \lceil \frac{|D|}{km} \rceil,$$

On the other hand, at most $\delta(s)$ lightpaths can start from source s and at most k destinations can be dropped in a k-path. Hence, we obtain the lower bound

$$\lceil \frac{|D|}{\delta(s)k} \rceil \leq W_p(G, s, D, k).$$

The proof is then finished. ■

In the above theorem, the upper bound is expressed in terms of the edge-connectivity of given graph, which can be determined in time $O(|V||E|)$ by employing the algorithm proposed in [1, 12]. If the edge-connectivity of graph G is equal to the degree of source node s, i.e., $\delta(s) = m$, then the upper bound and lower bound meet and they give $W_p(G, s, D, k)$. In addition, the proof of Theorem 9.1 also suggests the following algorithm for constrained multicast routing and wavelength assignment problem 9.1.

ALGORITHM 9.1 *Multicast Routing under Multi-drop Lightpath*

Step 1 Find m edge-disjoint spanning trees of graph G of maximal m.
Step 2 Find $\lceil |D|/km \rceil$ k-paths in each spanning tree by traversing each edge in the tree at most twice so that all destinations are designated in a k-path.
Step 3 Assign wavelengths to k-paths in an optimal way.

THEOREM 9.3 *Algorithm 9.1 has approximation performance ratio*

$$\lceil \frac{|D|}{km} \rceil / \lceil \frac{|D|}{\delta(s)k} \rceil \approx \frac{\delta(s)}{m}.$$

PROOF Due to Theorem 9.2, the optimal multicast k-routing needs at least $\lceil \frac{|D|}{\delta(s)k} \rceil$ wavelengths, and the proposed algorithm requires at most $\lceil \frac{|D|}{km} \rceil$ wavelengths. ■

1.1.2 Special networks

We will present some algorithms for solving the constrained multicast routing and wavelength assignment problem in some widely used local area networks. We first consider the networks whose physical topology is a complete graph. In this case, the source s is adjacent to every destination in D, and the link between them is a k-path for any k. Thus the optimal multicast routing consists of $|D|$ edge disjoint k-path which can share one wavelength. Thus we have the following straightforward result.

THEOREM 9.4 *For any multicast request $r(s, D)$ in the complete graph K_n of n vertices, $W_p(K_n, s, D, k) = 1$.*

Next we consider two more difficult cases studied in [11], toruses and hypercube. The technique used is to partition them into edge-disjoint Hamiltonian cycles.

THEOREM 9.5 *For any multicast request $r(s, D)$ in torus $T_{p,q}$ of p rows and q columns, $W_p(T_{p,q}, s, D, k) = \lceil |D|/4k \rceil$.*

PROOF Due to the lower bound given in Theorem 9.2, we have $W_p(T_{p,q}, s, D, k) \geq \lceil |D|/4k \rceil$. Thus to prove the theorem it suffices to show that this is also an upper bound.

A multicast request $r(s, D)$ can be implemented by partitioning D into some groups of size k, at most one of them has size less than k. Since a Hamiltonian cycle contains all vertices in G, data can be delivered from source s to any such group using any of the edge-disjoint Hamiltonian cycles. Thus, h edge-disjoint Hamiltonian cycles may be used to deliver the data to hk destinations using a single wavelength. Since the undirected torus can be decomposed into two Hamiltonian cycles (refer to [2, 5]), the symmetric directed torus can be decomposed into four directed Hamiltonian cycles. Therefore, multicast request $r(s, D)$ can be set up by using at most $\lceil |D|/4k \rceil$ wavelengths. ■

THEOREM 9.6 *For any multicast request $r(s, D)$ in the hypercube H_n of order n, if n is even, then*

$$W_p(H_n, s, D, k) = \lceil \frac{|D|}{nk} \rceil;$$

If n is odd, then

$$\lceil \frac{|D|}{nk} \rceil \leq W_p(H_n, s, D, k) \leq \lceil \frac{|D|}{(n-1)k} \rceil.$$

PROOF It is known that an undirected hypercube H_n contains $\lfloor n/2 \rfloor$ edge-disjoint Hamiltonian cycles (refer to [2, 5]). Thus when n is even, the symmetric directed hypercube H_n can be partitioned into n directed Hamiltonian cycles. By the same argument as in the proof of Theorem 9.5, we have $W_p(H_n, s, D, k) = \lceil |D|/nk \rceil$. When n is odd, $\lfloor n/2 \rfloor = (n-1)/2$, the symmetric directed hypercube H_n contains $(n-1)$ edge-disjoint directed Hamiltonian cycles. Therefore, we have $W_p(H_n, s, D, k) = \lceil |D|/(n-1)k \rceil$, and the lower bound is from Theorem 9.2. ■

In the end, we consider the most complicated case, mesh M_{pq} of p rows and q columns. In this case, the value $W_p(M_{pq}, s, D, k)$ is dependent on the location of source s as well as the cardinality of D and k. We divide the locations of source s into the following four cases, and then we consider them separately.

Case 1. Source s is at a corner of the mesh, as shown in Fig. 9.1(a-b).

Case 2. Source s is a node on the border but not at any corner of the mesh, as shown in Fig. 9.2(a-f).

Case 3. Source s is an inside node of the mesh, and it is adjacent to a node on the border of the mesh, as shown in Fig. 9.3(d).

Case 4. All other cases that are not included in Cases 1-3, as shown in Fig. 9.3(a-c).

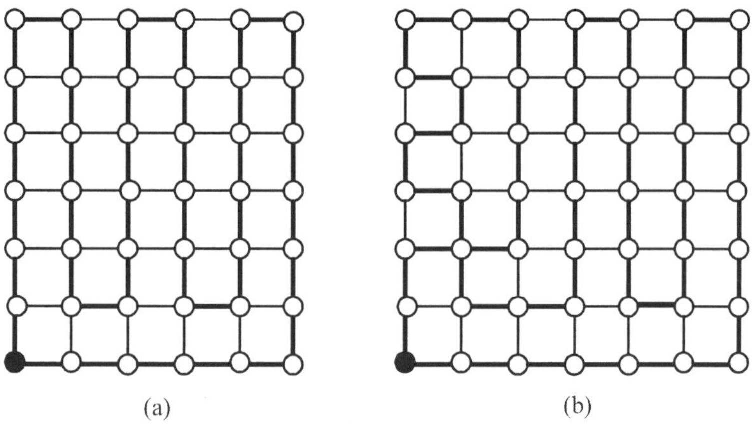

Figure 9.1. Source s is at a corner.

THEOREM 9.7 *For any multicast request $r(s, D)$ in mesh $M_{p,q}$,*
(1) *in Case 1,* $W_p(M_{p,q}, s, D, k) = \lceil |D|/2k \rceil$.

(2) *in Case 2,* $W_p(M_{p,q}, s, D, k) = \lceil |D|/3k \rceil$.
(3) *in Case 3,* $\lceil |D|/4k \rceil \leq W_p(M_{p,q}, s, D, k) \leq \lceil |D|/4k \rceil + 1$.
(4) *in Case 4,* $W_p(M_{p,q}, s, D, k) = \lceil |D|/4k \rceil$.

PROOF We consider these four cases one by one as below.

In Case 1, data can be sent along two trail as shown in Fig. 9.1. Fig. 9.1(a) and Fig. 9.1(b) demonstrate the situations where the number p of columns and the number q of rows are both odd and the rest of cases of p and q, respectively. At each transmission, data travels along those two lightpaths until it visits k destinations or they meet at some node. The latter occurs only when no enough destinations remained. During each transmission, data can be dropped at $2k$ destinations except the last time. The result thus follows.

In Case 2, we assume, without loss of generality, that source s is on the left border. Fig. 9.2(a-b) display the case of odd number of columns, Fig. 9.2(c-d) display the case of even number of columns. Moreover, Fig. 9.2(a) and Fig. 9.2(d) show the cases that there are odd number of rows above the row where source s is located; Fig. 9.2(b) and (c) show the cases where there are even number of rows above the row in which source s is located. In each of these cases, data can be sent along the trails as shown in Fig. 9.2(a-d) until it meets $\lceil |D|/3k \rceil$ destinations. Denote the path that data travels by P_1. It then can be easily checked that two other edge-disjoint paths P_2 and P_3 (also edge-disjoint to P_1) can be found in order such that they include the rest of destinations regardless of where P_1 terminates and they respectively contain $\lceil |D|/3k \rceil$ or $\lfloor |D|/3k \rfloor$ destinations as demonstrated in Fig. 9.2(e-f), where Fig. 9.2(e) shows the case where there are even number of columns and Fig. 9.2(f) shows the case where there are odd number of columns. Data can be sent along the three paths in a parallel way. At each transmission, data is sent along them until it meets k destinations or it reaches the ends (there are no k destinations left). Therefore, the data can be dropped at all destinations through $\lceil |D|/3k \rceil$ times of such transmissions.

In Case 4, we assume, without loss of generality, that the source s is in the region of down-left of mesh. Fig. 9.3(a-c) show three possible cases where source s is one, two, and more than two rows away from bottom line. It can be verified that there always exist four edge-disjoint trails that include all nodes in the mesh as shown in Fig. 9.3(a-c). Thus any multicast connection can be set up on them. Each of four lightpaths can include k destinations except when there are less than k destinations left, and it is easy to see that $\lceil |D|/4k \rceil$ wavelengths are enough for any multicast connection.

In Case 3, the situation is similar to Case 4, especially when there is no destination in D located at the corner, four lightpaths can be found as in Case 4 and $\lceil |D|/4k \rceil$ wavelengths are enough for such multicast connections. But one extra wavelength may be required when there is a destination in D located at the corner. For instance, suppose $k = 3$ and D contains 12 nodes (indicated in gray as shown in Fig. 9.3(d)), then $\lceil |D|/4k \rceil = 1$, but, 2 wavelengths are required. The proof is then finished. ■

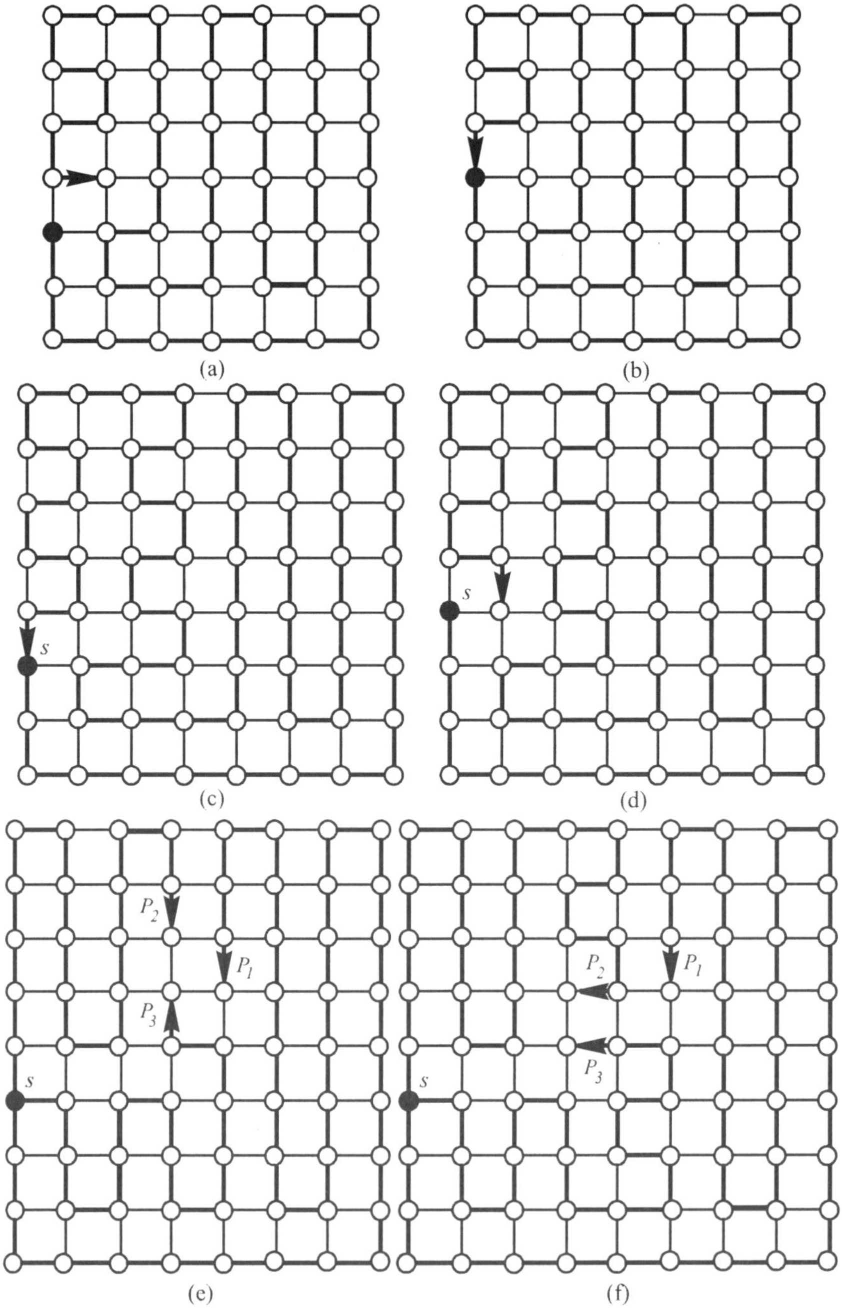

Figure 9.2. Source s is at the border.

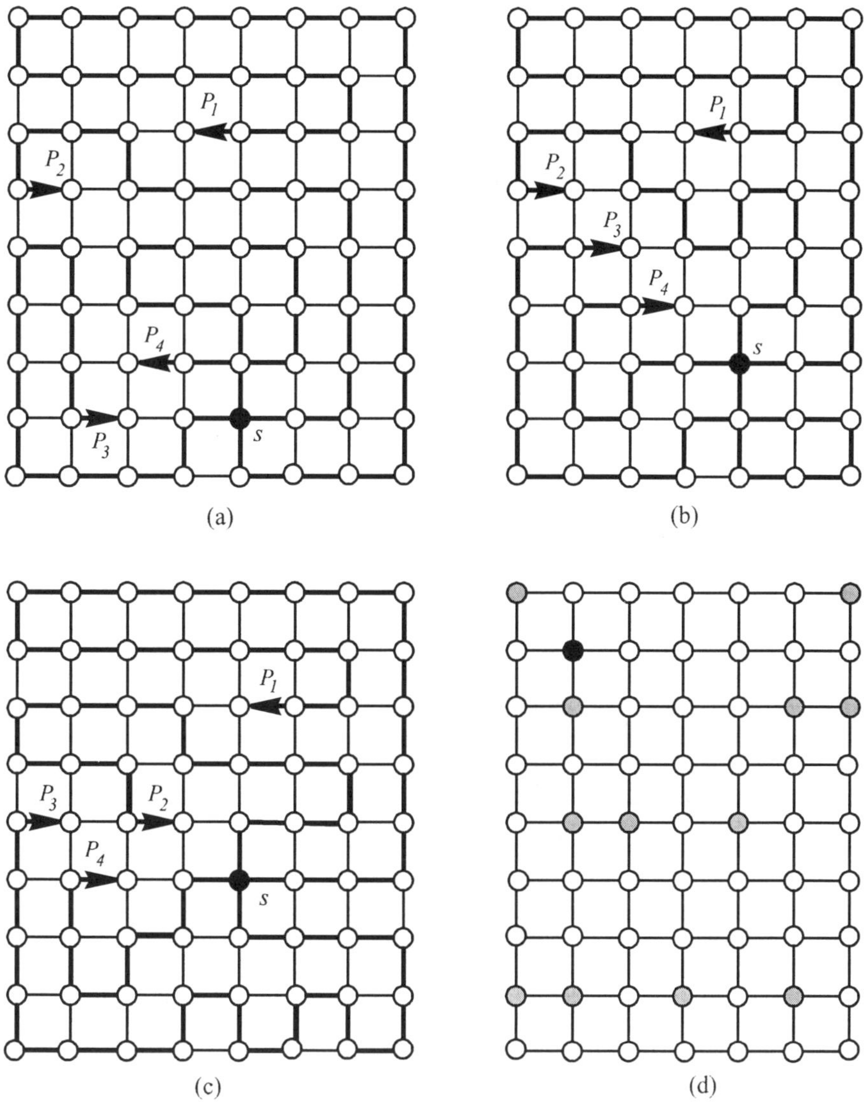

Figure 9.3. Source s is at the inside.

The argument used in the proof of Theorem 9.7 is constructive, so it suggests a routing and wavelength assignment algorithm for establishing a given multicast connection in mesh M_{pq}.

1.2 Multi-Drop Light-tree Model

In the recent development, multicast capable switches become available. A multicast capable node can split an incoming signal into multiple copies and put on multiple output ports. This leads to the light-tree model. A light-tree is a tree rooted at the source node, on which all edges must use the same wavelength to transfer data. Each time an optical signal is split, a splitting loss is incurred which reduces the power of the signal at the outputs [14]. Moreover, when a message is dropped off at a destination in the light-tree, it also incurs dropping loss. Thus, the number of destinations allowed to drop off in a light-tree is limited. In this section, we assume that the data can be dropped at most k destinations in a light-tree, where k is dependent on the power of light transmission. In the multi-drop light-tree model, the establishment of a multicast connection may need more than one light-trees and wavelengths.

The network model and notations used in this subsection are the same as in the preceding subsection. In a similar way, we define a *k-tree* as directed tree in$\vec{G}$ rooted at source s such that at most k nodes in the tree are allowed to receive the data (all other nodes in the tree, including destination nodes, can only forward the data to their neighbors in the tree), where an arc can appear in the tree at most once (but a node may appear more than once). The *multicast k-routing* under multi-drop light-tree model, denoted by $R_t(s, D, k)$, is to find a set of k-trees such every destination node in D is designated in exactly one light-tree to receive the data. The constrained multicast routing and wavelength assignment problem under multi-drop light-tree model can be formulated as follows.

PROBLEM 9.2 *Multicast Routing and Wavelength Assignment Problem*

Instance A digraph $\vec{G}(V, A)$, a multicast connection request $r(s, D)$, and an integer $k \geq 1$.
Solution A k-routing of $R_t(s, D, k)$ for $r(s, D)$ with assigned wavelengths to k-paths in $R_t(s, D, k)$.
Objective Minimizing the number of wavelengths used.

Since k-routing under multi-drop light-tree model becomes k-routing under multi-drop lightpath model when $k = 1$, the following theorem comes directly from Theorem 9.1.

THEOREM 9.8 *Problem 9.2 is $\mathcal{NP}$-hard.*

Let $W_t(G, s, D, k)$ denote the minimal number of wavelengths required for multicast connection request $r(s, D)$ in network G under multi-drop light-tree model. In the following discussion, we will first give an upper bound and a lower bound on $W_t(G, s, D, k)$ in general networks, and then present algorithms for solving the constrained multicast routing and wavelength assignment problem in some widely used local area networks.

1.2.1 General Networks

It is easy to see that $W_t(G, s, D, k) \leq W_p(G, s, D, k)$ since k-routing under multi-drop lightpath model is also k-routing under multi-drop lightpath model. To determine $W_t(G, s, D, k)$, we need to introduce some terms and lemmas about digraphs. Given a digraph $\vec{G}(V, A)$, and $u, v \in V$, an edge-separator of (u, v) is a set $A' \subseteq A$ of arcs such that every directed path from u to v uses at least one edge in A'. Let $M(u, v)$ be the least cardinality of an edge-separator of (u, v). Then the edge-connectivity of $\vec{G}(V, A)$ is $m = \min\{M(u, v) \mid (u, v) \in V \times V\}$. The following lemma was proved in [4].

LEMMA 9.1 *Let $s \in V$ be a vertex of digraph $\vec{G}(V, A)$, and $m \equiv \min\{M(s, v) \mid v \in V \setminus \{s\}\}$. Then there are m edge-disjoint directed spanning trees of $\vec{G}$ rooted at s.*

The following lemma can be immediately deduced from Lemma 9.1

LEMMA 9.2 *If $G(V, A)$ is an m-edge-connected graph, then for any $s \in V$, there are m edge-disjoint directed spanning trees of $\vec{G}$ rooted at s.*

PROOF Note that $m \leq \min\{M(s, v) \mid v \in V \setminus \{s\}\}$. ■

The following theorem gives both upper and lower bounds on $W_t(G, s, D, k)$.

THEOREM 9.9 *Let $G(V, A)$ be an m-edge- connected graph, then for any multicast request $r(s, D)$ and a positive integer k,*

$$\lceil \frac{|D|}{\delta(s)k} \rceil \leq W_t(G, s, D, k) \leq \lceil \frac{|D|}{km} \rceil,$$

where $\delta(s)$ is the degree of source s in graph G.

PROOF It is obvious that there are at most $\delta(s)$ edge-disjoint light-trees rooted at source s, which can share one wavelength and include at most $\delta(s)k$ destinations. This yields the lower bound on $W_t(G, s, D, k)$. Due to Lemma 9.2, there are m edge-disjoint directed spanning trees of rooted at source s, each of them being used as a light-tree rooted s can include k destinations. This implies that data can be dropped at km destinations on these m edge-disjoint light-trees that share one wavelength (except when there are less than k destinations left). This yields the upper bound on $W_t(G, s, D, k)$. ■

The following theorem suggests an approximation algorithm for Problem 9.2.

THEOREM 9.10 *There is a polynomial-time algorithm that, given any multicast request $r(s, D)$ in an m-edge-connected graph G and a positive integer k, produces a k-routing which requires at most $\delta(s)/m$ times wavelengths that the optimal k-routing uses.*

PROOF Note that there is an $O(m^2|V||A|^2)$-time algorithm, proposed in [4], for finding m edge-disjoint directed trees rooted at s. By Theorem 9.9 and its proof, this algorithm can be modified and used as an approximation algorithm to construct a k-routing. It requires at most $\lceil |D|/km \rceil$ wavelengths. Note that the optimal k-routing needs at least $\lceil |D|/\delta(s)k \rceil$ wavelengths. ∎

Two corollaries can be immediately deduced from the above two theorems.

COROLLARY 9.1 *If G is an m-edge-connected and k-regular graph, then*
(1) $W_t(G, s, D, k) = \lceil |D|/km \rceil = \lceil |D|/\delta(s)k \rceil$, for any multicast request $r(s, D)$;
(2) Problem 9.2 can be solved in polynomial-time.

COROLLARY 9.2 *Given graph G, if $G_p(G, s, D, k) = \lceil |D|/(\delta(s)k) \rceil$ for multicast request $r(s, D)$, then $G_t(G, s, D, k) = \lceil |D|/(\delta(s)k) \rceil$.*

1.2.2 Special Networks

For Problem 9.2 in special networks, k-routing and their performances analysis can be constructed and obtained in the same way as for Problem 9.1 in special networks. So in the following we just present the parallel results without proofs.

THEOREM 9.11 *For any multicast request $r(s, D)$ in the complete graph K_n of n vertices, $W_t(K_n, s, D, k) = 1$.*

THEOREM 9.12 *For any multicast request $r(s, D)$ in torus $T_{p,q}$ of p rows and q columns, $W_t(T_{p,q}, s, D, k) = \lceil |D|/4k \rceil$.*

THEOREM 9.13 *For any multicast request $r(s, D)$ in the hypercube H_n of order n, if n is even, then*

$$W_t(H_n, s, D, k) = \lceil \frac{|D|}{nk} \rceil;$$

If n is odd, then

$$\lceil \frac{|D|}{nk} \rceil \leq W_t(H_n, s, D, k) \leq \lceil \frac{|D|}{(n-1)k} \rceil.$$

THEOREM 9.14 *For any multicast request $r(s, D)$ in mesh $M_{p,q}$,*
(1) *in Case 1, $W_t(M_{p,q}, s, D, k) = \lceil |D|/2k \rceil$.*
(2) *in Case 2, $W_t(M_{p,q}, s, D, k) = \lceil |D|/3k \rceil$.*
(3) *in Cases 3-4, $W_t(M_{p,q}, s, D, k) = \lceil |D|/4k \rceil$.*

2. Routing for Minimum Cost

In the preceding section, we have studied the constrained multicast routing and wavelength assignment problem under multi-drop model. The objective of Problem 9.1-2 is to find k-routing so that the number of wavelengths required is minimized, where the cost issue of k-routing is not touched. Given a multicast connection

$r(s, D)$, the *network cost* of k-routing for establishing $r(s, D)$ is the total costs of lightpaths in $R_p(s, D, k)$ or light-trees $R_t(s, D, k)$. Observe that the produced k-routings in the preceding section have very high costs and long delays (see Fig. 9.1-3).

In this section, we will focus on how to, given a multicast connection $r(s, D)$ and a positive integer $k \geq 1$, construct a k-routing whose network cost is minimal. In addition, we will also consider the wavelength assignment problem under the optimal k-routing of minimum cost.

2.1 Multi-drop Lightpath Model

The network model is almost the same as the one we described in preceding section, except that a cost function on the links is introduced. The physical topology of WDM networks under consideration is modelled as an arc-weighted digraph $G(V, A, c)$, where vertex-set V is the set of nodes and arc-set A is the set of links between nodes. For arc $(u, v) \in A$, cost function $c : A \to \Re^+$ measures the desirability of using a particular link (a lower cost means more desirable). We assume that $G(V, A, c)$ is totally symmetric, that is, there is an arc $(u, v) \in A$ from u to v if and only if there is an arc $(v, u) \in A$ from v to u, and the costs on two arcs between u and v are the same, i.e., $c(u, v) = c(v, u)$. We also assume that the cost function c is additive over the links in a (directed) path $p(u, v)$ from u to v, i.e.,

$$c(p(u, v)) \equiv \sum_{a \in p(u,v)} c(a).$$

For the simplicity of presentation, we denote by $p_{G'}$ the shortest path from u to v in subdigraph G' of G.

A k-routing of multicast connection $r(s, D)$ is denoted by $R_p(s, D, k) = \{P_i \mid i\}$, where each P_i is a k-path. Clearly, $m \equiv \lceil |D|/k \rceil \leq |R_p(s, D, k)| \leq |D|$. Since, the data is transmitted through each arc in a k-path in $R_p(s, D, k)$ exactly once, the network cost of multicasting data is then defined as the total of costs of k-paths in $R_p(s, D, k)$, that is,

$$c(R_p(s, D, k)) \equiv \sum_{P_i \in R(s,D,k)} c(P_i). \qquad (9.1)$$

Under multi-drop lightpath model the multicast routing problem for minimal cost is then formulated as follows.

PROBLEM 9.3 *Minimum Multicast Routing Problem*

Instance A digraph $\vec{G}(V, A, c)$, a multicast connection $r(s, D)$, an integer $k \geq 1$.
Solution A multicast k-routing $R_p(s, D, k)$.
Objective Minimizing the network cost of $R_p(s, D, k)$ as defined in equation (9.1).

Observe that when $k = 1$, the optimal solution to the above problem consists of $|D|$ shortest paths from source s to each of $|D|$ destination nodes. Thus it can be found in polynomial-time. The following theorem shows that this is also true when $k = 2$.

THEOREM 9.15 *When $k = 2$ Problem 9.3 is polynomial-time solvable.*

PROOF We will show that in this case Problem 9.3 can be reduced in polynomial-time to the minimum weight matching problem, which can be solved in polynomial-time [13]. Given a multicast connection $r(s, D)$ on network $\vec{G}$, where $D = \{d_1, \cdots, d_{|D|}\}$, the reduction can be done as follows. See Fig. 9.4, where D contains six destination nodes in black.

Step 1. For $d_i, d_j \in D$ and $i \neq j$, compute the shortest path $p_G(d_i, d_j)$ between them, and the shortest path $p_G(s, d_i)$ between source s and d_i.

Step 2. Construct an auxiliary graph $G'(D \cup \{s_1, \cdots, s_{|D|}\}, E')$. There is an edge between d_i and d_j for $i \neq j$ that is given weight $w(d_i, d_j) = \min\{c(p_G(s, d_i)) + c(p_G(d_i, d_j)), c(p_G(s, d_j)) + c(p_G(d_j, d_i))\}$. There is an edge between s_i and s_j for $i \neq j$ that is given weight $w(s_i, s_j) = 0$. There is an edge between s_i and d_i for each i that is given weight $w(s_i, d_i) = c(p_G(s, d_i))$.

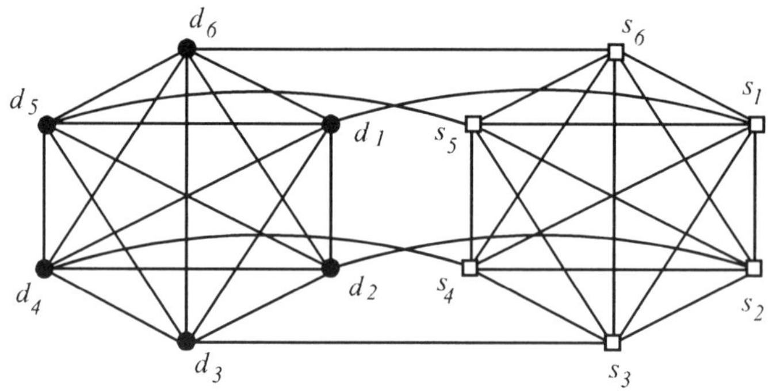

Figure 9.4. The reduction from Problem 9.3 to minimum weight matching problem.

Clearly, Step 1-2 can be done in polynomial-time. Moreover, given a *minimum weight matching* of G' that is a perfect matching M of G' whose weight (the total weight of edges in the matching) is minimal, we can produce an optimal k-routing for $r(s, D)$ on G in the following way.

(1) For each edge $(d_i, d_j) \in M$, produce a k-path from s to d_j via d_i that consists of $P_G(s, d_i)$and $P_G(d_i, d_j)$ if $w(d_i, d_j) = c(p_G(s, d_i)) + c(p_G(d_i, d_j))$ and a k-path from s to d_i via d_j that consists of $P_G(s, d_j)$ and $P_G(d_j, d_i)$ if $w(d_i, d_j) = c(p_G(s, d_j)) + c(p_G(d_i, d_j))$;

(2) For each edge $(s_i, d_i) \in M$, produce a k-path from s to d_i.
The correctness of the reduction follows from two facts, which can be easily verified. (a) Each of possible shortest k-paths is associated with exactly one edge in G, and its cost is the weight of the edge. (b) Each of destination nodes is incident to exactly one edge in M. That means, every destination node is designated in a produced k-path. The proof is then finished. ■

The following theorem, however, shows that the minimum multicast routing problem in general is not polynomial-time solvable unless $\mathcal{P} = \mathcal{NP}$.

THEOREM 9.16 *Problem 9.3 is $\mathcal{NP}$-hard.*

PROOF We will consider the decision version of Problem 9.3. Given a multicast connection $r(s, D)$ on network G, an integer $k > 2$ and a bound $B > 0$, the problem asks if there is a k-routing for $r(s, D)$ whose cost is at most B.

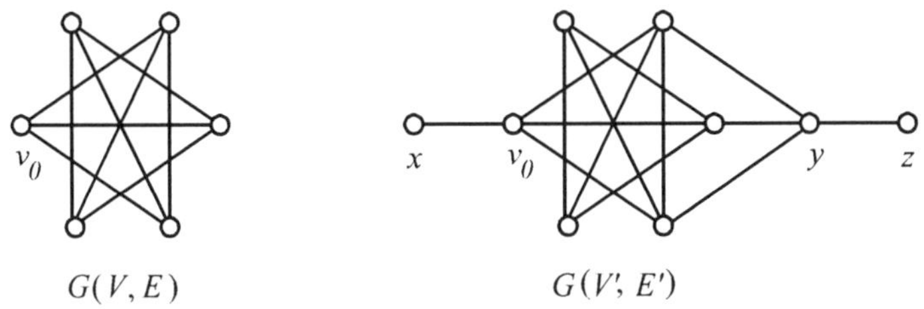

Figure 9.5. Constructing a new graph $G(V', E')$ from given 3-regular graph $G(V, E)$.

It was proved in [13] that Hamilton circuit problem for graphs with all nodes of degree three (that is a 3-regular graph) is $\mathcal{NP}$-complete. It was also proved there that Hamilton path problem is $\mathcal{NP}$-complete through a simple reduction as follows. Given a 3-regular graph $G(V, E)$, construct a new graph $G(V', E')$ where $V' = V \cup \{x, y, z\}$ and $E' = E \cup \{(y, z), (x, v_0)\} \cup \{(y, v) \mid (v, u_0) \in E\}$, for some fixed $v_0 \in V$. See Fig. 9.5. It can be verified that $G(V, E)$ has a Hamilton circuit if and only if $G(V', E')$ has a Hamilton path.

We now show that Hamilton path problem for graph $G(V', E')$ can be reduced in polynomial-time to Problem 9.3. First, construct $\vec{G}(V', A, c)$ by substituting each edge $(u, v) \in E'$ with a pair of arcs (u, v) and (v, u) whose costs c are equal to one. Secondly, set source node $s := x$, destination set $D := V' \setminus \{x\}$, bound $B := |V'| - 1$, and $k := |V'|$. In the end, it is easy to verify that $G(V', E')$ has a Hamilton path if and only if $\vec{G}(V', A, c)$ has a k-routing for $r(s, D)$ whose cost is at most B. ■

In the above we have proved that the minimum multicast routing problem 9.3 in general is $\mathcal{NP}$-hard, we will present two approximation algorithms with guaranteed performances. These two algorithms are based on the approximation of minimum Steiner tree.

The basic idea of our first algorithm is to transverse an approximation of minimum Steiner tree of $D \cup \{s\}$, which can be produced by a spanning tree based approximation algorithm (refer to Section 1 of Chapter 8).

ALGORITHM 9.2 *Constructing a k-routing*

Step 1 Construct an auxiliary complete edge-weighted graph G_a of $D \cup \{s\}$. For $u, v \in D \cup \{s\}$, the weight of edge (u, v) is the cost of the shortest path in G between u and v.

Step 2 Construct a minimum spanning tree T_{min} of G_a and substitute each edge in T_{min} by the shortest path between two endpoints in G. And then obtain a Steiner tree T_S of $D \cup \{s\}$ in G.

Step 3 Obtain a directed trail $\mathcal{T}$ of $D \cup \{s\}$ by traversing each vertex of $D \cup \{s\}$ along T_S whose edges are replaced by two arcs between two endpoints of the edges, that is, $\mathcal{T} = (s \to d_1 \to \cdots \to d_{|D|} \to s)$. See Fig. 9.6.

Step 4 Partition $\mathcal{T}$ into k subtrail $\mathcal{T}_i = \{d_{ik+1} \to d_{ik+2} \to \cdots \to d_{ik+k}\}$ for $i = 0, 1, \cdots k-1$. And then for each i find v_i that is the vertex in subtrail $\mathcal{T}_i$ which is closest to source s. See Fig. 9.7(a).

Step 5 Construct two k-paths via v_i including destination nodes in subtrail $\mathcal{T}_i$,
$P_i' = \{s \to d_{i'} \to d_{i'-1} \to \cdots \to d_{ik+1}\}$,
$P_i'' = \{s \to d_{i'+1} \to d_{i'+2} \to \cdots \to d_{ik+k}\}$,
where v_i lies between $d_{i'}$ and $d_{i'+1}$ on $\mathcal{T}_i$. See Fig. 9.7(b).

Step 6 Output k-routing $R_{st}(s, D, k) = \{P_i' \,|\, i\} \cup \{P_i'' \,|\, i\}$.

In fact, Step 1-2 of Algorithm 9.2 is Algorithm 8.1 for constructing a 2-approximation solution to the minimum Steiner tree problem (refer to Theorem 8.1). We now prove that the Steiner tree based Algorithm 9.2 has guaranteed performance under worst case analysis. To do this, we need the following lemma. Given a multicast connection demand $r(s, D)$, let R_{opt} be the optimal k-routing and $c(R_{opt})$ be its cost.

LEMMA 9.3 *Let $d_{i'}$ be the destination node in trail $\mathcal{T}_i$ that is closest to s. Then*

$$\sum_{i=1}^{k} c(p_G(s, d_{i'})) \leq c(R_{opt}).$$

PROOF Suppose that an optimal routing is $R_{opt} = \{P_i^* \,|\, i = 1, \cdots, N\}$, where $N \geq m$. Then we construct an auxiliary weighted bipartite graph $B(X, Y)$, where X is the set of subtrails $\mathcal{T}_i$ and $Y = \{P_i^* \,|\, i = 1, \cdots, N\}$. There exists an edge $(\mathcal{T}_i, P_j^*)$ in $B(X, Y)$ if and only if $\mathcal{T}_i$ and P_j^* include $\alpha \geq 1$ destination nodes in common and the weight of the edge is $w(\mathcal{T}_i, P_j^*) = \alpha$.

Now we prove, by using Hall's Theorem [9], that $B(X, Y)$ has a perfect matching such that each $\mathcal{T}_i$ is incident to an edge in the matching. Suppose, by contradiction, that there exists a subset $X_0 \subseteq X$ such that X_0's neighbor set $Y_0 \subseteq Y$, which consists

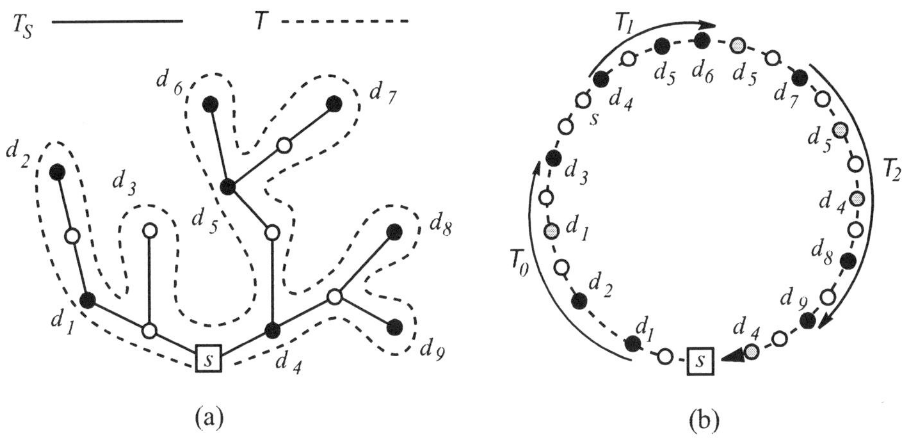

Figure 9.6. (a) Traversing all destination nodes in D along the Steiner tree T_s. (b) Partitioning the obtained trail $\mathcal{T}$ into subtrails $\mathcal{T}_i$.

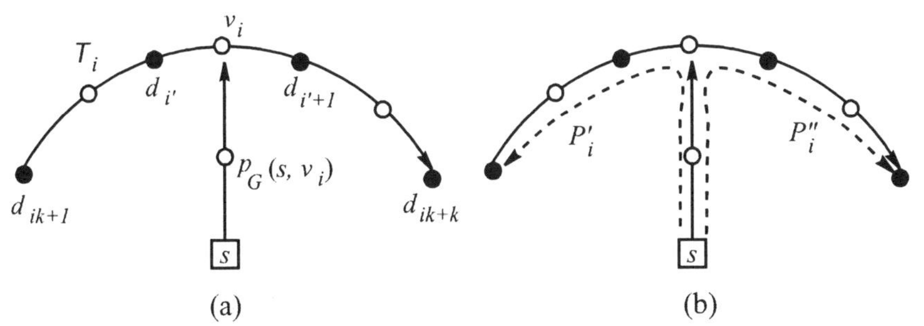

Figure 9.7. (a) Find node v_i in $\mathcal{T}_i$ which is closest to source s. (b) Produce two k-paths P'_i and P''_i.

of vertices adjacent with some vertex in X_0, satisfies $|Y_0| \leq |X_0| - 1$. Since each $\mathcal{T}_i$ includes at most k destinations and each of them is included in just one optimal k-path, then the total weight of edges incident to $\mathcal{T}_i$ is at most k. For each P_j^* we have the same result. Now for $X' \subseteq X$ and $Y' \subseteq Y$, let $w(X')$ and $w(Y')$ denote the total weights of edges incident to some $\mathcal{T}_i \in X'$ and $P_j^* \in Y'$, respectively. Then $w(Y') \leq k|Y'|$, this implies $w(X_0) \leq w(Y_0) \leq k|Y_0|$. In addition, we have

$$w(X \setminus X_0) \leq k|X \setminus X_0| = k(k - |X_0|).$$

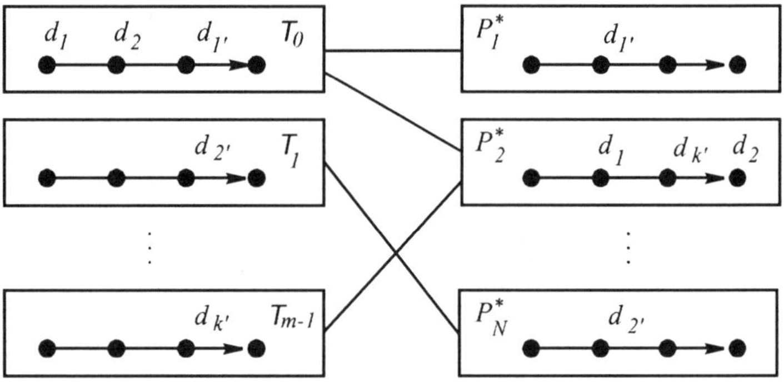

Figure 9.8. Constructing a bipartite graph $B(X, Y)$.

Hence we obtain the following contradiction.

$$|D| = w(X_0) + w(X \setminus X_0) \leq k|Y_0| + k(m - |X_0|) \leq k(m-1) < |D|.$$

Therefore, there exists a desired matching. Without loss of generality, we denote this matching by $M = \{(\mathcal{T}_i, P_i^*) \mid i\}$. This means that for each i there exists a destination node in both $\mathcal{T}_i$ and P_i^*. Thus the cost of P_i^* is not less than the cost of the shortest path from s to that common destination node, which, by the definition of $d_{i'}$, is not less than the cost of the shortest path from s to $d_{i'}$, i.e.,

$$c(P_i^*) \geq c(p_G(s, d_{i'})).$$

To sum up above inequality over i, we obtain the desired inequality. ■

THEOREM 9.17 *Given a multicast connection demand $r(s, D)$ and $k \geq 2$, Algorithm 9.2 produces a k-routing R_{st} in time $O(k|D|^k|V|^2)$ whose cost is at most four times that of an optimal k-routing R_{opt}.*

PROOF Notice that the optimal k-routing $R_{opt} = \{P_i^* \mid i = 1, \cdots, N\}$ can be transformed into a undirected spanning tree of G_a, that can be done by substituting each directed path in P_i^* from one destination node to another with the shortest path in G. Thus $c(R_{opt}) \geq c(T_{min})$. According to the rules of Algorithm 9.2 and Lemma 9.3, we have

$$\begin{aligned} c(R_{st}) &= \sum_{i=0}^{m-1} \left(c(P_i') + c(P_i'') \right) \\ &< 2c(T_S) + 2\sum_{i=0}^{m-1} c(p_G(s, v_i)) \\ &\leq 2c(T_{min}) + 2c(R_{opt}) \leq 4c(R_{opt}) \end{aligned}$$

Now consider the running time of Algorithm 9.2. First, notice that the shortest path between a pair of vertices in G can be found in time $O(|V|^2)$, thus the auxiliary graph G_a at Step 1 can be constructed in time $O(|D||V|^2)$. Secondly, notice that the minimal spanning tree T_{min} of G_a and Steiner tree T_S at Step 2 can be produced in time $O(|D|^2)$ and $O(|D||V|^2)$, respectively. Thirdly, Steiner tree T_S has at most $(|V|-1)$ edges, thus directed trail $\mathcal{T}$ at Step 3 and its partition into m subtrails at Step 4 can be obtained in time $O(|V|)$. In the end, every k-path can be produced in time $O(|V|^2)$. Therefore, Algorithm 9.2 can output a k-routing in time $O(k|D|^k|V|^k)$. The proof is then finished. ■

For some special cases, Algorithm 9.2 can be modified so that its approximation ratio 4 could be considerably reduced. Moreover, the produced k-routing has cost within a small constant times cost of minimum multicast routing under one light-tree model.

COROLLARY 9.3 *For the case of $m = 1$, there is a polynomial-time algorithm that produces a k-routing whose cost is at most two times that of an optimal k-routing and also at most four times that of minimum Steiner tree of $D \cup \{s\}$.*

PROOF In this case, a subtrail $\{s \to d_1 \to d_2 \to \cdots \to d_{|D|-1|} \to d_{|D|}\}$ can be a k-path. Its cost is less than that of trail $\mathcal{T}$, which is at most two times that of minimal spanning tree T_{min}. At the same time, the cost of T_{min} is not greater than that of an optimal k-routing and it is at most two times that of minimum Steiner tree of $D \cup \{s\}$ due to Theorem 8.1. ■

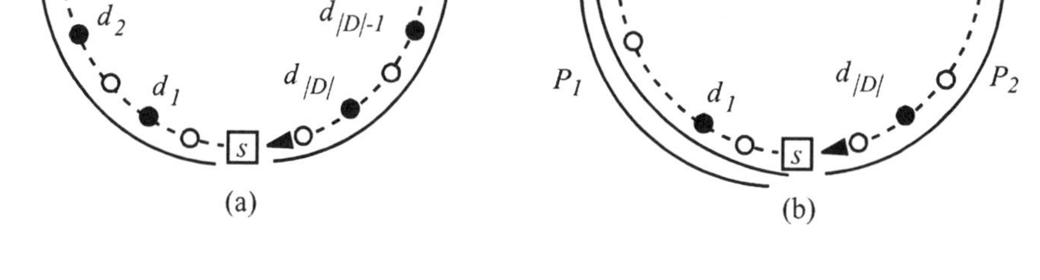

Figure 9.9. (a) k-routing for case of $m = 2$. (b) k-routing for case of $m = 3$.

COROLLARY 9.4 *For the case of $m = 2$, there is a polynomial-time algorithm that produces a k-routing whose cost is at most two times that of an optimal k-routing and also at most four times that of the minimum Steiner tree of $D \cup \{s\}$.*

PROOF After Step 4 of Algorithm 9.2, construct two k-paths (see Fig. 9.9(a)),

$$\begin{aligned} P_1 &= \{s \to d_1 \to d_2 \to \cdots \to d_{k-1} \to d_k\}, \\ P_2 &= \{s \to d_{|D|} \to d_{|D|-1} \to \cdots \to d_{k+2} \to d_{k+1}\}. \end{aligned}$$

Clearly, $c(P_1) + c(P_2) < c(T) \leq 2c(T_{min}) \leq 2c(R_{opt})$. In addition, by Theorem 8.1 again the cost of T_{min} is less than two times that of the minimum Steiner tree of $D \cup \{s\}$. ■

COROLLARY 9.5 *For the case of $m = 3$, there is a polynomial-time algorithm that produces a k-routing whose cost is at most three times that of an optimal k-routing and also at most six times that of the minimum Steiner tree of $D \cup \{s\}$.*

PROOF After Step 4 of Algorithm 9.2, construct two k-paths,

$$\begin{aligned} P_1 &= \{s \to d_1 \to d_2 \to \cdots \to d_{k-1} \to d_k\}, \\ P_2 &= \{s \to d_{|D|} \to d_{|D|-1} \to \cdots \to d_{2k+2} \to d_{2k+1}\}; \end{aligned}$$

And then make another k-path as follows (see Fig. 9.9(b)).

$$P_3 = \begin{cases} \{s \to d_{k+1} \to d_{k+2} \to \cdots \to d_{2k-1} \to d_{2k}\}, & \text{if } c(P_1) \leq c(P_2), \\ \{s \to d_{2k} \to d_{2k-1} \to \cdots \to d_{k+2} \to d_{k+1}\}, & \text{if } c(P_1) > c(P_2). \end{cases}$$

It is easy to verify $c(P_1) + c(P_2) + c(P_3) \leq \frac{3}{2}c(T) \leq 3c(T_{min}) \leq 3c(R_{opt})$. In addition, by Theorem 8.1 the cost of T_{min} is less than two times that of minimum Steiner tree of $D \cup \{s\}$. ■

The basic idea of the second algorithm is to transform Problem 9.3 into the minimum set cover problem, which can be solved approximately by a greedy algorithm [3]. Given a collection of subsets $C = \{S_1, \cdots, S_k\}$ of a finite set S, where $\cup_{i=1}^{k} S_i = S$ and each subset S_i is associated with a weight $w(S_i) > 0$. A subcollection $C' \subseteq C$ is called a *set cover* if each element in S is in at least one subset of C'. The problem is to find a set cover whose weight, the total weights of subsets in the set cover, is minimal.

Given $k \geq 1$, the algorithm first produces all possible j-paths for $1 \leq j \leq k$ in a nearly optimal way as in Algorithm 9.2. Each j-path can be considered as a subset of D that contains j destinations in D, and all possible j-paths constitute a collection C of subsets D. The algorithm then finds a subcollection C' to cover D in a greedy way. The obtained subcollection C' is a set of j-paths that constitute a k-routing since every destination in D is designated in one of j-paths.

ALGORITHM 9.3 *Constructing a k-routing*

Step 1 For each subset of j destination nodes $\{d_{i_1}, \cdots, d_{i_j}\} \subseteq D$, where $j = 1, \cdots, k$, produce a j-path as follows (as in the proof of Corollary 9.3).

Step 1.1 Construct an auxiliary graph $G(d_{i_1}, \cdots, d_{i_j})$ that is a complete edge-weighted graph of $\{d_{i_1}, \cdots, d_{i_j}\} \cup \{s\}$. For $u, v \in \{d_{i_1}, \cdots, d_{i_j}\} \cup \{s\}$ the weight of edge (u, v) is the cost of the shortest path in G between u and v.

Step 1.2 Construct a minimal spanning tree $T_{min}(d_{i_1}, \cdots, d_{i_j})$ of $G(d_{i_1}, \cdots, d_{i_j})$ and substitute each edge in $T_{min}(d_{i_1}, \cdots, d_{i_j})$ by the shortest path between two endpoints in G. Obtain a Steiner tree $T_S(d_{i_1}, \cdots, d_{i_j})$ of $\{d_{i_1}, \cdots, d_{i_j}\} \cup \{s\}$.

Step 1.3 Obtain a j-path $P(d_{i_1}, \cdots, d_{i_j})$ by traversing each vertex in $\{s\} \cup \{d_{i_1}, \cdots, d_{i_j}\}$ along $T_S(d_{i_1}, \cdots, d_{i_j})$ whose edges are replaced by two arcs between two endpoints of the edges.

Step 2 For each subset $\{d_{i_1}, \cdots, d_{i_j}\}$, assign weight $w(d_{i_1}, \cdots, d_{i_j}) \equiv c(P(d_{i_1}, \cdots, d_{i_j}))$. Find a set cover C' of D, by using a subcollection of $\mathcal{C} = \{\{d_{i_1}, \cdots, d_{i_j}\} \mid \text{all } \binom{|D|}{j}\ j\text{-tuples } \{d_{i_1}, \cdots, d_{i_j}\} \text{ and } j = 1, \cdots, k\}$, as follows.

Step 2.1 Choose a subset $S_g = \{d_{i_1}, \cdots, d_{i_g}\}$ that satisfies the equality $w(d_{i_1}, \cdots, d_{i_g})/g = \min\{w(d_{i_1}, \cdots, d_{i_j})/j \mid \{d_{i_1}, \cdots, d_{i_j}\} \in \mathcal{C}\}$.

Step 2.2 Put $\{d_{i_1}, \cdots, d_{i_g}\}$ into set cover C', and remove every element in $\{d_{i_1}, \cdots, d_{i_g}\}$ from $D, \mathcal{C}$ and each $\{d_{i_1}, \cdots, d_{i_j}\}$.

Step 2.3 Repeat (2.1-2) until D is empty.

Step 3 Output a k-routing $R_{sc} = \{P(d_{i_1}, \cdots, d_{i_j}) \mid \{d_{i_1}, \cdots, d_{i_j}\} \text{ in } C'\}$.

THEOREM 9.18 *Given a multicast connection $r(s, D)$ and $k \geq 2$, Algorithm 9.3 produces a k-routing R_{sc} in time $O(k|D|^k|V|^2)$ whose cost is $2\mathcal{H}(k)$ times that of an optimal k-routing R_{opt}, where $\mathcal{H}(k) = \sum_{i=1}^{k} 1/i$.*

PROOF Let C_{opt} be the optimal set cover of the set cover problem constructed at Step 2. Then it can be proved (refer to [3]) that

$$\begin{aligned} c(R_{sc}) &= \sum_{\{d_{i_1}, \cdots, d_{i_j}\} \in C'} c(P(d_{i_1}, \cdots, d_{i_j})) \\ &\leq \mathcal{H}(k) \sum_{\{d_{i_1}, \cdots, d_{i_j}\} \in C_{opt}} c(P(d_{i_1}, \cdots, d_{i_j})) \end{aligned}$$

Now suppose the optimal k-routing $R_{opt} = \{P^*(d_{i_1}, \cdots, d_{i_j}) \mid \{d_{i_1}, \cdots, d_{i_j}\} \in C^*\}$, where $P^*(d_{i_1}, \cdots, d_{i_j})$ is the optimal k-path including all destination nodes in $\{d_{i_1}, \cdots, d_{i_j}\}$. Since C^* is a set cover, then we have

$$\sum_{\{d_{i_1}, \cdots, d_{i_j}\} \in C_{opt}} c(P(d_{i_1}, \cdots, d_{i_j})) \leq \sum_{\{d_{i_1}, \cdots, d_{i_j}\} \in C^*} c(P(d_{i_1}, \cdots, d_{i_j})).$$

Notice that by using the same argument as in the proof of Theorem 9.17 and Corollary 9.3, we can prove that for each k-path $P(d_{i_1}, \cdots, d_{i_j})$ produced by Algorithm 9.3,

$$c(P(d_{i_1}, \cdots, d_{i_j})) \leq 2c(P^*(d_{i_1}, \cdots, d_{i_j})).$$

Therefore, to combine above three inequalities we have

$$c(R_{sc}) \leq 2\mathcal{H}(k) \sum_{\{d_{i_1},\cdots,d_{i_j}\}\in\mathcal{C}^*} c(P^*(d_{i_1}, \cdots, d_{i_j})) = 2\mathcal{H}(k)c(R_{opt}).$$

This is the desired bound on the performance ratio of Algorithm 9.3.

Now consider the running time of Algorithm 9.3. First, notice that the discussion in the proof of Theorem 9.17 shows that j-path can be produced in time $O(|V|^2)$, and total of $\sum_{j=1}^{k} \binom{|D|}{j}$ paths need to be produced. Thus the instance of set cover problem at Step 1 can be constructed in time $O(k|D|^k|V|^2)$. Secondly, since $|\mathcal{C}| = \sum_{j=1}^{k} \binom{|D|}{j}$, each time to find a subset S_g from $\mathcal{C}$ can be done in time $O(k|D|^m)$ and the set cover $\mathcal{C}'$ can be found in time $O(k|D|^{k+1})$ (since $\mathcal{C}'$ includes $O(|D|)$ subsets). Therefore, Algorithm 9.3 outputs a k-routing in time $O(k|D|^k|V|^2)$. The proof is then finished. ■

For the case of small k, Algorithm 9.3 can be modified so that its approximation ratio can be reduced to $\mathcal{H}(k)$.

COROLLARY 9.6 *When $k = 3$, there is a polynomial-time algorithm that produces k-routing whose cost is at most 11/6 times that of an optimal k-routing.*

PROOF Notice that at Step 1 of Algorithm 9.3 for each triple $\{d_{i_1}, d_{i_2}, d_{i_3}\}$, an optimal k-path including d_{i_1}, d_{i_2}, and d_{i_3} can be produced as follows: Find the shortest paths between each pair of them, which can be proved to make a closed trail. And then, compute respectively the shortest paths between source node s and d_{i_1}, d_{i_2}, and d_{i_3}. See Fig. 9.10(a). It is not difficulty to verify that the optimal k-path is one of the following six k-paths. See Fig. 9.10(b).

$$p_G(s, d_{i_1}) \cup p_G(d_{i_1}, d_{i_2}) \cup p_G(d_{i_2}, d_{i_3}), \quad p_G(s, d_{i_1}) \cup p_G(d_{i_1}, d_{i_3}) \cup p_G(d_{i_3}, d_{i_2}),$$

$$p_G(s, d_{i_2}) \cup p_G(d_{i_2}, d_{i_1}) \cup p_G(d_{i_1}, d_{i_2}), \quad p_G(s, d_{i_2}) \cup p_G(d_{i_2}, d_{i_3}) \cup p_G(d_{i_3}, d_{i_1}),$$

$$p_G(s, d_{i_3}) \cup p_G(d_{i_3}, d_{i_2}) \cup p_G(d_{i_2}, d_{i_1}), \quad p_G(s, d_{i_3}) \cup p_G(d_{i_3}, d_{i_1}) \cup p_G(d_{i_1}, d_{i_2}).$$

Therefore, it follows from the proof of Theorem 9.18 a k-routing whose cost is at most $\mathcal{H}(3)$ times that of an optimal k-routing can be found in polynomial-time. ■

2.2 Multi-drop Light-tree Model

In the subsection we will focus on the minimum multicast routing problem under multi-drop light-tree model. The network model and notations used in this subsection are the same as in the preceding subsections. Some of the ideas and techniques used

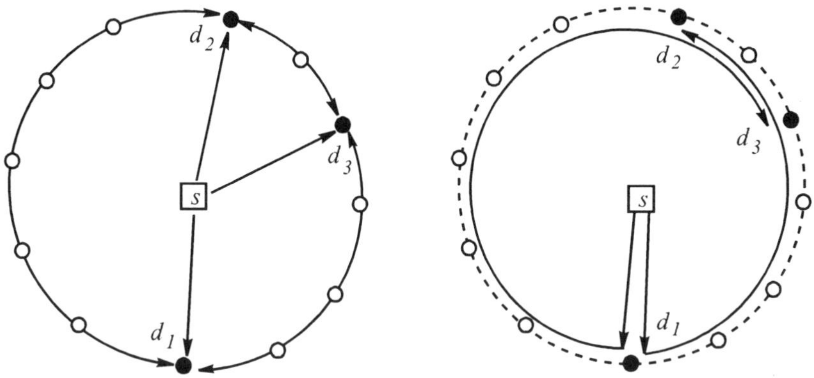

Figure 9.10. (a) The shortest paths between three destination nodes make a closed trail. (b) Two k-paths from source node s towards destination node d_1.

in the preceding subsection for multi-drop lightpath model can also be used in this subsection. A k-routing of multicast connection request $r(s, D)$ is denoted again by $R_t(s, D, k) = \{T_i \mid i\}$, where each T_i is a k-tree. Since the data is transmitted through each arc in k-tree T_i exactly once, the cost of multicasting data is then defined as the total of network costs of T_i in $R_t(s, D, k)$, that is,

$$c(R_t(s, D, k)) \equiv \sum_{T_i \in R_t(s,D,k)} c(T_i). \tag{9.2}$$

The minimum multicast routing problem under multi-drop light-tree model can be formulated as follows.

PROBLEM 9.4 *Minimum Multicast Routing Problem*

Instance A digraph $\vec{G}(V, A, c)$, a multicast connection $r(s, D)$, an integer $k \geq 1$.
Solution A multicast k-routing $R_t(s, D, k)$.
Objective Minimizing the network cost of $R_t(s, D, k)$ as defined in equation (9.2).

Observe that when $k = 1$, the above problem becomes Problem 9.3, thus it can be solved in polynomial-time. The following theorem claims that this is also true when $k = 2$. Observe that, however, when $k \geq 2$, the optimal solution to Problem 9.4 may be different from the optimal solution to Problem 9.3.

THEOREM 9.19 *The Problem 9.4 in case of $k = 2$ is polynomial-time solvable.*

PROOF As in the proof of Theorem 9.15 we will show again that Problem 9.4 can be reduced in polynomial-time to minimum weight matching problem, which is polynomial-time solvable. Given a multicast connection request $r(s, D)$ on network $G(V, A, c)$, where $D = \{d_1, \cdots, d_{|D|}\}$, the reduction can be done as follows.

Step 1. Computer the shortest path $p_G(u, v)$ each node pair u and v in V.

Step 2. Compute the minimum Steiner tree $T(d_i, d_j, s)$ of $\{d_i, d_j,\}$ for each node pair d_i and d_j in D. This can be done by finding $u \in V$ that minimizes $p_G(d_i, u) + p_G(d_j, u) + p_G(s, u)$.

Step 3. Construct an auxiliary graph $G'(D \cup \{s_1, \cdots, s_{|D|}\}, E')$. There is an edge between d_i and d_j for $i \neq j$ that is given weight $w(d_i, d_j) = c(T(d_i, d_j, s))$. There is an edge between s_i and s_j for $i \neq j$ that is given weight zero. There is an edge between s_i and d_i for each i that is given weight $w(s_i, d_i) = c(p_G(s_i, d_i))$. There is no edge between d_i and s_j for $i \neq j$.

Clearly, Step 1-3 can be done in polynomial-time. Moreover, given a minimum weight matching $M \subset E'$ of G', we can produce an optimal k-light-tree routing of $r(s, D)$ on G as follows.

(1) For each edge $(d_i, d_j) \in M$, produce a k-tree that is a minimum Steiner tree of $\{s, d_i, d_i\}$;

(2) For each edge $(s_i, d_i) \in M$, produce a k-tree that is the shortest path from s to d_i.

The proof is then finished. ■

THEOREM 9.20 *Problem 9.4 is $\mathcal{NP}$-hard.*

PROOF The minimum Steiner tree problem (Problem 8.1) can be reduced to Problem 9.4 by setting k any number no less than $|D|$, that is, a k-routing includes just one k-tree, which is a Steiner tree. ■

The two algorithms proposed in preceding subsections for Problem 9.3 under multi-drop lightpath model can be modified into two similar algorithms for Problem 9.4 under multi-drop light-tree model. Their approximation performance analysis can be obtained in a similar way.

We first examine Algorithm 9.3 for constructing a k-routing under multi-drop lightpath model and modify it into Algorithm 9.3' for constructing a k-routing under multi-drop light-tree model. For this purpose, in Step 5 of Algorithm 9.3, we can produce one k-tree T_i, instead of two k-paths, that consists of subtrail $\mathcal{T}_i$ and the shortest path $p_G(s, v_i)$ from source s to v_i (refer to Fig. 9.7).

THEOREM 9.21 *Given a multicast connection request $r(s, D)$ and $k \geq 2$, Algorithm 9.3' produces a k-routing R_{st} in time $O(k|D|^k|V|^2)$ whose cost is at most five times that of an optimal k-routing R_{opt}.*

PROOF The basic argument used in the proof of Theorem 9.17 is applicable here. The approximation ratio is increased from four to five, because under multi-drop light-tree model we can not obtain $c(T_{min}) \leq c(R_{opt})$ as in the case of multi-drop lightpath model. We have

$$c(T_{min}) \leq 2c(R_{opt}), \tag{9.3}$$

because the optimal k-routing under multi-drop light-tree model has cost no less than the cost of the optimal Steiner tree of $D \cup \{s\}$ which is not less than half times cast of T_{min} due to Theorem 8.1. Thus we have

$$\begin{aligned} c(R_{st}) &= \sum_{i=0}^{m-1} c(t_i) < 2c(T_s) + \sum_{i=0}^{m-1} c(p_G(s, v_i)) \\ &\leq 2c(T_{min}) + c(R_{opt}) \leq 5c(R_{opt}). \end{aligned}$$

The proof is then finished. ■

For the case of small m, Algorithm 9.3' can be modified so that its approximation ratio can be reduced.

COROLLARY 9.7 *When $m = 1$, there is a polynomial-time algorithm that produces a k-routing whose cost is at most two times that of an optimal k-routing and the minimum Steiner tree of $D \cup \{s\}$.*

PROOF Notice that when $m = 1$, Algorithm 9.3' becomes the spanning tree based approximation algorithm (Algorithm 8.1) for Steiner tree problem (Problem 8.1). Thus $c(R_{st}) = c(T_{min})$. The corollary immediately follows from the following inequality (9.3) and Theorem 8.1. ■

COROLLARY 9.8 *When $m = 2$, there is a polynomial-time algorithm that produces a k-routing whose cost is at most four times that of an optimal k-routing and the minimum Steiner tree of $D \cup \{s\}$.*

PROOF After operation Step 4 of Algorithm 9.3', construct two k-trees, T_1 including $\{s, d_1, \cdots, d_{k-1}, d_k\}$ and T_2 containing $\{s, d_{|D|}, \cdots, d_{k+2}, d_{k+1}\}$. Clearly, $c(T_1) + c(T_2) \leq 2c(T_{min})$. The corollary then follows from inequality (9.3) and Theorem 8.1. ■

We now examine Algorithm 9.4 for constructing a k-routing under multi-drop lightpath model and modify it into Algorithm 9.4' for constructing a k-routing under multi-drop light-tree model. For this purpose, we can remove Step 1.3 of Algorithm 9.4 since Steiner tree $T_S(d_{i_1}, \cdots, d_{i_j})$ of $\{d_{i_1}, \cdots, d_{i_j}\}$ obtained at the end of Step 1.2 can be used as a j-tree rooted at s that designates destinations in $\{d_{i_1}, \cdots, d_{i_j}\}$ to receive the data.

THEOREM 9.22 *Given a multicast connection request $r(s, D)$ and $k \geq 2$, Algorithm 9.4' produces a k-routing R_{sc} in time $O(k|D|^k|V|^2)$ whose cost is at most $2\mathcal{H}(k)$ times that of an optimal k-routing R_{opt}.*

PROOF The argument used in the proof of Theorem 9.18 is applicable here. Under multi-drop light-tree model, we have

$$c(R_{sc}) = \sum_{\{d_{i_1}, \cdots, d_{i_j}\} \in C'} c(T_S(d_{i_1}, \cdots, d_{i_j}))$$

$$\leq \mathcal{H}(m) \sum_{\{d_{i_1},\cdots,d_{i_j}\}\in\mathcal{C}_{opt}} c(T_S(d_{i_1},\cdots,d_{i_j}))$$
$$\leq \mathcal{H}(m) \sum_{\{d_{i_1},\cdots,d_{i_j}\}\in\mathcal{C}^*} c(T_S(d_{i_1},\cdots,d_{i_j})).$$

Let $T^*(d_{i_1},\cdots,d_{i_j})$ denote the optimal k-tree that designates destinations in $\{d_{i_1},\cdots,d_{i_j}\}$. Then due to Theorem 8.1, we have

$$c(T_S(d_{i_1},\cdots,d_{i_j})) \leq 2c(T^*(d_{i_1},\cdots,d_{i_j})).$$

Therefore, to combine above two inequalities we have

$$c(R_{sc}) \leq 2\mathcal{H}(m) \sum_{\{d_{i_1},\cdots,d_{i_j}\}\in\mathcal{C}^*} c(T^*(d_{i_1},\cdots,d_{i_j})) = 2\mathcal{H}(k)c(R_{opt}).$$

This is the desired bound on the performance ratio of Algorithm 9.4'. ∎

2.3 Simulation Study

In the preceding subsections, we have proposed some algorithms for constructing k-routings under multi-drop lightpath and light-tree models with objective to minimize the network cost of k-routing. At the same time we have made theoretical analysis of their performances, respectively. In this subsection, we will investigate how many wavelengths the produced k-routing require through simulation study. To allocate wavelengths to the produced k-routing, we will use Algorithm 8.3. It is expected that a k-routing with less network cost requires fewer wavelengths.

The objective of the simulation work is to determine how much the multi-drop k-routing can save in network cost and wavelengths over the shortest path routing (i.e., point-to-point connection) under various network environments. In addition, in order to see the effective of the proposed routing algorithm, we use the network cost of minimum routing under Steiner tree model as the performance benchmark.

In the simulations we use two network topologies. One is the backbone of NSFnet (refer to Chapter 1). It consists of 14 nodes representing 14 states in the USA. The cost of a link joining two states is the driving distance between them. Another one is randomly generated again by using the method described in Chapter 1. The generated graphs have 100 nodes whose degrees are 6.5 on average. Cost function c on an edge in the generated graphs is the distance between its two end nodes on the rectangular coordinated grid. The multicast connection requests are generated randomly. Source node s and destination nodes D are randomly picked up from the nodes in generated network graph.

In the simulations, we simulate the network cost and the number of wavelengths against two parameters: $|D|$ the number of destination nodes varying from 5 to 50 and k the number of drops in a lightpath varying from 2 to 5. The number of wavelengths presented in the figures below are the mean values of 50 simulation runs.

In the simulation, when $k = 2$, k-routings under multi-drop lightpath and light-tree models are produced by using methods described in the proofs of Theorem 9.15 and Theorem 9.16 (they have minimal costs), respectively. When $k \geq 3$, k-routings under multi-drop lightpath and light-tree models are produced by applying Algorithm 9.3 and Algorithm 9.3', respectively.

Fig. 9.11 and Fig. 9.13 show the costs and the number of wavelengths used by shortest path, Steiner tree, and multi-drop routing methods against the number of drops allowed in NSFnet, respectively. Two sets of results are displayed. One in solid line is for the case of 13 destinations and the other in dashed line is for the case of 7 destinations. Notice that the cost and the number of wavelengths used by the shortest path and Steiner tree routing do not vary with the number of drops, i.e., they are constant (shown as horizontal lines). Fig. 9.12 and Fig. 9.14 show the costs and the number of wavelengths used by these four routing methods against the number of destinations for with multi-drop number 4, respectively. Notice that the costs and the number of wavelengths used by all routing methods increase proportionately as the number of the destinations increases. Fig. 9.15-18 show the parallel results in the randomly generated network.

Observe that in Fig. 9.11, Fig. 9.13, Fig. 9.15, and Fig. 9.17, we do not join the results of $k = 2$ with the results of $k = 3$, since they are produced by different algorithms and the former are the optimal solution while the latter are the approximate solutions. From those eight figures we can draw the following conclusions.

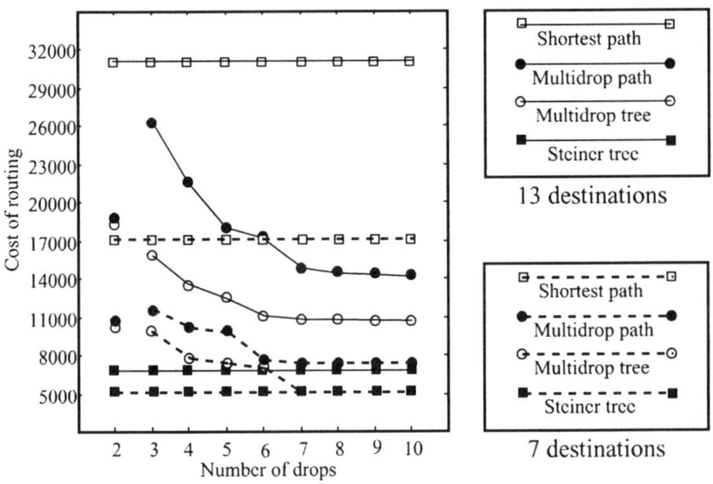

Figure 9.11. The network cost against the number of drops allowed in NSFnet.

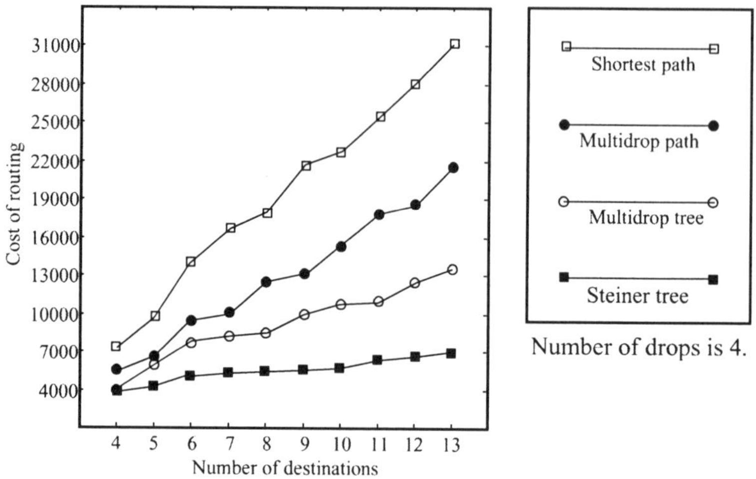

Figure 9.12. The network cost against the number of destinations in NSFnet.

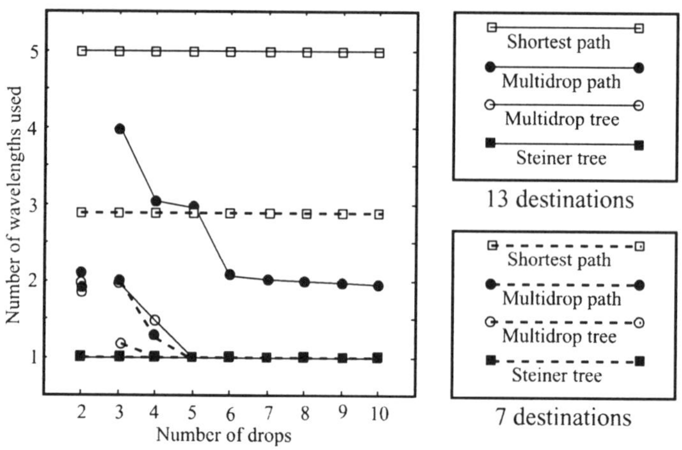

Figure 9.13. The number of wavelengths required against the number of drops allowed in NSFnet.

(1) The network cost of the shortest path routing is about one and two times that of multi-drop lightpath and light-tree routings, respectively. These ratios are independent of the sizes of multicast connection and they are very stable. This can be observed from Fig. 9.11-12 and Fig. 9.15-16.

(2) The network costs of multi-drop k-routing decrease as the number of multi-drop number k increases. However, increasing k is not very effective in decreasing the network cost. This can be seen from Fig. 9.11 and Fig. 9,15. The reason behind this interesting result is that when multi-drop number k becomes bigger, although k-routing will consist of less number of k-paths or k-trees, each k-path or k-tree will become bigger so that it includes more number of destinations. This will make k-paths or k-trees more costly.

(3) The network cost of the Steiner tree routing is about one and two times that of multi-drop lightpath and light-tree k-routings, respectively. In fact, when multi-drop number k becomes large enough, the ratios are much better than guaranteed performance ratio 4 and 5 of Algorithm 9.3 and Algorithm 9.3'. This can be observed from Fig. 9.11 and Fig. 9.15. The reason is that the performance ratio is obtained by using the worst case analysis.

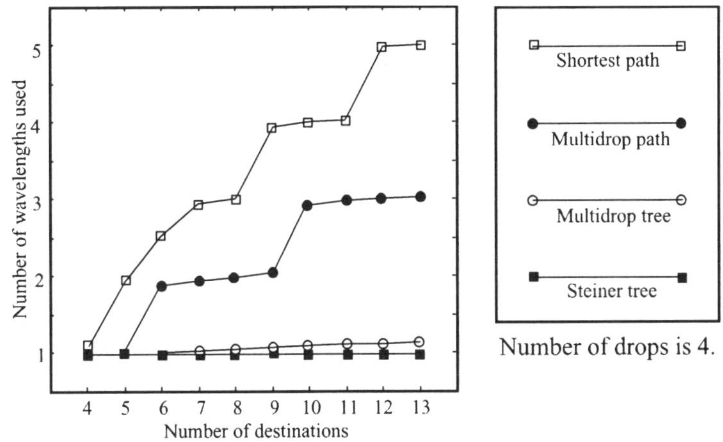

Figure 9.14. The number of wavelengths required against the number of destinations in NSFnet.

(4) The number of wavelengths used by the shortest path routing is about three and four times that of multi-drop lightpath and light-tree routings, respectively. These ratios are independent of the sizes of multicast connections and they are very stable. This can be observed from Fig. 9.14 and Fig. 9.18.

(5) The number of wavelengths used by multi-drop k-routings decreases as multi-drop number k increases. However, increasing k is not very effective in decreasing the network cost for multi-drop k-routings. This can be observed from Fig. 9.13 and Fig. 9.17. The reason behind this interesting result is that when multi-drop number k becomes bigger, although k-routing will consist of less number of k-paths or k-trees, each k-path or k-tree will contain more number of links so that it includes

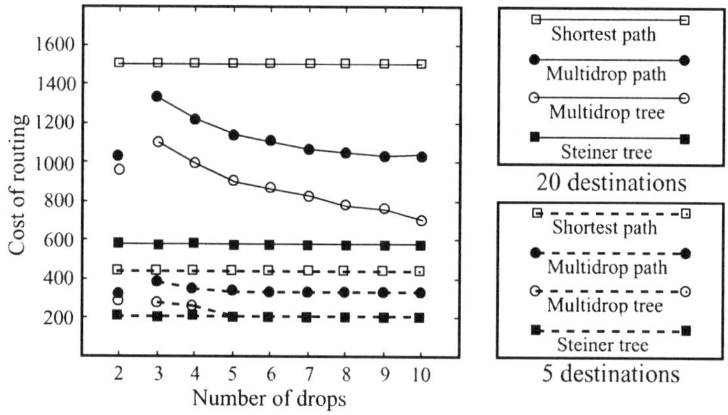

Figure 9.15. The network cost against the number of drops allowed in the general network.

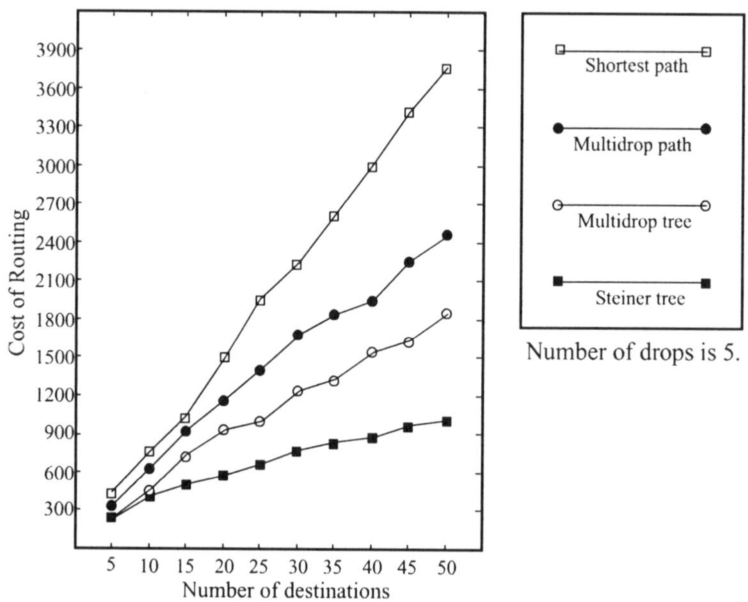

Figure 9.16. The network cost against the number of destinations in the general network.

more number of destinations. This will make k-trees to have more chances to share links among them, and thus prevent them from sharing a wavelength.

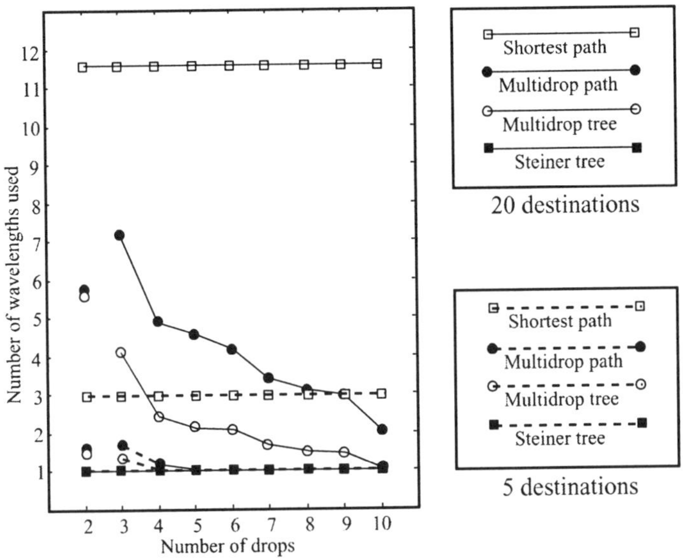

Figure 9.17. The number of wavelengths required against the number of drops allowed in the general network.

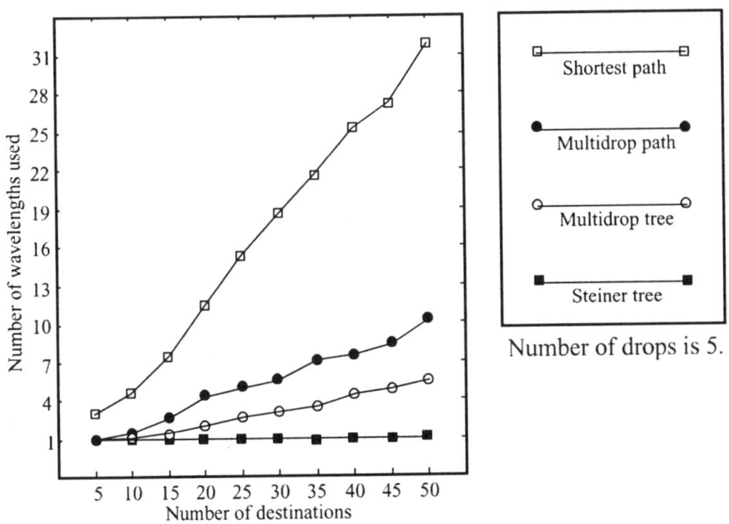

Figure 9.18. The number of wavelengths required against the number of destinations in the general network.

(6) In general, the multi-drop routing algorithms are more effective in saving the wavelengths than the network cost, although they are designed to construct minimum cost multi-drop routings. The reason is that a multi-drop k-routing of less cost consists of k-paths or k-trees of less costs. Thus these k-paths or k-trees tend to have less number of links, and as a result they have more chances to share a wavelength with others.

(7) The proposed algorithms have the same performance in NSFnet and the randomly generated networks. This shows that they are very effective.

3. Discussion

In this chapter we have studied the multicast routing and wavelength assignment problem under multi-drop lightpath and light-tree models. The obtained results show that compared with point-to-point connection model, multi-drop k-routings under both models can considerably reduce the network cost and save the number of wavelengths used even for small k. This is very significant from application point of view. In addition, using multi-drop k-routings is more effective in saving the number of wavelengths used than in reducing the network cost.

Ravi et al [15] studied a closely related network-design problems with two different objectives, the total cost of the edges and nodes in the network and the maximum degree of any node in the network. A prototypical example is the degree-constrained node-weighted Steiner tree problem as follows. Given an undirected graph $G(V, E)$, with a nonnegative integral function d on V that specifies an upper bound $d(v)$ on the degree of each vertex $v \in V$ in the Steiner tree to be constructed, nonnegative costs on the nodes, and a subset of k nodes called terminals. The goal is to construct a Steiner tree T containing all the terminals such that the degree of any node v in T is at most the specified upper bound $d(v)$ and the total cost of the nodes in T is minimum. They proposed a bicriteria approximation algorithm whose output is approximate in terms of both the degree and cost criteria. The degree of any node $v \in V$ in the output Steiner tree is $O(d(v) \log k)$ and the cost of the tree is $O(\log k)$ times that of a minimum-cost Steiner tree that obeys the degree bound $d(v)$ for each node v.

In an earlier work, Fürer and Raghavachari [6] consider another related problem, the *minimum-degree Steiner tree problem*. Given a graph $G(V, E)$ and a set of distinguished vertices $D \subseteq V$, the problem is to find a Steiner tree of D whose maximal degree is minimal. They proposed an iterative polynomial-time algorithm that produces a tree whose maximal degree is at most one more than the maximal degree of an optimal Steiner tree.

In a recent work, Zhang et al [17] proposed and studied four algorithms for constructing source-based multicast light-forest, which consist of one or more multicast trees, for each multicast connection. Their performances were studied through simulation and compared in terms of the average number of wavelengths used, average number of branches involved and average number of hops encountered from a source to a destination.

References

[1] R. K. Ahuja, T. L. Magnanti, and J. B. Orlin, *Network Flows: Theory, Algorithms, and Applications*, Prentice Hall, Englewood Cliffs, New Jersey, 1993, 273-288.

[2] B. Alspach, J.-C. Bermond, and D. Sotteau, Decompositions into cycles I: Hamiltonian decompositions, in *Cycles and Rays*, edited by G. Hahn, G. Sabidussi, and R. E. Woodrow, Kluwer Academic Publishers, 1989.

[3] V. Chvátal, A greedy heuristic for the set-covering problem, *Mathematics of Operations Research*, 4 (3) (1979), 233-235.

[4] S. Even, *Graph Algorithms*, Computer Science Press, Inc., 1979.

[5] M. F. Forreger, Hamiltonian decompositions of products and cycles, *Discrete Mathematics*, 24 (1978), 251-260.

[6] M. Fürer and B. Raghavachari, Approximating the minimum-degree Steiner tree to within one of optimal, *Journal of Algorithms*, 17 (3) (1994), 409-423.

[7] S.-X. Gao, X.-H. Jia, X.-D. Hu, D.-Y. Li, Wavelength requirements and routing for multicast connections in lightpath and light-tree models of WDM networks with limited drops, *IEE Proceedings on Communications*, 148 (6) (2001), 363-367.

[8] D. Gusfield, Connectivity and edge-disjoint spanning trees, *Information Processing Letters*, 6 (1983), 87-89.

[9] P. Hall, On representatives of subsets, *Journal of London Mathematics Society*, 10 (1935), 26-30.

[10] S. Kundu, Bounds on the number of disjoint spanning trees, *Journal of Combinatoric Theory* (Ser.B), 17 (1974), 199-203.

[11] R. Libeskind-Hadas, Efficient collective communication in WDM networks with a power budget, in *Proceedings of the 9th IEEE International Conference on Computer Communications and Networks*, (2000), 612-616.

[12] D. W. Matula, Determining edge connectivity in $O(mn)$, in *Proceedings of the 28th Symposium on Foundations of Computer Science* (FOCS), 1987, 249-251.

[13] C. H. Papadimitriou and K. Steiglitz, *Combinatorial Optimization: Algorithms and Complexity*, Prentice-Hall, Englewood Cliffs, NJ, 1982.

[14] R. Ramaswami, Multi-Wavelength lightwave networks for computer communication, *IEEE Communication Magazine*, 31 (1993), 78-88.

[15] R. Ravi, M. V. Marathe, S. S. Ravi, D. J. Rosenkrantz, H. B. Hunt III, Approximation Algorithms for Degree-Constrained Minimum-Cost Network-Design Problems, *Algorithmica*, 31 (2001), 58-78.

[16] L. H. Sahasrabuddhe and B. Mukherjee, Multicast routing algorithms and protocols: a tutorial, *IEEE Network*, 14 (1) (2000), 90-102.

[17] X. Zhang, J. Wei, and C. Qiao, Constrained multicast routing in WDM networks with sparse light splitting, *Proceedings of IEEE Conference on Computer Communications* (INFOCOM), 2000.

Index

Author Index

Network Theory and Applications

1. P.-J. Wan: *Multichannel Optical Networks.* 2000 ISBN 0-7923-5776-0
2. B. Fortz: *Design of Survivable Networks with Bounded Rings.* 2000 ISBN 0-7923-6414-7
3. B. Liu and H.J. Lai: *Matrices in Combinatorics and Graph Theory.* 2000 ISBN 0-7923-6469-4
4. G.J. Chang, L. Cui and F.K. Hwang: *Reliabilities of Consecutive-k Systems.* 2000 ISBN 0-7923-6661-1
5. D.-Z. Du and H.Q. Ngo (eds.): *Switching Networks: Recent Advances.* 2001 ISBN 0-7923-7153-X
6. L. Ruan and D.-Z. Du (eds.): *Optical Networks - Recent Advances.* 2001 ISBN 0-7923-7166-6
7. J. Xu: *Topological Structure and Analysis of Interconnection Networks.* 2001 ISBN 1-4020-0020-0
8. B. Lu et al. (eds.): *Layout Optimization in VLSI Design.* 2001 ISBN 1-4020-0089-8
9. X. Jia, X-D. Hu and D-Z. Du: *Multiwavelength Optical Networks.* 2002 ISBN 1-4020-0804-X

Kluwer Academic Publishers – Dordrecht / Boston / London